Source Books in the History of the Sciences

Edward H. Madden, *General Editor*

A Source Book in Geography

Edited by GEORGE KISH

HARVARD UNIVERSITY PRESS
Cambridge, Massachusetts, and London, England   1978

G
80
S63

Library of Congress Cataloging in Publication Data

Main entry under title:

A source book in geography.
    (Source books in the history of the sciences)
    Includes index.
    1. Geography—History—Sources.   I. Kish,
George, 1914–   II. Series.
G80.S63   910′.9   77–25972
ISBN 0-674-82270-6

To the memory of Paul, Count Teleki,
geographer and historian of geography

General Editor's Preface

The *Source Books* in this series are collections of classical papers that have shaped the structure of the various sciences. Some of these classics are not readily available and many of them have never been translated into English, thus being lost to the general reader and in many cases to the scientist himself. The point of this series is to make these texts easily accessible and to provide good translations of the ones that have not been translated at all, or only poorly.

The series was planned to include volumes in all the major sciences from the Renaissance through the nineteenth century. It has been extended to include ancient and medieval science and the development of the sciences in the first half of the present century. Many of these books have been published already and others are in various stages of preparation.

The Carnegie Corporation originally financed the series by a grant to the American Philosophical Association. The History of Science Society and the American Association for the Advancement of Science have approved the project and are represented on the Editorial Advisory Board. This Board at present consists of the following members:

Marshall Clagett, History of Science, Institute for Advanced Study, Princeton
I. Bernard Cohen, History of Science, Harvard University
Thomas A. Goudge, Philosophy, University of Toronto
Gerald Holton, Physics, Harvard University
Ernst Mayr, Zoology, Harvard University
Ernest Nagel, Philosophy, Columbia University
Dorothy Needham, Chemistry, Cambridge University

Harry Woolf, History of Science, Institute for Advanced Study, Princeton

I am indebted to the members of the Advisory Board and to William Bennett, Editor for Science and Medicine, Harvard University Press, for their indispensable aid in guiding the course of the *Source Books*.

Edward H. Madden

Department of Philosophy
State University of New York at Buffalo

Preface

Quum oceanus movetur, totus movetur.
Bernardus Varenius, *Geographia Generalis* (1650)

Geography is as old as man's first search for a bit of soil to dig for plantings, for a path that leads to water, for a trail to a place where hard rock for arrowheads may be found. Geography is as new as man's current search for ways to relieve urban congestion, to establish well-marked international boundaries, to describe and analyze vegetation patterns in remote parts of the earth.

The first work entitled *Geography* was written in Alexandria, in the third century B.C., by Eratosthenes. But writings dealing with the shape of the earth and the nature of the environment date from the days of Ionian science, as early as the sixth century B.C. From those beginnings the study of the earth as the abode of man, of earth processes, and of the distribution of terrestrial phenomena has continued to our day; this is the domain of general geography.

But people have always been interested in knowing more about the earth than generalities. They have always been curious about what lies beyond their horizons, what kinds of plants, animals, and men live beyond their own limits. Descriptions of the lands and peoples of various parts of the earth represent another long-established branch of geographical writings: regional geography.

To acquire knowledge of other lands and people, travelers since earliest times have set out beyond their own frontiers and have reported their findings to those who stayed at home. These reports of travel—undertaken for profit, out of curiosity, to establish relations with people in distant countries, for the sake of religious experience—form a third and very popular type of geographical writing.

This volume is composed of examples of all three types of writing that may be called geographical. Chronologically, they range from biblical texts and early Greek explanations of our earth to the first formal statements on geography as a science, written between 1650 and 1850. Although the majority of the sources quoted are from the European tradition, Islam's intense interest in geography is well represented by the writings of Moslem geographers, and examples of geographical writings from classical China are also included.

The principles of geographical writing were formulated in the days of Greek and Hellenistic science. It was on that foundation that Christian and Moslem practitioners based their contributions. Bridging the transition from medieval to modern times, a series of selections from the Age of Discovery epitomizes the mood, methods, and results of that time. The final segment of the book represents the growth of the new, scientific methodology of geography during the Age of Enlightenment and the first half of the nineteenth century.

I am deeply in debt to the work of my predecessors and my contemporaries, historians of geography: C. Raymond Beazley, Edward Bunbury, Siegmund Günther, Joachim Lelewel, Oscar Peschel, and Sophus Ruge, in the period prior to 1914; Hanno Beck, Robert Dickinson, Richard Hartshorne, Joseph Schmithüsen, William Stahl, Eric Warmington, and John K. Wright, in our own time. I have also benefited from the anthology "Selected Readings in the Development of Geographic Thought," prepared at the University of Hawaii in mimeographed form by my late friend, Curtiss A. Manchester.

A generous grant from the Faculty Research Fund of the Horace H. Rackham School of Graduate Studies at the University of Michigan provided support for the project. I also wish to express my gratitude to my research assistants: Gail Lebeau Howell, Michael Watts, and Katherine Baker Drysdale.

George Kish
Ann Arbor, at the House of the Seven Plum Trees
November 1977

Contents

Revival of Geography in the West   236

Enlarging Horizons by Travel   249

Physical Geography in the Later Middle Ages   275

Geographical Writings of the Age of Discovery   289

German Geographers of the Sixteenth Century   348

The Beginnings of Modern Geography:
The Seventeenth Century   364

Eighteenth Century Concepts of Geography   378

A Source Book in Geography

Ever since Almighty God commanded Adam to subdue the earth, there have not wanted in all ages some heroical spirits, which, in obedience to that high mandate, either from manifest reasons alluring them, or secret instinct inforcing them, have expended their wealth, employed their times, and adventured their persons, to find the true circuit thereof.

Francis Drake the Younger

The Beginnings

The first selections in this book—from the Bible and from Hesiod—are among the earliest of written statements on the problems and methods of geography. The Bible, the source of Judaeo-Christian religious and ethical tradition, is an important source of historical and geographical traditions as well. The Voice out of the Whirlwind speaking to Job asks questions that continue to concern the practitioner of modern geography, just as the description of the commerce of the great Phoenician city of Tyre is a detailed and accurate statement of the subject matter of what we now call regional or economic geography.

Hesiod's poems represent the earliest written traditions of our Greek heritage. Hesiod, who probably lived during the eighth century B.C., was a farmer and poet who wrote about the yearly cycle in the life of Greek country folk of his day. *Works and Days* has been called a shepherd's calendar, and describes the march of the seasons and the changes in the environment as seen by a farmer. In the poem *Theogony,* Hesiod describes creation and the history of divine dynasties; the excerpt quoted here depicts the power of the winds over the lives of men.

1. The Lord speaks to Job on man and his world

VOICE OUT OF THE WHIRLWIND. Who is this that darkeneth counsel
By words without knowledge?

Job 38:1–41, from *The Bible Designed to Be Read as Living Literature,* ed. E. S. Bates (New York: Simon & Schuster, 1936), by permission of the publisher. Copyright 1936, 1964 by Simon & Schuster, Inc.

Gird up now thy loins like a man;
For I will demand of thee, and declare thou unto me.
Where wast thou when I laid the foundations of the earth?
Declare, if thou hast understanding.
Who determined the measures thereof, if thou knowest?
Or who stretched the line upon it?
Whereupon were the foundations thereof fastened?
Or who laid the corner stone thereof;
When the morning stars sang together,
And all the sons of God shouted for joy?
Or who shut up the sea with doors,
When it broke forth, as if it had issued out of the womb;
When I made the cloud the garment thereof,
And thick darkness a swaddlingband for it,
And prescribed for it my decree,
And set bars and doors,
And said, "Hitherto shalt thou come, but no further;
And here shall thy proud waves be stayed"?
Hast thou commanded the morning since thy days began,
And caused the dayspring to know its place;
That it might take hold of the ends of the earth,
And the wicked be shaken out of it?
It is changed as clay under the seal;
And all things stand forth as a garment:
And from the wicked their light is withheld,
And the high arm is broken.
Hast thou entered into the springs of the sea?
Or hast thou walked in the recesses of the deep?
Have the gates of death been revealed unto thee?
Or hast thou seen the gates of the shadow of death?
Hast thou comprehended the breadth of the earth?
Declare, if thou knowest it all.
Where is the way to the dwelling of light,
And as for darkness, where is the place thereof;
That thou shouldest take it to the bound thereof,
And that thou shouldest discern the paths to the house thereof?
Doubtless, thou knowest, for thou wast then born,
And the number of thy days is great!
Hast thou entered the treasuries of the snow,
Or hast thou seen the treasuries of the hail,
Which I have reserved against the time of trouble,
Against the day of battle and war?
By what way is the light parted,
Or the east wind scattered upon the earth?
Who hath cleft a channel for the waterflood,

Or a way for the lightning of the thunder;
To cause it to rain on a land where no man is;
On the wilderness, wherein there is no man;
To satisfy the waste and desolate ground;
And to cause the tender grass to spring forth?
Hath the rain a father?
Or who hath begotten the drops of dew?
Out of whose womb came the ice?
And the hoary frost of heaven, who hath gendered it?
The waters are hidden as with stone,
And the face of the deep is frozen.
Canst thou bind the cluster of the Pleiades,
Or loose the bands of Orion?
Canst thou lead forth the Mazzaroth in their season?
Or canst thou guide the Bear with her train?
Knowest thou the ordinances of the heavens?
Canst thou establish the dominion thereof in the earth?
Canst thou lift up thy voice to the clouds,
That abundance of waters may cover thee?
Canst thou send forth lightnings, that they may go,
And say unto thee, "Here we are"?
Who hath put wisdom in the inward parts?
Or who hath given understanding to the mind?
Who can number the clouds by wisdom?
Or who can pour out the bottles of heaven,
When the dust runneth into a mass,
And the clods cleave fast together?
Wilt thou hunt the prey for the lioness?
Or satisfy the appetite of the young lions,
When they couch in their dens,
And abide in the covert to lie in wait?
Who provideth for the raven his food,
When his young ones cry unto God,
And wander for lack of meat?

2. Ezekiel describes the commerce of Tyre

The word of the Lord came again unto me, saying, "Now, thou Son of Man, take up a lamentation for Tyrus; and say unto Tyrus, 'O thou that art

Ezekiel, 27:1–25, from *The Bible Designed to Be Read as Living Literature,* ed. E. S. Bates (New York: Simon & Schuster, 1936), by permission of the publisher. Copyright 1936, 1964 by Simon & Schuster, Inc.

situate at the entry of the sea, which art a merchant of the people for many isles, thus saith the Lord God: "O Tyrus, thou hast said, 'I am of perfect beauty.' Thy borders are in the midst of the seas, thy builders have perfected thy beauty. They have made all thy ship boards of fir trees of Senir: they have taken cedars from Lebanon to make masts for thee. Of the oaks of Bashan have they made thine oars; the company of the Ashurites have made thy benches of ivory, brought out of the isles of Chittim. Fine linen with broidered work from Egypt was that which thou spreadest forth to be thy sail; blue and purple from the isles of Elishah was that which covered thee. The inhabitants of Sidon and Arvad were thy mariners: thy wise men, O Tyrus, that were in thee, were thy pilots. The ancients of Gebal and the wise men thereof were in thee thy calkers: all the ships of the sea with their mariners were in thee to occupy thy merchandise. They of Persia and of Lud and of Phut were in thine army, thy men of war: they hanged the shield and helmet in thee; they set forth thy comeliness. The men of Arvad with thine army were upon thy walls round about, and the Gammadims were in thy towers: they hanged their shields upon thy walls round about; they have made thy beauty perfect. Tarshish was thy merchant by reason of the multitude of all kind of riches; with silver, iron, tin, and lead, they traded in thy fairs. Javan, Tubal, and Meshech, they were thy merchants: they traded the persons of men and vessels of brass in thy market. They of the house of Togarmah traded in thy fairs with horses and horsemen and mules. The men of Dedan were thy merchants; many isles were the merchandise of thine hand: they brought thee for a present horns of ivory and ebony. Syria was thy merchant by reason of the multitude of the wares of thy making: they occupied in thy fairs with emeralds, purple, and broidered work, and fine linen, and coral, and agate. Judah, and the land of Israel, they were thy merchants: they traded in thy market wheat of Minnith, and Pannag, and honey, and oil, and balm. Damascus was thy merchant in the multitude of the wares of thy making, for the multitude of all riches; in the wine of Helbon, and white wool. Dan also and Javan going to and fro occupied in thy fairs: bright iron, cassia, and calamus, were in thy market. Dedan was thy merchant in precious clothes for chariots. Arabia, and all the princes of Kedar, they occupied with thee in lambs, and rams, and goats: in these were they thy merchants. The merchants of Sheba and Raamah, they were thy merchants: they occupied in thy fairs with chief of all spices, and with all precious stones, and gold. Haran, and Canneh, and Eden, the merchants of Sheba, Asshur, and Chilmad, were thy merchants. These were thy merchants in all sorts of things, in blue clothes, and broidered work, and in chests of rich apparel, bound with cords, and made of cedar, among thy merchandise. The ships of Tarshish did sing of thee in thy market: and thou wast replenished, and made very glorious in the midst of the seas.

3. Hesiod on the seasons

At the time when the Pleiades, the daughters
    of Atlas, are rising,
begin your harvest, and plow again when they
    are setting.
. . . . .
The Pleiades are hidden for forty nights and forty
days, and then, as the turn of the year reaches
    that point
they show again, at the time you first sharpen
    your iron.
This is the usage, whether you live in the plains,
    or whether   ·
close by the sea, or again in the corners
    of the mountains
far away from the sea and its tossing water,
    you have your rich land;
wherever you live: strip down to sow, and strip
    for plowing,
and strip for reaping, if you wish to bring in
    the yields of Demeter
all in their season, and so that each crop
    in its time will increase
for you; so that in aftertime you may not be in need
and go begging to other people's houses,
    and get nothing;
. . . . .
At the time when you hear the cry of the crane
    going over, that annual
voice from high in the clouds, you should take notice
    and make plans.
She brings the signal for the beginning of planting,
    the winter
season of rains, but she bites the heart
    of the man without oxen.
At this time, keep your horn-curved oxen indoors,
    and feed them.
. . . . .

Hesiod, *The Works and Days,* trans. Richmond Lattimore (Ann Arbor: The University of Michigan Press, 1959), pp. 63, 65, 71, 99, by permission of the publisher. Copyright © by the University of Michigan 1959.

Even if you plant late, here is one thing
      that might save you:
at that time when the cuckoo first makes his song
      in the oak leaves,
and across immeasurable earth makes glad
      the hearts of mortals,
if at that time Zeus should rain three days
      without stopping,
and it neither falls short of, nor goes over,
      the height of an ox hoof;
then the late planter might come out even
      with the early planter.

. . . . .

Beware of the month Lenaion, bad days,
      that would take the skin off
an ox; beware of it, and the frosts, which,
      as Boreas,
the north wind, blows over the land, cruelly develop;
he gets his breath and rises on the open water
      by horse-breeding
Thrace, and blows, and the earth
      and the forest groan, as many
oaks with sweeping foliage, many solid fir trees
along the slopes of the mountains his force bends
      against the prospering
earth, and all the innumerable forest
      is loud with him.
The beasts shiver and put their tails
      between their legs, even
those with thick furry coats to cover their hides,
      the cold winds
blow through the furs of even these, for all
      their thickness.
The wind goes through the hide of an ox,
      it will not stop him;
it goes through a goatskin, that is fine-haired;
      but not even Boreas'
force can blow through a sheepskin to any degree,
      for the thick fleece
holds him out. It does bend the old man
      like a wheel's timber.
For fifty days, after the turn
      of the summer solstice,
when the wearisome season of the hot weather

    goes to its conclusion
then is the timely season for men to voyage.
    You will not
break up your ship, nor will the sea drown
    its people, unless
Poseidon, the shaker of the earth,
    of his own volition,
or Zeus, the king of the immortals, wishes
    to destroy it,
for with these rests authority for all outcomes,
    good or evil.
At that time the breezes can be judged,
    and the sea is untroubled.
At that time, trusting your swift ship to the winds,
    you can draw her
down to the sea at will, and load all your cargo
    inside her;
but make haste still, for the sake of
    an earlier homecoming,
and do not wait for the season of new wine,
    and the autumn
rain, and the winter coming on,
    and the hard-blowing southwind
who comes up behind the heavy rains that Zeus sends
    in autumn
and upheaves the sea and makes the open water
    difficult.
    There is one other sailing season for men,
in spring time.
At that point, when you first make out
    on the topmost branches
of the fig tree, a leaf as big as the print
    that a crow makes
when he walks; at that time also the sea is navigable
and this is called the spring sailing season.
    I for my part
do not like it. There is nothing about it
    that I find pleasant.
It's snatched. You will find it hard
    to escape coming to grief. Yet still
and even so, men in their short-sightedness
    do undertake it;
for acquisition means life to miserable mortals;
but it is an awful thing to die among the waves.

4. Hesiod on the winds

And from Typhoeus comes the force of winds
  blowing wetly:
all but Notos, Boreas, and clearing Zephyros,
for their generation is of the gods,
  they are a great blessing
to men, but the rest of them blow wildly
  across the water
and burst upon the misty face
  of the open sea, bringing
heavy distress to mortal men,
  and rage in malignant
storm, and blow from veering directions,
  and scatter the shipping
and drown the sailors,
  and there is no remedy against this evil
for men who run into such winds
  as these on the open water.
And then again, across the limitless
  and flowering
earth, they ruin the beloved works
  of ground-dwelling people
by overwhelming them with dust
  and hard tornadoes.

Hesiod, *Theogony*, trans. Richmond Lattimore (Ann Arbor: The University of Michigan Press, 1959), pp. 175, 176, by permission of the publisher. Copyright © by the University of Michigan 1959.

Early Greek Geography

The study of Greek geography has always been hampered by a scarcity of texts. Since the entire opus of a classical geographer does not become available for study until the last two centuries of the pre-Christian era, the reader can only progress slowly from hearsay to fragments to chapters. Thus it is only a transmitted tradition that Eratosthenes, the Alexandrine, writing in the third century B.C., considered the Ionians to be the first geographers. Eratosthenes believed that Anaximander of Miletus was first among the Ionians, having written a work on the nature of the earth, and having also drawn a world map. Today we assume that the place of *primus inter pares* belongs to Thales, another citizen of Miletus, whose life spanned the seventh and sixth centuries B.C.

5. Thales' views of a floating earth

An interest in their environment and in the shape and mechanics of the cosmos are among the characteristic features of early Greek writers on science. Several of these men were born and spent much of their lives along the shores of the Aegean Sea, principally on the eastern one, in Asia Minor. Others lived in what was later called Magna Graecia, or Greater Greece, which incorporated southernmost Italy and Sicily.

The first selection is from Seneca, *Naturales Quaestiones,* trans. T. H. Corcoran (Cambridge, Mass.: Harvard University Press, 1971–1972), p. 147. The other three selections are from E. H. Warmington, *Greek Geography*, The Library of Greek Thought Series (New York, 1934), p. 1. Published in the United States by E. P. Dutton and reprinted by their permission.

Thales of Miletus (625 B.C.?–? 547 B.C.) was considered by Aristotle to be the founder of the Ionian school of philosophy: he is usually referred to as being the first of the Greek philosophers to concern himself with the earth, its position, and its characteristics. His contemporaries believed him to be a wizard, mainly because, according to tradition, he accurately predicted the occurrence of the solar eclipse of May 27, 585 B.C. Thales was a mathematician, an engineer, an astronomer, and a practical statesman who freely acknowledged his indebtedness to Babylonian and Egyptian science. His writings on the earth survive only in fragments, as quotations used by his successors.

Seneca

Thales of Miletus judges that the whole earth is buoyed up and floats upon liquid that lies underneath, whether you call it the ocean, the great sea, or consider it the as yet elementary water of a different character and call it merely a humid element. The disc is supported by this water, he says, just as some big heavy ship is supported by the water which it presses down upon.

Aristotle

Thales says that the first principle is water, whence he declared that the earth is floating on water. Thales of Miletus . . . maintained that the earth stays still because it floats like wood . . . as though the same theory had not been put forward of the water also on which the earth rides . . .

Aetius

Some, following Thales, say that the earth is in the centre (of our universe).

Diodorus

Thales . . . says that it is the etesian winds blowing against the mouths of the river (*Nile*) that prevent the stream from pouring into the sea, and that the river being filled because of this floods over Egypt which is low and plainland.

6. Anaximander considers the earth; he offers an explanation for wind and rain, thunder and lightning

The ideas of Anaximander of Miletus (c. 610 B.C.–c. 545 B.C.), considered by later Greek writers to be the true founder of geography, come to us through fragments quoted by his successors. In the following discussion by Charles Kahn A. refers to Aetius, C. to Cicero, H. to Hippolytus, and P. to the Pseudo-Plutarch. In addition to his views on the shape and nature of the earth, Anaximander is credited with the invention of the gnomon, the simple yet efficient Greek instrument that allowed the early measurement of latitude. Like his fellow citizen and older contemporary Thales, Anaximander is also credited with the prediction of eclipses.

Position of the Earth

H. The earth is aloft, not dominated by anything; it remains in place because of the similar distance from all points [of the celestial circumference].

Arist. There are some who say that the earth remains in place because of similarity [or symmetry], as did Anaximander among the ancients; for a thing established in the middle, with a similar relationship to the extremes, has no reason to move up rather than down or laterally; but since it cannot proceed in opposite directions at the same time, it will necessarily remain where it is.

Anaximander's view of the earth as resting in equipoise at the center of the heavens is perhaps the most significant single piece of information which has reached us concerning the development of scientific thought in sixth-century Miletus. Thales' prediction of a solar eclipse, which symbolized to the ancients the scientific attainments of this period, is less impressive for us who know that such a *tour de force* could only have been achieved on the basis of century-long Babylonian observations.

For the history of ideas, Anaximander's theory of the earth's position is of an entirely different order of importance. Even if we knew nothing else concerning its author, this alone would guarantee him a place among the creators of a rational science of the natural world.

What is most striking in this doctrine is its specifically mathematical character. No matter what terms were used for its formulation, it must, in substance, presuppose the standard definition of the circle as "that which is in every way equidistant from the middle to the extremes." That this clear geometric concept was itself the work of Anaximander is unlikely, and is in

---

From Charles H. Kahn, *Anaximander and the Origins of Greek Cosmology* (New York: Columbia University Press, 1960), pp. 76–78, 81–82, 100–102, by permission of the publisher.

fact contradicted by the ancient tradition, which ascribes to Thales a proof that the diameter of a circle divides it into two equal parts. In all probability the rudiments of geometry were an essential part of Anaximander's formation in the "school" of Thales. The definition of a circle in terms of its equal radii might have been suggested by the spokes of a wheel; for the word κύκλος originally had this concrete sense. But the fact that such a notion had been formulated in a precise way is in itself worthy of note.

More important for us is Anaximander's own use of this geometric idea, as a general expression for the principle of symmetry or indifference. It is indeed the same notion which was glorified in modern times by Leibniz as his Principle of Sufficient Reason, according to which everything which is true or real implies a reason why it is so and not otherwise. Such considerations of symmetry have by no means ceased to play a role in modern mathematical thought. If the geometric sphere imposed itself with such power on the ancient scientific imagination (and indeed still on that of Galileo), it must be due above all to the intellectual prestige of this figure as the image par excellence of regularity, order, and rational proportion.

Form of the Earth

H. The form of the earth is moist, rounded like [the drum of] a stone column.

P. Its form is cylindrical, with a depth one third of its width.

The geometrical turn of mind confronts us with equal distinctness in Anaximander's description of the earth, as a cylinder whose altitude is one third the diameter of its base. Whether or not he made use of the classical term κύλινδρος or "roller" (which was familiar to Democritus, B 155) is not known, but the idea is perfectly expressed by a comparison of the earth to the stocky, rounded stones of which a Greek column is composed. That this image is due to Anaximander himself, and not simply to Theophrastus, is a very likely conjecture of Diels . . .

The originality of Anaximander's conception of the earth is by no means limited to his precise numerical ratio for the dimensions of the cylinder. The epic, it is true, also conceives the surface of the earth as flat, and even as circular in shape, bounded by the circumjacent Ocean. As one might expect, this view is a good deal older than Homer. But what the early poets lack is any distinct notion of the subterranean regions. The *Iliad* speaks of the Titans seated in the darkness of Tartarus "at the limits of earth and sea" . . . This "deepest pit under the earth" is as far below Hades as heaven is from earth . . . The Hesiodic *Theogony* contains several elaborate descriptions of this place, where the "sources and limits" of Earth, Tartarus, Sea, and Heaven converge, while "above grow the roots of Earth and Sea" . . .

It would be hopeless to draw a diagram to accompany such a description. The poetic Tartarus is vividly and dramatically conceived. A diagram, however, requires not drama but a precise geometric arrangement, and nothing could be more alien to the poet's state of mind when describing such mysterious regions. It is, on the other hand, the characteristic feature of Anaximander's view of the earth that it lends itself directly to geometric representation. We can scarcely doubt that the Milesians were in fact accustomed to discuss such matters with the aid of diagrams or of simple models. And, in Ionia at any rate, the standard model for the earth remained that of Anaximander until the time of Democritus.

The doxographical description of Anaximander's earth as a rather low cylinder can be supplemented by data from a different source. When the Greek geographers looked back to the origins of their science, they recognized the same Milesian as the first to have produced a πίναξ or chart of the inhabited earth. And what little we know about the details of that first Greek map shows the same geometric spirit reigning here as in the rest of the cosmos. We have no description of Anaximander's chart as such, but F. Jacoby has shown that the geographical ideas which Herodotus ascribes to the "Ionians" are essentially those of Hecataeus of Miletus, and he has rightly pointed out that the general lines of Hecataeus' view must already have been those of Anaximander. The fullest statement of the relation between Hecataeus and Anaximander is given by the late geographer Agathemerus, probably on the authority of Eratosthenes:

Anaximander of Miletus, the pupil of Thales, was the first to depict the inhabited earth on a chart (ἐν πίνακι γράψαι). After him Hecataeus of Miletus, a much-traveled man, made it more precise so as to be a thing of wonder . . . Now the ancients drew the inhabited earth as round, with Hellas in the middle, and Delphi in the middle of Hellas, since it holds the navel (τόν ὀμφαλόν ἔχειν) of the earth. Democritus, a man of great experience, was the first to recognize that the earth is oblong, with its length one-and-a-half times its width.

Wind

H. Rainfall arises from the vapor emitted by the earth under the action of and move together in one mass.

A. Wind is a flow of air, when the finest parts in it are set in motion by the sun, and the wettest portions are melted [i.e., liquefied].

Rain

H. Rainfall arises from the vapor emitted by the earth under the action of the sun.

Lightning and Thunder

H. Lightning occurs when a wind (ἄνεμος) leaping forth separates the clouds.

A. Concerning thunder, lightning, thunderbolt, fire-winds (?), whirlwinds: All these arise from wind (πνεῦμα). For, when it is enclosed in a thick cloud and bursts out violently because of its lightness and the fineness of its parts, then the tearing of the cloud produces the noise, the dilatation [or explosion] causes the flash, by contrast with the darkness of the cloud.

Winds for Anaximander are a result of "separating-off" (ἀποκρίνε-σθαι), the same process by which Hot and Cold, the circles of the stars, and the heavens as a whole are produced. The finest vapors are separated from the grosser bulk of ἀήρ, under the influence of the sun's heat. They are thus set in motion and, when a sufficient quantity have been gathered together, constitute a wind. This is described by Aëtius as a "flow of air," although in fact only the finest portions of the air are involved. We are tempted to say, "the finest *particles*," although the connotations of this word are anachronistic.

Our authorities do not make clear what is left over in the ἀήρ when the finest vapors have been removed. However, this residue would naturally be a thick cloud on its way to precipitation in raindrops or some similar form ... Wind and water are thus produced by ἀπόκρίοις from air under the influence of the sun's heat. To this corresponds the statement from another source ... that "at first the entire region about the earth was moist, but, as it was dried up by the sun, what was evaporated produced the winds (πνεύματα) ... while the remainder is the sea." In cosmogony, also, the action of celestial fire is thus to separate wind from water. The emergence of atmosphere and sea from primitive moisture is strictly comparable to the production of wind and rain cloud out of the air. The process in both cases is the same: "separating-out."

It may easily be seen that Anaximander's ἀήρ is not quite what we call "air." If, as a matter of convenience, we may translate the word by its direct descendant in our own tongue, it must be remembered that ἀήρ in Homer—and still in some fifth-century texts—means "mist" or "haze." This will be seen in detail in the next chapter. That Anaximander's notion of air is still close to this old sense of the word is clear from the fact that fine vapors (and, presumably, "gross" or thick ones as well) can be separated out from it. Ἀήρ, then, is a gaseous substance of a certain consistency, like smoke or steam. Its resemblance to the primeval moisture which lay about the earth before the sea was formed is probably so great as to fade almost into identity. One text actually speaks of the first appearance of fire above "the air around the earth" ... The roots of the doctrine of Anaximenes are here apparent: both Milesian cosmogonies begin with "an aery fluid charged with water vapor."

Rain was, of course, derived by Anaximander from the vapor raised by the sun and concentrated in the clouds ... We have inferred that this atmospheric moisture which tends to condense as rain is identical with the wettest portions of air, whose "melting," i.e., liquefaction, is mentioned by one text ... Rain and wind then represent the two opposite products of the ἀήρ. The air we breathe is not expressly assigned to either form, but in the nature of the case it must be closely connected with the wind. The very word πνεῦμα applies to both forms of "blowing." Clouds, on the other hand, are naturally associated with the moist ἀπόκρισις leading to rain. Accordingly, we find here in virtual form a theory of the entire atmosphere as ἀήρ, although what is emphasized is not the unity of this medium, but its polar analysis into fine and gross, wet and dry. The primary Hot and Cold . . . do not reappear in our brief meteorological summaries, but their presence here must be taken for granted in connection with solar heat.

It is of thunder and lightning that the most detailed description has reached us . . . The active agent here is wind (ἄνεμος), i.e., a movement of fine air vapors under the influence of the sun. The passive condition or foil is a thick surrounding cloud, the corresponding residue of ἀήρ after wind has been separated out. When some wind is enclosed within this thick envelope, its reaction is to escape outwards in virtue of its fineness and lightness. The consequent ripping and rending of the cloud is cause of the thunder roar; the flash is produced from "the explosion by contrast with the darkness of the cloud." Unfortunately this statement of Aëtius is by no means clear. Seneca speaks in this context of "fire" and "flame." Now the connection between the earthly fire and the bolt of Zeus is age-old. It is presupposed by the theft of Prometheus. Since the burning flame of a thunderbolt reappears as an important element in the classical theory, we can be sure that its absence in the Greek doxographies for Anaximander is due only to the meagerness of the excerpts which have reached us. To contrast with the darkness of a storm cloud, a wind must have acquired the brilliance of a flame. This wind is then an inflammable air current, trapped in the cloud like air in a balloon. When its heat or innate motion causes it to expand, the envelope is torn, the wind shoots outwards and bursts into flame.

It is impossible to mistake the parallel between this meteorological fire and that of the celestial rings. The envelope of obscure cloud or "air," the fire or hot wind within, the flaming emission, all justify the remark of Tannery that the great heavenly bodies resemble a lightning flash which could last indefinitely. In both these cases, as in that of evaporation and the production of winds, the Hot operates as the active cause. In the description of lightning we see that other opposite pairs—fine and thick, bright and dark—also play a decisive role.

7. Anaxagoras on the shape of the earth, eclipses, and atmospheric phenomena

Anaxagoras of Clazomenae (500–?? 428 B.C.) is considered to be the most prominent of the later Ionians; his views on the earth and on terrestrial phenomena constitute further elaboration of those ascribed to Anaximander. The first selection is from Hippolytus' *Refutatio Omnium Heresium;* the second from Alexander of Aphrodisias' *De Fatis;* the third from Plutarch's *De Placitis Philosophorum.*

Hippolytus

The earth is flat in shape. It stays up because of its size, because there is no void, and because the air, which is very resistant, supports the earth, which rests on it. Now we turn to the liquids on the earth: The sea existed all along, but the water in it became the way it is because it suffered evaporation, and it is also added to from the rivers which flow into it. Rivers originate from rains and also from subterranean water; for the earth is hollow and has water in its hollows.

The moon does not shine with its own light, but receives its light from the sun. The stars go under the earth in their course. Eclipses of the moon occur when the earth cuts off the light, and sometimes when the bodies below the moon cut off the light. Eclipses of the sun take place at new moon, when the moon cuts off the light. The sun and moon reach turning points in their courses when they are forced back by the air. Often the moon turns when it can not prevail over the cold. Anaxagoras was the first to describe the circumstances under which eclipses occur and the way light is reflected by the moon. He said that the moon is made of earth and has plains and gullies on it.

Alexander of Aphrodisias

It is evident that hail is ice, that is, water in a solid state. As we have said, it seems inconsistent with this fact that it hails in warm weather rather than in cold weather and it is difficult to see how so much water can remain suspended in the air long enough to freeze. For these reasons, some people—among them Anaxagoras—say that hail falls, and freezes the way it does, because clouds are thrust to colder, higher altitudes by the warmth surrounding the earth below.

Plutarch

Concerning thunder, lightning, thunderbolts, hurricanes attended with lightning, and typhoons ... Anaxagoras says that these phenomena occur when heat falls into the cold below, i.e. when a part of the aether falls into the atmosphere. Thunder is produced by the noise, lightning by the contrast between the color of the aether and the black of the clouds, thunderbolts by a large quantity of light. Typhoons are produced by heat substance composed of yet more particles, and hurricanes attended with lightning are produced by heat substance mixed with clouds.

8. The Pythagoreans: Philolaus and Parmenides

Pythagoras of Samos (c. 560–480 B.C.) founded one of the most influential schools of philosophy in the Greek world. Its principal centers were on the shores of the Ionian sea, in the Greek cities of what is now southernmost Italy: Krotoṉ (present-day Crotone), Tarentum (now Taranto), and Metapontum, a now-abandoned city west of Taranto.

The theories of Pythagoras and the Pythagoreans involved mathematics, astronomy, and music as well as geography. Their influence extended through several centuries, even to the formulation of Euclid's theorems of geometry. Their geographical theories included concepts of a revolving earth and even of a heliocentric universe.

Parmenides of Elea (c. 515–450 B.C.) and Philolaus (fl. 450–400 B.C.) are regarded as writers in the Pythagorean tradition. Besides their formulation of a world system having a central fire at its core, they may be credited with the introduction of one of the most durable models in geography: the division of the earth into zones, habitable and inhabitable. These include the torrid zone, extending north and south from the equator; the two temperate zones, northern and southern, best suited for men; and the two frigid zones, northern and southern.

It remains to speak of the earth, where it is, whether it should be classed among things at rest or things in motion, and of its shape.

The first of these selections is from Aristotle, *On the Heavens*, trans. W. K. C. Guthrie, Loeb Classical Library (Cambridge, Mass.: Harvard University Press, 1939), pp. 217 and 219. The second selection is from John Burnet, *Early Greek Philosophy*, 3rd ed. (London, A. & C. Black, 1920), pp. 296–297. The third selection is from T. L. Heath, *Aristarchus of Samos* (Oxford, Oxford University Press, 1913), pp. 63–64, 65–66. By permission of the Oxford University Press.

Concerning its position there is some divergence of opinion. Most of those who hold that the whole Universe is finite say that it lies at the centre, but this is contradicted by the Italian school called Pythagoreans. These affirm that the centre is occupied by fire, and that the earth is one of the stars, and creates night and day as it travels in a circle about the centre. In addition they invent another earth, lying opposite our own, which they call by the name of "counter-earth," not seeking accounts and explanations in conformity with the appearances, but trying by violence to bring the appearances into line with accounts and opinions of their own. There are many others too who might agree that it is wrong to assign the central position to the earth, men who see proof not in the appearances but rather in abstract theory. These reason that the most honourable body ought to occupy the most honourable place, that fire is more honourable than earth, that a limit is a more honourable place than what lies between limits, and that the centre and outer boundary are the limits. Arguing from these premises, they say it must be not the earth, but rather fire, that is situated at the centre of the sphere.

The Pythagoreans make a further point. Because the most important part of the Universe—which is the centre—ought more than any to be guarded, they call the fire which occupies this place the Watch-tower of Zeus, as if it were the centre in an unambiguous sense, being at the same time the geometrical centre and the natural centre of the thing itself. But we should rather suppose the same to be true of the whole world as is true of animals, namely that the centre of the animal and the centre of its body are not the same thing. For this reason there is no need for them to be alarmed about the Universe, nor to call in a guard for its mathematical centre; they ought rather to consider what sort of thing the true centre is, and what is its natural place. For it is that centre which should be held in honour as a starting-point; the local centre would seem to be rather an end than a starting-point, for that which is defined is the local centre, that which defines it is the boundary: but that which encompasses and sets bounds is of more worth than that which is bounded, for the one is matter, the other the substance of the structure.

The planetary system which Aristotle attributes to "the Pythagoreans" and Aetios to Philolaos is sufficiently remarkable. The earth is no longer in the middle of the world; its place is taken by a central fire, which is not to be identified with the sun. Round this fire revolve ten bodies. First comes the Antichthon or Counterearth, and next the earth, which thus becomes one of the planets. After the earth comes the moon, then the sun, the planets, and the heaven of the fixed stars. We do not see the central fire and the antichthon because the side of the earth on which we live is always turned away from them. This is to be explained by the analogy of the moon, which always presents the same face to us, so that men living on the other side of it would never see the earth. This implies, of course, from our point of view, that these bodies rotate on their axes in the same time as they revolve round the central fire, and that the antichthon revolves round the central fire in the same time as the earth, so that it is always in opposition to it.

The main difference between the cosmologies of Parmenides and the Pythagoreans appears to be this. It seems almost certain that Pythagoras himself conceived the universe to be a sphere, and attributed to it daily rotation round an axis (though this was denied by Philolaus afterwards); this involved the assumption that it is itself finite but that something exists round it; the Pythagoreans, therefore, were bound to hold that, beyond the finite rotating sphere, there was limitless void or empty space; this agrees with their notion that the universe breathes, a supposition which Tannery attributes to the Master himself because Xenophanes is said to have denied it. Parmenides, on the other hand, denied the existence of the infinite void, and was therefore obliged to make his finite sphere motionless, and to hold that its apparent rotation is only an illusion . . .

Secondly, Parmenides is said to have been the first to 'define the habitable regions of the earth under the two tropic zones'; on the other hand we are told that Pythagoras and his school declared that the sphere of the whole heaven was divided into five circles which they called "zones". Hultsch bids us reject the attribution to Pythagoras on the ground that these zones would only be possible on a system in which the axis of the universe about which it revolves passes through the centre of the earth; the zones are therefore incompatible with the Pythagorean system, according to which the earth moves round the central fire. Hultsch admits, however, that this argument does not hold if the hypothesis of the central fire was not thought of by any one before Philolaus; and there is no evidence that it was. As soon as Pythagoras had satisfied himself that the universe and the earth were concentric spheres, the centre of both being the centre of the earth, the definite portion of the heaven marked out by the extreme deviations of the sun in latitude (north and south) might easily present itself to him as a zone on the heavenly sphere. The Arctic Circle, already known in the sense of the circle including within it the stars which never set, would make another division, while a corresponding Antarctic Circle would naturally be postulated by one who had realized the existence of antipodes. With the intervening two zones, five divisions of the heaven were ready to hand. It would next be seen that straight lines drawn from the centre of the earth to all points on all the dividing circles in the heaven would cut the surface of the earth in points lying on exactly corresponding circles, and the zone-theory would thus be transferred to the earth. We are told, however, that Parmenides' division of the earth into zones was different from the division which would be arrived at in this way, in that he made his torrid zone about twice as broad as the zone intercepted between the tropic circles, so that it spread over each of those circles into the temperate zones. This seems to be the first appearance of zones viewed from the standpoint of physical geography.

9. Xenophanes on the origin of fossils

From an early date the Greeks perceived the environment as existing within a continuum of time. Xenophanes of Colophon (c. 580/570–c. 478 B.C.) offered an explanation of fossil remains as well as an early diluvian theory of the evolution of the earth—reported here by Hippolytus.

Xenophanes thinks that a mixture of the earth with the sea is taking place, and that the earth in the course of time is being dissolved by the moisture; he states that he has the following proofs: in interior parts of the land and in the mountains are found shells, and he says that in the stone-quarries at Syracuse were found imprints of a fish and of sea-weed, and in Paros the imprint of an anchovy in the depth of the stone, and in Melite (*Malta or its Liburnian namesake*) flat outlines of all and sundry sea-creatures. And he says that this came about when everything was "mudded" long ago, and the imprint was fixed dry in the mud; all mankind is destroyed whenever the earth is brought down into the sea and has become mud, and then begins the process of birth all over again.

From E. H. Warmington, *Greek Geography*, The Library of Greek Thought Series (New York, 1934), p. 6. Published in the United States by E. P. Dutton, and reprinted by their permission.

Periplus and Periegesis: Greek Maritime Writings

Living on the shores of the Mediterranean, the Greeks turned to the sea at an early date. The travels of Greek merchants and seamen, and of their keen competitors, the Punic navigators, produced two of the earliest genres of Greek geographical writing. The *periplus*, the more formal of the two, was developed to provide detailed, concrete information for seafaring folk about distances at sea and anchorages near land; the *periegesis* was a more general view of the major bodies of water. The close concordance between the *periploi* and the data recorded on the earliest surviving navigators' charts tends to support the theory that these sea charts, the first relatively accurate maps known in the Western world, may well have been first developed during Greco-Roman times in the Mediterranean.

The following excerpts illustrate the typical features of *periploi* and *periegeses*.

10. Hanno reports on West Africa, Himilco on the Atlantic

Carthage was the greatest competitor of the Greek city-states in maritime commerce, and it is generally accepted that Carthaginian ships engaged in extra-Mediterranean trade, sailing regularly to Atlantic destinations both on the western coasts of Europe and the west coast

The first selection is from *The Periplus of Hanno—A Voyage of Discovery down the West African Coast by a Carthaginian Admiral of the Fifth Century B.C.* trans. Wilfred H. Schoff. Philadelphia, Commercial Museum, 1912; the second is from E. H. Warmington, *Greek Geography*, The Library of Greek Thought Series (New York, 1934), pp. 76–77. Published in the United States by E. P. Dutton, and reprinted by their permission.

of North Africa. Hanno's coasting voyage on the African coast, which according to tradition took place around 500 B.C., may well have taken him and his men as far as Sierra Leone; tradition also states that a report of this voyage, engraved upon a tablet, was put up in the main temple of Carthage.

The fragments included here on Himilco's voyage—another Carthaginian commander—are taken from the writings of Rufus Festus Avienus (fl. 380–420 A.D.), a popularizer of the late Roman Empire.

The Voyage of Hanno, King of the Carthaginians

To the Libyan regions of the earth beyond the Pillars of Hercules, which he dedicated also in the Temple of Baal, affixing this

1. It pleased the Carthaginians that Hanno should voyage outside the Pillars of Hercules, and found cities of the Libyphœnicians. And he set forth with sixty ships of fifty oars, and a multitude of men and women, to the number of thirty thousand, and with wheat and other provisions.

2. After passing through the Pillars we went on and sailed for two days' journey beyond, where we founded the first city, which we called Thymiaterium; it lay in the midst of a great plain.

3. Sailing thence toward the west we came to Solois, a promontory of Libya, bristling with trees.

4. Having set up an altar here to Neptune, we proceeded again, going toward the east for half the day, until we reached a marsh lying no great way from the sea, thickly grown with tall reeds. Here also were elephants and other wild beasts feeding, in great numbers.

5. Going beyond the marsh a day's journey, we settled cities by the sea, which we called Caricus Murus, Gytta, Acra, Melitta and Arambys.

6. Sailing thence we came to the Lixus, a great river flowing from Libya. By it a wandering people, the Lixitæ, were pasturing their flocks; with whom we remained some time, becoming friends.

7. Above these folk lived unfriendly Æthiopians, dwelling in a land full of wild beasts, and shut off by great mountains, from which they say the Lixus flows, and on the mountains live men of various shapes, cave-dwellers, who, so the Lixitæ say, are fleeter of foot than horses.

8. Taking interpreters from them, we sailed twelve days toward the south along a desert, turning thence toward the east one day's sail. There, within the recess of a bay we found a small island, having a circuit of fifteen stadia; which we settled, and called it Cerne. From our journey we judged it to be situated opposite Carthage; for the voyage from Carthage to the Pillars and thence to Cerne was the same.

9. Thence, sailing by a great river whose name was Chretes, we came

to a lake, which had three islands, larger than Cerne. Running a day's sail beyond these, we came to the end of the lake, above which rose great mountains, peopled by savage men wearing skins of wild beasts, who threw stones at us and prevented us from landing from our ships.

10. Sailing thence, we came to another river, very great and broad, which was full of crocodiles and hippopotami. And then we turned about and went back to Cerne.

11. Thence we sailed toward the south twelve days, following the shore, which was peopled by Æthiopians who fled from us and would not wait. And their speech the Lixitæ who were with us could not understand.

12. But on the last day we came to great wooded mountains. The wood of the trees was fragrant, and of various kinds.

13. Sailing around these mountains for two days, we came to an immense opening of the sea, from either side of which there was level ground inland; from which at night we saw fire leaping up on every side at intervals, now greater, now less.

14. Having taken in water there, we sailed along the shore for five days, until we came to a great bay, which our interpreters said was called Horn of the West. In it there was a large island, and within the island a lake of the sea, in which there was another island. Landing there during the day, we saw nothing but forests, but by night many burning fires, and we heard the sound of pipes and cymbals, and the noise of drums and a great uproar. Then fear possessed us, and the soothsayers commanded us to leave the island.

15. And then quickly sailing forth, we passed by a burning country full of fragrance, from which great torrents of fire flowed down to the sea. But the land could not be come at for the heat.

16. And we sailed along with all speed, being stricken by fear. After a journey of four days, we saw the land at night covered with flames. And in the midst there was one lofty fire, greater than the rest, which seemed to touch the stars. By day this was seen to be a very high mountain, called Chariot of the Gods.

17. Thence, sailing along by the fiery torrents for three days, we came to a bay, called Horn of the South.

18. In the recess of this bay there was an island, like the former one, having a lake, in which there was another island, full of savage men. There were women, too, in even greater number. They had hairy bodies, and the interpreters called them *Gorillæ*. When we pursued them we were unable to take any of the men; for they all escaped, by climbing the steep places and defending themselves with stones; but we took three of the women, who bit and scratched their leaders, and would not follow us. So we killed them and flayed them, and brought their skins to Carthage. For we did not voyage further, provisions failing us.

From Avienus

The Atlantic

Himilco records that in the region far towards the west there are interminable swirling waters, where the sea and surge spread open far and wide. "Nobody has visited these waters hitherto; nobody has brought ships into that wide stretch, for there no driving blasts of winds are felt upon the deep, no breath of heaven helps on a vessel; moreover dark mist shrouds the sky as with a cloak; fog at all times hides the swirling waters, and clouds last all day long in thickest gloom."

The Sargasso Sea?

He adds further that amidst the swirl of waters sea-weed rises up straight and often holds back a ship as brushwood might. Nevertheless, he says, here the body of the sea does not go deeply down, and the sea-bed is scarcely covered over by the shallow water; you constantly encounter sea-beasts roving hither and thither; the vessels creep along slowly and sluggishly, and among them the monsters swim. (*Himilco saw all this*).

11. A periplus of the Mediterranean: Greek sailing directions

The following excerpt from one of the most complete *periploi* was attributed in the nineteenth century to the Greek merchant Scylax of Caryanda, known to have traded with settlements not only along the Red Sea but also on the shores of the Indian Ocean. This attribution is no longer held, but the authenticity of this itinerary at sea, written probably in Hellenistic times, is not questioned.

Libya begins beyond the Canopic mouth of the Nile. Adyrmachidae. The first people in Libya are the Adyrmachidae. From Thonis the voyage to Pharos, a desert island (good harbourage but no drinking-water) is 150 stadia. In Pharos are many harbours. But ships water at the Marian mere, for it is drinkable. It is a short sail from Pharos to the mere. Here is also Chersonesus (peninsula) and harbour: the coasting thither is 200 stadia. Beyond Chersonesus is the bay of Plinthine. The mouth of the bay of Plinthine to Leuce Acte (the white beach), is a day and night's sail; but sailing round by the head of the bay of Plinthine is twice as long. The shores of the bay are inhabited. From Leuce Acte to the harbor of Laodamantium to the harbour of Paraetonium is a half-day's sail. Then comes the city of Apis. As far as this point is governed by the Egyptians. . . .

From A. E. Nordenskiöld, *Periplus* (Stockholm: Norstedt, 1897), pp. 6–8.

From the harbour of Cyrene to the harbour near Barcae to the Hesperides is 620 stadia. But between Cyrene and the Hesperides are harbours and the shores are indented; such is the bay of Phycus; above this is the garden of the Hesperides. The place is 18 fathom deep, broken sheer all round, having a descent nowhere, and it is two stadia everyway, not less, in breadth and length and width. The garden is thickly shaded by trees twined in one another in the thickest manner possible. These trees are: lotus-trees, all kinds of apples, pomegranates, pears, arbutus-fruits, mulberry-trees, vines, myrtles, laurels, ivy, olives, oleasters, almond- and walnut-trees . . .

From Carthage hither to the pillars of Hercules with most favourable navigation is a seven days and seven nights' sail. Gadeira. These islands are near Europe, and of them the one has a city. Opposite them are the pillars of Hercules, the one in Libya being low, but that in Europe being lofty. Now these are capes opposite each other, and they are separated from each other by a day's sail. The sailing along Libya from the Canopic mouth in Egypt to the pillars of Hercules, reckoning being made according to that described in Asia and in Europe, takes 74 days if one coast round the bays. Such towns and ports as have been described in Libya, from the Syrtis that is by the Hesperides as far as the pillars of Hercules in Libya, all belong to the Carthaginians.

Beyond the pillars of Hercules, sailing into the outer sea, having Libya to the left, one comes to a large bay as far as the Hermaean cape; for here too there is a Hermaean cape. At the middle of the bay lies the place and city of Pontion. Round the city lies a great mere, and in this mere lie many islands. Around the mere grow cane, cyprus, reeds, (ρλέως), and rushes. There are also a kind of peacock, which are not to be found elsewhere unless they are imported from here. This mere is named Cephesias, but the bay Cotes. It lies midway between the pillars of Hercules and the cape of Hermaea. From the cape of Hermaea extend great reefs, that is, from Libya towards Europe, not rising above the sea; it washes over them at times. This reef extends towards another cape of Europe right opposite. This cape is named the Holy Promontory. Beyond the cape of Hermaea is a river Anides; this runs out into a large mere. After Anides is another great river—Lixus, and a city of the Phoenicians—Lixus, and another city of the Libyans is on the other side of the river with a harbour. After Lixus the river Crabis, with a harbour and a city of the Phoenicians, Thymiateria by name. From Thymiateria one sails to cape Soloes, which juts far into the sea. But all this district of Libya is very famous and very sacred. And on the point of the cape there is a great altar consecrated to Poseidon, and on the altar are engraved figures of men, lions, dolphins. It is said to be the work of Daedalus. Beyond cape Soloes is a river by name Xion. Around this river dwell the sacred (Hesperian) Ethiopians. Outside there is an island by name Cerne. Coasting from the pillars of Hercules to cape Hermaea is two days; but from cape Hermaea to cape Soloes coasting is three days; and from Soloes to Cerne, seven days' coasting. This whole coasting from the pillars of Hercules to Cerne Island takes twelve days. The parts beyond the isle of Cerne are no longer navigable because of shoals, mud, and sea-weed. This sea-weed has the width of a palm, and is sharp towards the points, so as to prick. The traders here are Phoenicians. When they arrive at the island of Cerne, they anchor their cargo-boats, and pitch tents for themselves on Cerne. Then they unload and ferry their merchandise in small boats to the mainland. They are Ethiopians on the mainland; and it is with these Ethiopians that they trade. They exchange their wares for the skins of deer, and lions, and leopards, and skins and teeth of elephants, and of tame cattle. The Ethiopians use as an ornament speckled skins, and drinking-cups of ivory as goblets, and their women use as an ornament armlets of ivory; even the horses are adorned with ivory. And these Ethiopians are the tallest of all people we know, greater than four cubits; some are even five cubits; and they wear a beard and long hair and are the most handsome of all people. And he rules them who happens to be the tallest. They are also trained horse-men, javelin-throwers, and archers, and use missiles hardened by means of burning. The Phoenician merchants import to them ointment, Egyptian stone, Attic earthware, and vessels for measuring ... These Ethiopians are flesh-eaters and milk-drinkers, and they make much wine from grapes, and this the Phoenicians export. They have also a great city, to which the Phoenician merchants sail, and some say that these Ethiopians stretch right along inhabiting the country thence to Egypt, and that this sea is continuous, and that Libya is a peninsula.

12. A periegesis: Dionysius on the Mediterranean

The following passage, translated by Priscian, is taken from the writings of Dionysius, called "Periegetes." It is an excellent example of the Greek *periegesis* or maritime survey. Dionysius worked during the first century A.D.; his translator, Priscian, during the fifth. (fl. 468–475 A.D.).

The vast abyss of the ocean, however, surrounds the earth on every side; but the ocean, although there is only one, takes many names. In the western

----

From a translation of Priscian's translation of Dionysius, found in Martin Waldseemüller, *Cosmographiae Introductio*, trans. Joseph Fischer, and Franz Von Wieser, ed. Charles George Hebermann (New York: The United States Catholic Historical Society, 1907), pp. 70–75. By permission.

countries it is called the Atlantic Ocean, but in the north, where the Arimaspi are ever warring, it is called the sluggish sea, the Saturnian Sea, and by others the Dead Sea . . . Where, however, the sun rises with its first light, they call it the Eastern or the Indian Sea. But where the inclined pole receives the burning south wind, it is called the Ethiopian or the Red Sea . . . Thus the great ocean, known under various names, encircles the whole world . . .

Of its arms the first that stretches out breaks through Spain with its waves, and extends from the shores of Libya to the coast of Pamphylia. This is smaller than the rest. A larger gulf is the one that enters into the Caspian land, which receives it from the vast waters of the north. The arm of the sea which Tethys (the ocean) rules as the Saturnian Sea is called the Caspian or the Hyrcanian. But of the two gulfs that come from the south sea, one, the Persian, running northward, forms a deep sea, lying opposite the country where the Caspian waves roll; while the other rolls and beats the shores of Panchæa and extends to the south opposite to the Euxine Sea . . .

Let us begin in regular order with the waters of the Atlantic, which Cadiz makes famous by Hercules' gift of the pillar, where Atlas, standing on a mountain, holds up the columns that support the heavens. The first sea is the Iberian, which separates Europe from Libya, washing the shores of both. On either side are the pillars. Both face the shores, the one looking toward Libya, the other toward Europe. Then comes the Gallic Sea, which beats the Celtic shores. After this the sea, called by the name of the Ligurians, where the masters of the world grew up on Latin soil, extends from the north to Leucopetra; where the island of Sicily with its curving shore forms a strait. Cyrnos (modern Corsica) is washed by the waters that bear its name and flow between the Sardinian Sea and the Celtic. Then rolls the surging tide of the Tyrrhenian Sea, turning toward the south; it enters the sea of Sicily, which turns toward the east and spreading far from the shores of Pachynum extends to Crete, a steep rock, which stands out of the sea, where powerful Gortyna and Phæstum are situated in the midst of the fields. This rock, resembling with its peak the forehead of a ram, the Greeks have justly called Κριοῦ μέτωπον (ram's forehead). The sea of Sicily ends at Mt. Garganus on the coast of Apulia.

Beginning there the vast Adriatic extends toward the northwest. There also is the Ionian Sea, famous throughout the world. It separates two shores, which, however, meet in one point. On the right fertile Illyria extends, and next to this the land of the warlike Dalmatians. But its left is bounded by the Ausonian peninsula, whose curving shores the three seas, the Tyrrhenian, the Sicilian, and the vast Adriatic, encircle on all sides. Each of these seas within its limits has a wind peculiar to itself. The west wind lashes the Tyrrhenian, the south wind the Sicilian, while the east wind breaks the waters of the Adriatic which roll beneath its blasts.

Leaving Sicily the sea spreads its deep expanse to the greater Syrtis which the coast of Libya encircles. After the greater Syrtis passes into the lesser, the two seas beat far and wide upon the re-echoing shores. From Sicily the Cretan Sea stretches out toward the east as far as Salmonis, which is said to be the eastern end of Crete.

Next come two vast seas with dark waves, lashed by the north wind coming from Ismarus, which rushes straight down from the regions of the north. The first, called the Pharian Sea, washes the base of a steep mountain. The second is the Sidonian Sea, which turns toward the north, where the gulf of Issus joins it. This sea does not continue far in a straight line; for it is broken by the shores of Cilicia. Then bending westward it winds like a dragon because, forcing its way through the mountains, it devastates the hills and worries the forests. Its end bounds Pamphylia and surrounds the Chelidonian rocks. Far off to the west it ends near the heights of Patara.

Next look again toward the north and behold the Ægean Sea, whose waves exceed those of all other seas, and whose vast waters surround the scattered Cyclades. It ends near Imbros and Tenedos, near the narrow strait through which the waters of the Propontis issue, beyond which Asia with its great peoples extends to the south, where the wide peninsula stretches out. Then comes the Thracian Bosporus, the mouth of the Black Sea. In the whole world they say there is no strait narrower than this. There are found the Symplegades, close together. There to the east the Black Sea spreads out, situated in a northeasterly direction. From either side a promontory stands out in the middle of the waters; one, coming from Asia on the south, is called Carambis; the other on the opposite side juts out from the confines of Europe and is called Κριοῦ μέωπον (ram's forehead.) They face each other, therefore, separated by a sea so wide that a ship can cross it only in three days. Thus you may see the Black Sea looking like a double sea, resembling the curve of a bow, which is bent when the string is drawn tight. The right side resembles the string, for it forms a straight line, outside of which line is found Carambis only, which projects toward the north. But the coast that encloses the sea on the left side, making two turns, describes the arc of the bow. Into this sea toward the north Lake Mæotis (modern Sea of Azov) enters, enclosed on all sides by the land of the Scythians, who call Lake Mæotis the mother of the Black Sea. Indeed, here the violent sea bursts forth in a great stream, rushing across the Cimmerian Bosporus (modern Crimea), in those cold regions where the Cimmerians dwell at the foot of Taurus. Such is the picture of the ocean; such the glittering appearance of the deep.

From the Geographical
Writings of Plato and Aristotle

Early Greek science comes to full bloom in the works of Plato and Aristotle. It was Aristotle, in particular, who was to exert a lasting influence on Western thought; hence the extent to which his works are quoted here.

13. Socrates explains the nature of the earth

Plato (427?–347 B.C.) reported in his *Phaedo* a capsule version of Socrates' ideas about our earth, including that oft-quoted phrase about the Greeks living around the Mediterranean as frogs live around a pool.

"Well then," said Socrates, "I at least am convinced first that if the earth is in the middle of the heavens and is round, it has no need of air or any other constraint to prevent it from falling; but the indifference of the heavens with themselves in all directions and the equipoise of the earth itself are enough to hold it fast. For a thing in equipoise placed in the middle of something uniform will not be capable of deflection either more or less in any direction.... Again, I am convinced that it is something altogether big, and that we who stretch from the Phasis (*Rion*) to the Pillars of Heracles (*Straits of Gibraltar*) dwell in a small part of it, living round the sea like ants or frogs round a pool; and that many other men dwell in many

---

Plato's *Phaedo*, from E. H. Warmington, *Greek Geography*, The Library of Greek Thought Series (New York, 1934), p. 22. Published in the United States by E. P. Dutton and reprinted by their permission.

other parts in many regions such as ours. For everywhere round the earth there are many hollows of all kinds of shapes and sizes into which water and mist and air have flowed together. The earth is clear and is placed in the heavens which are clear, in which are the stars, and which is called 'aether' by most of those who are accustomed to speak about such matters. The substances (*water, mist, air*) mentioned above are sediment of it, and constantly flow together into the hollows of the earth. We then dwelling in its hollows are unaware that we do so and believe that we are dwelling above on the earth."

14. Plato on the fate of Atlantis

In the *Timaeus* Plato gives us his own version of the disappearance of the fabled island.

———

Many great and wonderful deeds are recorded of your State in our histories. But one of them exceeds all the rest in greatness and valor. For these histories tell of a mighty power which was aggressing wantonly against the whole of Europe and Asia, and to which your city put an end. This power came forth out of the Atlantic Ocean, for in those days the Atlantic was navigable; and there was an island situated in front of the straits which you call the columns of Heracles; the island was larger than Libya and Asia put together, and was the way to other islands, and from the islands you might pass through the whole of the opposite continent which surrounded the true ocean; for this sea which is within the Straits of Heracles is only a harbor, having a narrow entrance, but that other is a real sea, and the surrounding land may be most truly called a continent. Now in this island of Atlantis there was a great and wonderful empire which had rule over the whole island and several others, as well as over parts of the continent, and, besides these, they subjected the parts of Libya within the columns of Heracles as far as Egypt, and of Europe as far as Tyrrhenia. The vast power thus gathered into one, endeavored to subdue at one blow our country and yours and the whole of the land which was within the straits; and then, Solon, your country shone forth, in the excellence of her virtue and strength, among all mankind; for she was the first in courage and military skill, and was the leader of the Hellenes. And when the rest fell off from her, being compelled to stand alone, after having undergone the very extremity of danger, she defeated and triumphed over the invaders, and preserved from slavery those who were not yet subjected, and freely liberated all the others who dwell within the limits of Heracles. But afterwards there occurred violent earth-

———

From *The Dialogues of Plato*, trans. B. Jowett (New York, 1874), II, 520–521.

quakes and floods; and in a single day and night of rain all your warlike men in a body sank into the earth, and the island of Atlantis in like manner disappeared, and was sunk beneath the sea. And that is the reason why the sea in those parts is impassable and impenetrable, because there is such a quantity of shallow mud in the way; and this was caused by the subsidence of the island.

15. Aristotle on the cosmos and the oikumene

The following passages from Aristotle (384–322 B.C.) were taken from *On the Heavens* and *On the Cosmos*. They illustrate Aristotle's views on the universe, as well as those regarding the *oikumene*—the habitable world known to the Greeks—and some of the phenomena observed in it.

On the heavens

Now since there are some who say that there is a right and a left side to the heaven, e.g. the school known as the Pythagoreans (for in fact this contention is theirs), we must consider whether, if we must attribute these principles to the body of the whole, it works as they suppose or otherwise. At the very outset, if it has a right and a left, it must be supposed a fortiori to contain the principles which are prior to those two. This is something which has been worked out in the treatises on the movements of animals, being properly a part of biology, for in living creatures it is plain and obvious that some have all these features—right and left and so forth—and others some, whereas plants have only above and below. If then we are to attribute anything of this sort to the heaven, it is natural, as I was saying, that that which is present in the lowest stage of animal life is present also in it; for each of the three pairs in question is of the nature of a principle. The three I mean are above and below, the front and its opposite, and right and left. These three dimensional differences may reasonably be supposed to be possessed by all complete bodies. Above is the principle of length, right of breadth, and front of depth; or alternatively their nature as principles may be defined with reference to motions, for I mean by principles the starting-points of the motions of the bodies possessing them. Growth is from above, locomotion from the right, the motion which follows sensation from in front

The first selection is from Aristotle, *On the Heavens*, trans. W. K. C. Guthrie, Loeb Classical Library (Cambridge, Mass.: Harvard University Press, 1939), pp. 139–147. The second is from *On the Cosmos*, trans. D. J. Furley, Loeb Classical Library (Cambridge, Mass.: Harvard University Press, 1955), pp. 347–349, 353–357, 357–363, 363–367.

(since the meaning of "front" is that towards which the sensations are directed).

It follows that above and below, right and left, front and back, are not to be looked for in all bodies alike, but only in those which, because living, contain within themselves a principle of motion; for in no part of an inanimate object can we trace the principle of its motion. Some do not move at all, whereas others, though they move, do not move in every direction alike. Fire, for instance, moves upwards only, earth to the centre. It is in relation to ourselves that we speak of above and below, or right and left, in these objects. We name them either as they correspond with our own right hands (as in augury), or by analogy with our own (as with the right-hand side of a statue), or from the parts which lie on opposite sides to our own, i.e. calling right in the object that part which is on our left-hand side, and vice versa (and back that which in relation to ourselves is the front). But in the objects themselves we detect no difference: if they are turned round, we call the opposite points right and left, above and below, or front and back.

In face of this, one would wonder why it is that the Pythagoreans posited only two of these principles, right and left, to the neglect of the other four, which are no less important; for above and below, front and back, are as distinctive from each other in all animals as right is from left. Sometimes the difference is in function only, sometimes in shape as well. Above and below belong to all living things, plants as well as animals, but right and left are absent from plants. Moreover, length is prior to breadth, so that if above is the principle of length, and right the principle of breadth, and the principle of that which is prior is itself prior, then above must be prior to right (prior, that is, in order of generation; the word "prior" has several senses). Finally, if above is the starting-point of the motion of growth, right the place where locomotion originates, and front the goal of appetitive motion, this too gives above the standing of a first principle in relation to the other forms.

One reason, then, why we may justly censure the Pythagoreans is that they omitted the more important principles, and another is that they supposed the principles which they did recognize to be inherent in all things alike. As for our own position, we have already decided that these functions are found in whatever contains a principle of motion, and that the heaven is alive and contains a principle of motion, so it is clear that the heaven possesses both upper and lower parts, and right and left. We need not be deterred by the fact that the world is spherical, nor wonder how one side of it can be the right and another the left if all its parts are similar and in eternal motion. We must suppose it to resemble a being of the class whose right differs from their left in shape as well as in function, but which has been enclosed in a sphere: the right will still have a distinctive function, though this will not be apparent owing to the uniformity of shape. The same reasoning applies to the origin of its motion. Even if the motion never had a beginning, nevertheless it must have a principle, from which it would have begun had it had a beginning, and would be started again should it stop.

By the length of the heaven I mean the distance between its poles, and I hold that one pole is the upper and one the lower; for only two of the possible hemispheres have a distinguishing mark, and that is the immobility of the poles. In ordinary speech too, when we speak of "across" the world we mean, not the line of the poles, but what lies around it, implying that the line of the poles represents above and below; for "across" means that which lies around the line of above and below.

Of the poles, the one which we see above us is the lowest part, and the one which is invisible to us the uppermost. For we give the name of right-hand to that side of a thing whence its motion through space starts. Now the beginning of the heaven's revolution is the side from which the stars rise, so that must be its right, and where they set must be its left. If this is true, that it begins from the right and moves round to the right again, its upper pole must be the invisible one, since if it were the visible, the motion would be leftward, which we deny. Clearly therefore the invisible pole is the upper, and those who live in the region of it are in the upper hemisphere and to the right, whereas we are in the lower and to the left. It is the contrary of the Pythagorean view, for they put us above and on the right, and the others below and on the left. The truth is just the reverse. Nevertheless in relation to the secondary revolution, i.e. that of the planets, we are in the upper and right-hand part, and they in the lower and left; for the place from which these bodies start is on the opposite side—since they move in the opposite direction—so that we are at the beginning and they at the end. So much for the dimensional parts of the world and those which are defined by their position.

On the cosmos, the elements, and the inhabited world

Cosmos, then, means a system composed of heaven and earth and the elements contained in them. In another sense, cosmos is used to signify the orderly arrangement of the universe, which is preserved by God and through God. The centre of the cosmos, which is unmoved and fixed, is occupied by "life-bearing earth," the home and mother of living beings of all kinds. The region above it, a single whole with a finite upper limit everywhere, the dwelling of the gods, is called heaven. It is full of divine bodies which we call stars; it moves eternally, and revolves in solemn choral dance with all the stars in the same circular orbit unceasingly for all time. The whole of the heaven, the whole cosmos, is spherical, and moves continuously, as I have said; but there are necessarily two points which are unmoved, opposite one another, just as in the case of a ball being turned in a lathe; they remain fixed, holding the sphere in a position, and the whole mass revolves in a circle round them; these are called poles. If we think of a straight line joining these two together (some call this the axis), it will be a diameter of the cosmos, having the earth at its centre and the two poles at its extremities. One of these two stationary poles is always visible, above our heads in the North: it is called the Arctic pole. The other is always hidden under the earth, in the South: it is called the Antarctic pole . . .

After the aetherial and divine element, which is arranged in a fixed order, as we have declared, and is also unchangeable, unalterable and impassive, there comes next the element that is through the whole of its extent liable to change and alteration, and is, in short, destructible and perishable. The first part of this is the fine and fiery substance that is set aflame by the aether because of the latter's great size and the swiftness of its motion. In this fiery and disorderly element, as it is called, meteors and flames shoot across, and often planks and pits and comets, as they are called, stand motionless and then expire.

Next under this is spread the air, opaque and icy by nature, but when it is brightened and heated by movement, it becomes bright and warm. In the air, which itself also has the power to change, and alters in every kind of way, clouds are formed and rain falls in torrents; there is snow, frost and hail, and gales and whirlwinds; thunder and lightning, too, and falling thunderbolts, and the clash of innumerable storm-clouds.

Next to the element of air comes the fixed mass of earth and sea, full of plants and animals, and streams and rivers, some winding about the surface of the earth, others discharging themselves into the sea. This region is adorned with innumerable green plants, high mountains, deep-shaded woodland, and cities established by the wise creature, man; and with islands in the sea, and continents. The inhabited world is divided by the usual account into islands and continents, since it is not recognized that the whole of it is really one island, surrounded by the sea which is called Atlantic. Far away from this one, on the opposite side of the intervening seas, there are probably many other inhabited worlds, some greater than this, some smaller, though none is visible to us except this one; for the islands we know stand in the same relation to our seas as the whole inhabited world to the Atlantic Ocean, and many other inhabited worlds to the whole ocean; for these are great islands washed round by great seas. The whole mass of the wet element lies on the surface of the earth, allowing the so-called inhabited worlds to show through where there are projections of the earth; it is this element that would properly be next in order to the air. After this, set in the depths at the centre of the cosmos, densely packed and compressed, is the whole mass of the earth, unmoved and unshaken. And this is the whole of that part of the cosmos that we call the lower part. So these five elements, occupying five spherical regions, the larger sphere always embracing the smaller—earth in water, water in air, air in fire, fire in aether—make up the whole cosmos; the upper part as a whole is distinguished as the abode of the gods, and the lower part as that of mortal creatures. Of the latter, some is wet, and this part we call rivers and springs and seas; the rest is dry, and this part we name land and continents and islands . . .

The ocean that is outside the inhabited world is called the Atlantic, or Ocean, and surrounds us. To the West of the inhabited world, this ocean makes a passage through a narrow strait called the Pillars of Heracles, and so makes an entry into the interior sea, as if into a harbour; gradually it broadens and spreads out, embracing large bays joined up to each other, here contracting into narrow necks of water, there broadening out again. They say that the first of these bays that the sea forms, to starboard, if you sail in through the Pillars of Heracles, are two, called the Syrtes, of which one is called the Major, the other the Minor; on the other side it does not form gulfs at first in the same way, but makes three seas, the Sardinian, Galatian and Adriatic; next to these, and across the line of them, is the Sicilian sea; after this, the Cretan; and continuing this on one side are the Egyptian and Pamphylian and Syrian seas, on the other the Aegean and Myrtian. Lying opposite these that I have described, in another direction, is the Pontus, and this has very many parts: the innermost part is called Maeotis, and the outermost part, towards the Hellespont, is joined by a strait to the sea called Propontis.

In the East, the Ocean again penetrates (the inhabited world); it opens out the gulf of India and Persia and without a break reveals the Red Sea, embracing these as parts of itself. Towards the other promontory (of Asia), passing through a long narrow strait and then broadening out again, it makes the Hyrcanian or Caspian sea; beyond this, it occupies a deep hollow beyond Lake Maeotis. Then little by little, beyond the land of the Scythians and Celts, it confines the inhabited world as it passes towards the Galatian Gulf and the Pillars of Heracles, already described, on the farther side of which the Ocean flows round the earth. There are two very large islands in it, called the British Isles, Albion and Ierne; they are larger than those already mentioned, and lie beyond the land of the Celts. No smaller than these are Taprobane (Ceylon) beyond the Indians, which lies obliquely to the inhabited world, and the island known as Phebol, by the Arabian Gulf. There is quite a number of other small islands round the British Isles and Spain, set in a ring round this inhabited world, which as we have said is itself an island; its breadth, at the deepest point of the continent, is a little short of 40,000 stades, in the opinion of good geographers, and its length is approximately 70,000 stades. It is divided into Europe, Asia and Libya.

Europe is the area which is bounded in a circle by the Pillars of Heracles and the inner parts of the Pontus and the Hyrcanian Sea, where a very narrow isthmus passes between it and the Pontus; but some have said the river Tanais, instead of this isthmus. Asia is the region from this isthmus, which lies between the Arabian Gulf and the Mediterranean; it is surrounded by the Mediterranean and the encircling stream of the Ocean; but some say that Asia stretches from the Tanais to the mouths of the Nile. Libya lies between the Arabian isthmus and the Pillars of Heracles (but some say from the Nile to the Pillars). Egypt, which is encompassed by the mouths of the Nile, is attached by some to Asia, and by others to Libya, and some make the islands separate, others attribute them to their nearest region of mainland . . .

Now let us turn to the most notable phenomena in and about the inhabited world, summarizing only the most essential points.

There are two exhalations from it, which pass continually into the air above us, composed of small particles and entirely invisible, except that in the early mornings some can be observed rising along rivers and streams. One of these is dry and like smoke, since it emanates from the earth; the other is damp and vaporous, since it is exhaled from the wet element. From the latter come mists, dews, the various kinds of frost, clouds, rain, snow and hail; from the dry exhalation come the winds and various breezes, thunder and lightning, fiery bolts (ηρηστῆρες) and thunderbolts and all the other things of the same class. Mist is a vaporous exhalation which does not produce water, denser than air but less dense than cloud; it comes into being either from a cloud in the first stage of formation or from the remnant of a cloud. The condition contrary to this is rightly called a clear sky, for it is simply air, with no cloud or mist. Dew is moisture that falls out of a clear sky in a light condensation; ice is solidified water, frozen in a clear sky: hoar-frost is frozen dew, and dew-frost is half-frozen dew. Cloud is a dense, vaporous formation, productive of water: rain comes from the compression of a well-compacted cloud, and varies in character according to the pressure on the cloud: if the pressure is light it scatters gentle drops of rain, but if it is heavy the drops are fuller; and we call this latter condition a downpour, for it is larger than a shower of rain and pours continuous drops of rain upon the earth. Snow occurs when well-condensed clouds break up and split before the formation of water: the split causes the foamy and brilliantly white condition of the snow, and its coldness is caused by the coagulation of the moisture contained in it, which has not had time to be either fused or rarefied. If there is a thick and heavy fall of snow, we call it a snowstorm. Hail occurs when a snowstorm is solidified and gathers weight because of its increased density so as to fall more rapidly; the hailstones increase in size and their movement increases in violence according to the size of the fragments that are broken off the cloud. These then are the natural products of the wet exhalation.

From the dry exhalation, when it is forced to flow by the cold, wind is produced: for this is nothing but air moving in quantity and in a mass. It is also called breath. In another sense "breath" means that substance found in plants and animals and pervading everything, that brings life and generation; but about that there is no need to speak now. The breath that breathes in the air we call wind, and the breath that comes from moisture we call breeze. Of the winds, some blow from the earth when it is wet and are called land-winds; some arise from gulfs of the sea and are called gulf-winds. There is a similarity between these winds and those which come from rivers and lakes. Those which arise at the breaking up of a cloud and resolve its density against themselves are called cloud-winds: those which burst out all at once accompanied by water are called rain-winds . . .

The earth contains in itself many sources, not only of water, but also of wind and fire. Some of these are subterranean and invisible, but many have vents and blow-holes, like Lipara and Etna and the volcanoes in the Aeolian islands. These often flow like rivers and throw up fiery, red-hot lumps. Some of the subterranean sources, which are near springs of water, impart heat to these: some of the streams they make merely lukewarm, some boiling, and some moderately and pleasantly hot.

16. Aristotle considers the city-state

In his *Politics,* Aristotle takes a close look at the environment, site, situation, and resources of the Greek polis: the city-state.

———

With reference to the position of the city itself we must pray that it turn out to be on a slope, having regard to four things. First we must have regard to health, for cities which have a slope towards the east and the breezes which blow from the sunrise are healthier than others, the second place being held by those which slope away from the north, for these have milder winters. Of the remaining considerations, a city so placed is well situated for administration and military activities. For military activities, then, it is necessary that it should have an easy exit for the people themselves, and be difficult for adversaries to approach or besiege, and as far as possible have its own plentiful supply of waters and springs, and if it does not, such a supply has been invented by building a large number of big receptacles for rain-water . . . Again, since we must take thought of the health of the inhabitants, and this depends on the locality being situated in a healthy district, and well placed as regards health, and secondly on using healthy waters, we must bestow attention on the following also, and not as on a matter of secondary importance. For it is the things which we

———

Aristotle's *Politics,* from E. H. Warmington, *Greek Geography,* The Library of Greek Thought Series (New York, 1934), p. 65. Published in the United States by E. P. Dutton and reprinted by their permission.

use most of all and most often for our bodies that make the greatest contribution to our own health; and the influence of waters and of air has
such a nature. Wherefore in cities of sound wisdom, if all the springs
are not alike in purity and there is not a large number of them, water-
supplies used for drinking and those put to other uses must be kept
separate. With regard to fortified sites not all constitutions alike find
the same conditions advantageous for instance, a citadel-hill suits an
oligarchy and a monarchy, while level ground suits a democracy; but
an aristocracy likes neither, bur rather several strong places.

17. Aristotle discusses water and dry land, world views and maps, earthquakes and their causes

In *Meteorologica* Aristotle first presents some general observations
on terrestrial phenomena and then discusses in detail the causes of
earthquakes. His preoccupation with these seismic phenomena is
attributable to their relatively frequent occurrence in that zone of
instability, the Mediterranean.

The same parts of the earth are not always moist or dry, but change
their character according to the appearance or failure of rivers. So also
mainland and sea change places and one area does not remain earth,
another sea, for all time, but sea replaces what was once dry land, and
where there is now sea there is at another time land. This process must,
however, be supposed to take place in an orderly cycle. Its originating cause
is that the interior parts of the earth, like the bodies of plants and animals,
have their maturity and age. Only whereas the parts of plants and animals
are not affected separately but the whole creature must grow to maturity
and decay at the same time, the parts of the earth are affected separately,
the cause of the process being cold and heat. Cold and heat increase and
decrease owing to the sun's course, and because of them the different
parts of the earth acquire different potentialities; some are able to remain
moist up to a certain point and then dry up and become old again, while
others come to life and become moist in their turn. As places become drier
the springs necessarily disappear, and when this happens the rivers at
first dwindle from their former size and finally dry up; and when the
rivers are removed and disappear in one place, but come into existence
correspondingly in another, the sea too must change. For wherever it has

------

From Aristotle, *Meteorologica*, trans. H. D. P. Lee, Loeb Classical Library (Cambridge, Mass.: Harvard University Press, 1952), pp. 107–109, 111, 179–183,
205–223.

encroached on the land because the rivers have pushed it out, it must when it recedes leave behind it dry land: while wherever it has been filled and silted up by rivers and formed dry land, this must again be flooded.

But these changes escape our observation because the whole natural process of the earth's growth takes place by slow degrees and over periods of time which are vast compared to the length of our life, and whole peoples are destroyed and perish before they can record the process from beginning to end . . .

This has happened in Egypt. This is a land which is obviously in the process of getting drier, and the whole country is clearly a deposit of the Nile: but because the adjacent peoples have only encroached on the marshes gradually as they dried up, the beginning of the process has been lost in the lapse of time. We can see, however, that all the mouths of the Nile, except the one at Canopus, are artificial and not formed by the action of the river itself; and the old name of Egypt was Thebes. Homer's evidence proves this last point, though in relation to such changes he is comparatively modern; for he mentions the country as though Memphis either did not exist as yet at all or at any rate were not a place of its present importance. And it is quite likely that this was in fact so. For the higher lands were inhabited before the lower-lying, because the nearer a place is to the point where silt is being deposited the longer it must remain marshy, as the land last formed is always more water-logged . . .

For there are two habitable sectiors of the earth's surface, one, in which we live, towards the upper pole, the other towards the other, that is the south pole. These sectors are drum-shaped—for lines running from the centre of the earth cut out this shaped figure on its surface: they form two cones, one having the tropic as its base, the other the ever-visible circle, while their vertex is the centre of the earth; and two cones constructed in the same way towards the lower pole cut out corresponding segments on the earth's surface.

These are the only habitable regions; for the lands beyond the tropics are uninhabitable, as there the shadow would not fall towards the north, and we know that the earth ceases to be habitable before the shadow disappears or falls towards the south, while the lands beneath the Bear are uninhabitable because of the cold . . .

The way in which present maps of the world are drawn is therefore absurd. For they represent the inhabited earth as circular, which is impossible both on factual and theoretical grounds. For theoretical calculation shows that it is limited in breadth but could, as far as climate is concerned, extend round the earth in a continuous belt: for it is not difference of longitude but of latitude that brings great variations of temperature, and if it were not for the ocean which prevents it, the complete circuit could be made. And the facts known to us from journeys by sea and land also confirm the conclusion that its length is much greater than its breadth.

For if one reckons up these voyages and journeys, so far as they are capable of yielding any accurate information, the distance from the Pillars of Heracles to India exceeds that from Aethiopia to Lake Maeotis and the farthest parts of Scythia by a ratio greater than that of 5 to 3. Yet we know the whole breadth of the habitable world up to the unhabitable regions which bound it, where habitation ceases on the one side because of the cold, on the other because of the heat; while beyond India and the Pillars of Heracles it is the ocean which severs the habitable land and prevents it forming a continuous belt round the globe.

On the causes of earthquakes

Now it is clear, as we have already said, that there must be exhalation both from moist and dry, and earthquakes are a necessary result of the existence of these exhalations. For the earth is in itself dry but contains much moisture because of the rain that falls on it; with the result that when it is heated by the sun and its own internal fire, a considerable amount of wind is generated both outside it and inside, and this sometimes all flows out, sometimes all flows in, while sometimes it is split up.

This process is inevitable. Our next step should therefore be to consider what substance has the greatest motive power. This must necessarily be the substance whose natural motion is most prolonged and whose action is most violent. The substance most violent in action must be that which has the greatest velocity, as its velocity makes its impact most forcible. The farthest mover must be the most penetrating, that is, the finest. If, therefore, the natural constitution of wind is of this kind, it must be the substance whose motive power is the greatest. For even fire when conjoined with wind is blown to flame and moves quickly. So the cause of earth tremors is neither water nor earth but wind, which causes them when the external exhalation flows inwards.

This is why the majority of earthquakes and the greatest occur in calm weather. For the exhalation being continuous in general follows its initial impulse and tends either all to flow inwards at once or all outwards. There is, however, nothing inexplicable in the fact that some earthquakes occur when a wind is blowing; for we sometimes see several winds blowing at the same time, and when one of these plunges into the earth the resultant earthquake is accompanied by wind. But these earthquakes are less violent, because the energy of their original cause is divided. Most major earthquakes occur at night, and those that occur in daytime at midday, this being as a rule the calmest time of day, because when the sun is at its strongest it confines the exhalation within the earth, and it is at its strongest about midday; and the night again is calmer than the day because of the sun's absence. So at these times the flow turns inwards again, like an ebb as opposed to the outward flood. This happens especially towards dawn, for it is then that winds normally begin to blow. If, then, the original impulse of the exhalation changes direction, like the Euripus, and turns inwards, it causes a more violent earthquake because of its quantity.

Again, the severest earthquakes occur in places where the sea is full of currents or the earth is porous and hollow. So they occur in the Hellespont and Achaea and Sicily, and in the districts in Euboea where the sea is supposed to run in channels beneath the earth. The hot springs at Aedepsus are due to a similar cause. In the places mentioned earthquakes occur mostly because of the constricted space. For when a violent wind arises the volume of the inflowing sea drives it back into the earth, when it would naturally be exhaled from it. And places whose subsoil is porous are shaken more because of the large amount of wind they absorb.

For the same reason earthquakes occur most often in spring and autumn and during rains and droughts, since these periods produce most wind. For summer and winter both bring calm weather, the one because of its frosts, the other because of its warmth, the one thus being too cold, the other being too dry to produce winds. But in times of drought the air is full of wind, drought simply being an excess of dry over moist exhalation. In times of rain the exhalation is produced within the earth in greater quantity, and when what has been so produced is caught in a constricted space and forcibly compressed as the hollows within the earth fill with water, the impact of the stream of the wind on the earth causes a severe shock, once the compression of a large quantity of it into a small space begins to have its effect. For we must suppose that the wind in the earth has effects similar to those of the wind in our bodies whose force when it is pent up inside us can cause tremors and throbbings, some earthquakes being like a tremor, some like a throbbing. We must suppose, again, that the earth is affected as we often are after making water, when a sort of tremor runs through the body as a body of wind turns inwards again from without. For the force that wind has can be seen not only by studying its effects in the air, when one would expect it to be able to produce them because of its volume, but also in the bodies of living things. Tetanus and spasms are movements caused by wind, and are so strong that the combined strength and efforts of a number of men is unable to master the movements of their victims. And if we may compare great things with small, we must suppose that the same sort of thing happens to the earth.

As evidence we may cite occurrences which have been observed in many places. For in some places there has been an earthquake which has not ceased until the wind which was its motive force has broken out like a hurricane and risen into the upper region. This happened recently, for instance, in Heracleia in Pontus, and before that in Hiera, one of the so-called Aeolian islands. For in this island part of the earth swelled up and rose with a noise in a crest-shaped lump; this finally exploded and a large quantity of wind broke out, blowing up cinders and ash which smothered the neighbouring city of Lipara, and even reached as far as some of the cities in Italy. The place where this eruption took place can still be seen. (This too must be regarded as the cause of the fire that there is in the earth; for when the air is broken up into small particles, percussion then causes it to catch fire.)

And there is a proof that winds circulate beneath the earth in something else that happens in these islands. For when a south wind is going to blow it is heralded by noises from the places from which eruptions occur. This is because the sea, which is being driven forward from far off, thrusts the wind that is erupting out of the earth back again when it meets it. This causes a noise but no earthquake because there is plenty of room for the wind, of which there is only a small quantity and which can overflow into the void outside.

Further evidence that our account of the cause of earthquakes is correct is afforded by the facts that before them the sun becomes misty and dimmer though there is no cloud, and that before earthquakes that occur at dawn there is often a calm and a hard frost. The sun is necessarily misty and dim when the wind which dissolves and breaks up the air begins to retreat into the earth. Calm and cold towards sunrise and dawn are also necessary concomitants. Calm must usually fall, as we have explained, because the wind drains back as it were into the earth, and the greater the earthquake the more this happens; for the earthquake is bound to be more severe if the wind is not dispersed, some outside and some in, but moves in a mass. The reason for the cold is that the exhalation, which is by nature essentially warm, is directed inwards. (Winds are not usually supposed to be warm because they set the air in motion and the air contains large quantities of cold vapour. This can be seen when wind is blown out of the mouth: close by it is warm, as when we breathe with open mouth, though there is too little of it to be very noticeable, while farther off it is cool for the same reason as the winds.) So the warm element disappears into the earth, and wherever this happens, the vaporous exhalation being moist condenses and causes cold. The cause of a sign which often heralds earthquakes is the same. In clear weather, either by day or a little after sunset, a fine long streak of cloud appears, like a long straight line carefully drawn, the reason being that the wind is dying down and running away. Something like it happens on the seashore too. For when the sea runs high the breakers are large and uneven, but when there is a calm they are fine and straight [because the amount of exhalation is small]. The wind produces the same effects on the cloud in the sky as the sea on the shore, so that when there is a calm the clouds that are left are all straight and fine like breakers in the air.

For the same reason an earthquake sometimes occurs at an eclipse of the moon. For when the interposition is approaching but the light and warmth from the sun, though already fading, have not entirely disappeared from the air, a calm falls when the wind runs back into the earth. And this causes the earthquake before the eclipse. For there are often winds also before eclipses, at nightfall before a midnight eclipse, at midnight before an eclipse at dawn. The reason for this is the failure of the heat from the moon when its course approaches the point at which the eclipse will take place. Thus when the cause which held it quiet ceases to operate the air is set in motion again and a wind rises, and the later the eclipse, the later this happens.

When an earthquake is severe the shocks do not cease immediately or at once, but frequently go on for forty days or so in the first instance, and symptoms appear subsequently for one or two years in the same district. The cause of the severity is the amount of the wind and the shape of the passages through which it has to flow. When it meets with resistance and cannot easily get through, the shocks are severest and air is bound to be left in the narrow places, like water that cannot get out of a vessel. Therefore, just as throbbings in the body do not stop at once or quickly, but gradually as the affliction which is their cause dies away, so the originating cause of the exhalation and the source of the wind clearly do not expend all at once the material which produces the wind which we call an earthquake. Until, therefore, the rest of it is expended shocks must continue, their force decreasing until there is too little exhalation to cause a shock that is noticeable.

Wind is also the cause of noises beneath the earth, among them the noises that precede earthquakes, though they have also been known to occur without an earthquake following. For as the air when struck gives out all sorts of noises, so also it does when it is itself the striker; the effect is the same in either case, since every striker is itself also struck. The sound precedes the shock because the sound is of finer texture and so more penetrating than the wind itself. When the wind is too fine to communicate any impulse to the earth, being unable to do so because of the ease with which it filters through it, nevertheless when it strikes hard or hollow masses of all shapes it gives out all sorts of noises, so that sometimes the earth seems to bellow as they say it does in fairy stories.

Water has sometimes burst out of the earth when there has been an earthquake. But this does not mean that the water was the cause of the shock. It is the wind which is the cause, whether it exerts its force on the surface or from beneath—just as the winds are the cause of waves and not the waves of winds. Indeed one might as well suppose that the earth is the cause of the shock as that the water is; for in an earthquake it is overturned like water, and upsetting water is a form of overturning. But in fact both earth and water are material causes, being passive not active, but wind the motive cause.

When a tidal wave coincides with an earthquake the cause is an opposition of winds. This happens when the wind which is causing the earthquake

is unable quite to drive out the sea which is being driven in by another wind, but pushes it back and piles it together till a large mass has
collected. Then if the first wind gives way the whole mass is driven in
by the opposing wind and breaks on the land and causes a flood. This is
what happened in Achaea. For in Achaea there was a south wind, outside
a north wind; this was followed by a calm when the wind plunged into
the earth, and so there was a tidal wave at the same time as the earthquake
—an earthquake which was all the more violent because the sea gave no
vent to the wind that had run into the earth, but blocked its passage. So
in their mutual struggle the wind caused the earthquake, the wave by its
subsidence the flood.

Earthquakes are confined to one locality, often quite a small one, but
winds are not. They are localized when the exhalations of a particular
locality and its neighbour combine, which was what we said happens in
local droughts and rainy seasons. Earthquakes are produced in this way,
but not winds. For rain, droughts and earthquakes originate in the earth,
and so their constituent exhalations tend to move all in one direction; the
sun has less power over them than it has with the exhalations in the air
which therefore flow on in one direction when the sun's movement gives
them an impulse, differing according to the difference of its position.

So then, when the quantity of wind is large it causes an earthquake shock
which runs horizontally, like a shudder: occasionally in some places the
shock runs up from below, like a throb. The latter type of shock is therefore the rarer, for sufficient force to cause it does not easily collect since
there is many times as much of the exhalation that causes shocks horizontally as of that which causes them from below. But whenever this type
of earthquake does occur, large quantities of stones come to the surface,
like the chaff in a winnowing sieve. This kind of earthquake it was that
devastated the country round Sipylos, the so-called Phlegraean plain and
the districts of Liguria.

Earthquakes are rarer in islands that are far out at sea than in those
close to the mainland. For the quantity of the sea cools the exhalations and
its weight crushes them and prevents their forming; and the force of the
winds causes waves and not shocks in the sea. Again, its extent is so
great that the exhalations do not run into it but are produced from it and
joined by those from the land. On the other hand, islands close to the
mainland are for all practical purposes part of it, the interval between
them being too small to be effective. And islands out at sea can feel no
shock that is not felt by the whole of the sea by which they are surrounded.

This completes our explanation of the nature and cause of earthquakes,
and of their most important attendant circumstances.

Hippocrates of Cos:
An Early Environmentalist

Hippocrates of Cos (460–? 360 B.C.) is best known for his contributions to medicine. He was one of the first writers to concern himself with the relationship between the environment and the health and general character of men. In *On Airs, Waters, and Places* he considers environmental influences in general, emphasizing the importance of these factors in city life; he also describes specific environments in Asia.

18. Hippocrates on the effects of the environment

He who arrives in a strange city must observe its site, how it lies with respect to the winds and the sunrise, for the effects are not the same when it is oriented to the north or to the south, toward the rising or the setting sun. All of this information must be acquired, as well as notions on the water supply, as to whether the people use marshy and soft water, or hard water from high and rocky places, or brackish and contaminated water. The soil, too, must be considered; whether it is bare and lacking water, or wooded and well watered; whether it is low lying and hot, or cold and frigid. The eating habits of the people, too, must be considered, what do they partake of most: are they drunkards and gluttons given to indolence or are they hard working, eating more than drinking . . .

The first selection is from Hippocrates, *On Airs, Waters, and Places*, Littré's edition and French translation of the Greek text (London, 1881); editor's translation. The selection on Asia is from the same work by Hippocrates and is taken from E. H. Warmington, *Greek Geography*, The Library of Greek Thought Series (New York, 1934), pp. 54–59. Published in the United States by E. P. Dutton and reprinted by their permission.

Let us assume that a city is so situated as to be exposed to hot winds, which I will place [originating from directions] between winter sunrise and winter sunset; that these winds are typical of the locality, but that the city is protected from the northerly winds. In such a place waters will be plentiful yet somewhat brackish, coming from shallow sources, and therefore warm in the summer, cold in winter . . .

But cities that have the opposite exposure, being sheltered from the warm, southerly winds and exposed to the cold winds, those originating from directions between the summer sunrise and summer sunset, display the following characteristics . . . the waters are cold and hard, and difficult to soften. Men in these places must be robust and slender . . .

As for cities facing in the direction between summer sunrise and winter sunrise (northeast to southeast), and those facing the opposite way (northwest to southwest), this is their nature. Those facing the rising sun are naturally more healthy than those facing north and those facing south, even though the distance between them be only a single stadium. First of all, in these places heat and cold are moderate; further, those waters having their source toward sunrise are necessarily clear, fragrant, soft and pleasant, since the sun upon rising and shining upon them disperses the morning vapors. The people in these places have a better and more florid complexion, unless counteracted by a disease. Their voices are clear, and in temper and intellect they are clearly superior to those living in northern lands; in fact all that grows there is of a superior quality. A city so situated enjoys a spring-like climate because of the moderate heat and cold that prevails there. Diseases are fewer and less virulent, and resemble those found in cities exposed to hot winds. Women are very fertile and deliver their children easily . . .

On Scythia

What is called the Scythian desert is a plain, with numerous meadows, rather high-lying, and reasonably well-watered, for there are large rivers into which the waters of the plain are drained.

There live in that country the Scythians, called nomads, because they do not have houses but live in wagons. The smallest of these have four wheels, others have six. These wagons are covered with felt and built in the manner of houses: some have but a single room, others as many as three. They are so strongly built that they resist rain, snow, and winds. The wagons are drawn by oxen, some by a pair, others by three, and these oxen do not have horns due to the cold climate of the land that prohibits their growth.

The women live in the wagons, while the men ride horses, driving their sheep, cattle and horses. They stay in one place as long as there is enough grazing for their livestock, when that fails, they move on to another district.

They eat boiled meat and drink mares' milk, and consume a dish called *hippace,* a cheese made from mares' milk.

On Asia

Asia differs widely from Europe with regard to natural conditions of things in general—of the wild products of the earth, and of mankind; for all things are much finer and bigger when they are produced in Asia. In Asia the country is civilized and its inhabitants in their different nations are milder and more patient of toil. The cause of this is temperature of the seasons—Asia lies towards the east midway between the two risings of the sun, and is situated farther from the cold *(than Europe is).* Increase and cultivation are produced above all when no condition violently predominates but a universal balance holds sway. Not that like conditions hold good throughout all Asia. All the territory which lies midway between heat and cold is the most productive in fruits and trees, has the clearest climate, and is endowed with the best waters both from the sky and from the earth . . . [*no extremes of cold or heat, drought or flood, or of snow*] . . . There is naturally in that region an abundance of things in season, both crops from hand-sown seeds, and all the wild plants which the earth itself sends up and of which mankind enjoys the fruits by reclamation and culture and by transplantation for his supply and use; the cattle that are bred there naturally flourish better than anywhere else; there they breed most plenteously and are the finest when reared; the inhabitants are naturally well nourished, handsomest in form and tallest in stature, and in form and stature one from another differ least of men. It is likely that this region approaches nearest to the conditions of springtime with regard to the moderate temperature of its seasons and the nature of its products. But manliness, hardiness, will to work, and courageous spirit could not be well engendered under such conditions of nature. [*A good deal, especially about Egypt and Libya, is lost here.*] Pleasure is inevitably master there. Hence it is that many forms are exhibited amongst the wild animals. Such then are my opinions about the conditions of the Egyptians and the Libyans. But with regard to the regions situated on the right of the summer rising of the sun as far as Lake Maeotis (this is the boundary between Europe and Asia) the facts are as follows. These nations differ from each other more than those I have already described because of the changes in their seasons and the nature of their territory. Like conditions would hold good for the earth and for mankind. For where the seasons suffer very great and very frequent changes, there the ground also is very wild and very uneven, and there you will find mountains most numerous and most thickly wooded; plains also and meadows. But where the seasons do not alter greatly, there the ground is most level . . . There are differences in the seasons which cause changes in the nature of man's physical shape, and if the seasons differ greatly from each other, then there are produced more differences in men's forms accordingly . . .

As regards the people in Phasis, the country is marshy, hot, humid, and well wooded. The rainfall there is during all seasons abundant and violent; the inhabitants pass their life in the marshes; and their dwellings made of wood and reeds are constructed in the water. They walk but little, that is to say, to town and market, and otherwise make journeys by water up and down in dug-out boats, for they have many canals. As for drink, they use waters that are warm and stagnant and tainted through the sun's influence, and increased constantly by rainfall, the Phasis itself being the most stagnant of all rivers and the gentlest in its flow. The fruits that are yielded for them are all sickly and soft and unripe because of the superabundance of damp; wherefore they do not grow mellow. Much mist from the water covers the land. It is surely from these reasons that the Phasians have forms diverse from the rest of mankind. In stature they are tall, and in bulk very fat; no joint or vein shows through their flesh. Their complexion is sallow, as though they were in the grip of jaundice. Their voices are the huskiest of all human voices, because the air they breathe is not clear, but foggy and wet. As regards hardiness they are physically and by nature somewhat lazy. Their seasons do not vary very much either towards stifling heat or cold. Their winds are mostly from the south save for a single local breeze; this sometimes blows violently and hot and is difficult to face . . . The north wind does not appear often; when it does blow it is feeble and faint . . . As for the faintheartedness and unmanliness of the inhabitants (sc. of Asia), in that the Asiatics are less warlike than the Europeans, and milder in character, the seasons are the chief cause of this, because they make little change either towards heat or cold, but are very similar. There are no sudden shocks of fear to assail their minds, nor any violent change to affect the body . . . The Asiatic race seems to be unvaliant because of its institutions also. For most nations of Asia are ruled by kings . . .

In Europe there is a Scythian nation which dwells round Lake Maeotis (Sea of Azov) and is different from all other nations. They are called the Sauromatae . . .

As for the similarity in form of the rest of the Scythians one to another, while differing from others, the same reasoning holds good as we gave about the Egyptians, except that the latter are oppressed by heat, the former by cold. The so-called 'Scythians' Desert' is a plain covered with meadows, bare of trees, and moderately watered, for there are big rivers which drain away in channels the water from the plains. Here also live the Scythians, who are called Pastoral because they have no dwellinghouses, but dwell in wagons . . . With regard to their seasons and physique, and the fact that the Scythian race differs widely from the rest of mankind, and yet preserves similarity within itself like the Egyptian, and is not prolific at all; and that its territory breeds the rarest and smallest animals, here are the reasons

—the country lies right under the Bears and the Rhipaean mountains whence the north wind blows; the sun is only at its nearest when at last it reaches the summer solstice, and then only is warm for a short time; and not to any extent do the clear fair winds, that blow from the hot regions, reach so far, except rarely and with feeble force. But from the north there blow constant cold winds that come from snow and ice and abundant water. Nowhere do mountains cease, and under the above conditions they are hardly habitable. Thick fog covers the plains in daytime, and the people live in damp and wet. Thus winter is there always, and summer for a few days, and very slight at that. For the plains are elevated and bare; they are not crowned with circles of mountains . . . The remainder of the European race differs individual from individual, in stature and form, because of the changes in the seasons in that these are great and frequent. There occur in turn violent heat-waves, hard winters, abundant rainfalls, and in turn long lasting droughts, and winds also. From these arise the many changes of all kinds . . . That is why the bodily forms of the Europeans vary more, I think, than those of the Asiatics, and their statures offer very great variations among themselves in every city . . . With regard to character the same reasoning holds good. Savagery, unsociability, and high-spiritedness are engendered in such a nature. For the occurrence of sudden shocks of fear, frequently assailing the mind, implants savagery and dulls the sense of gentleness and civilization. That is why I think the inhabitants of Europe are nobler in spirit than the inhabitants of Asia. For in a climate that is always unvarying in itself indolence is inherent, and, in a changing climate, hardiness in body and soul . . . But there are also in Europe tribes which differ one from another in stature, shape, and manliness . . . People who inhabit a country which is mountainous, rough, high, and watery, and where the changes which they experience in the seasons vary greatly, naturally possess a physical form which is large and well-fitted by nature for hardiness and manliness. Savagery too and brute ferocity are possessed by such natures most of all. But those who inhabit regions which are depressed, covered with meadows, and stiflingly hot, and experience a greater share of warm than of cold winds, and use warm waters, could not well be large or straight as a rod, but are bred with a tendency to be broad and fleshy, black-haired too, and swarthy rather than pale. They suffer from excess of phlegm rather than of bile. Manliness and hardiness would not alike be inherent in their souls, though the additional effect of institutions might produce this. Again, if we suppose there are rivers in the country, all those who drain away the stagnant water and the rain-water from the land will naturally be healthy and clear-skinned. But if there should be no rivers, and they drink spring-water, stagnant water, and water out of marshes, such people must of necessity be physically rather pot-bellied and splenetic. Those again who dwell in a country which is lofty, level, windy, and watery, will possess large forms individually alike; but their mentality will be rather unmanly and mild. Again, those who inhabit light, waterless, and bare regions, and where the changes in the seasons are not well tempered, in such a country their forms will naturally be slender, well strung, and blond rather than dark, and in their manners and passions they will be self-willed and self-opinionated. For where the changes in the seasons are most frequent and differ most widely one from another, there you will find the greatest difference in physique, manners, and natures alike. These then (sc. *those caused by variations in seasons*) are the greatest variations in man's nature. Next in importance comes the effect of the country in which a man is reared, and of the waters he uses. For you will find that for the most part both the physical and moral characteristics of mankind follow the nature of his country. For where the soil is rich, soft, and wet, and keeps its water very near the surface, so that it is warm in summer, and is well situated with regard to the seasons, there the inhabitants also are fleshy and have invisible joints; they are flabby, and not hardy, and mostly cowards at heart. Indolence and drowsiness are to be seen in them. With regard to arts and crafts they are obese and gross, and not clever. But where the soil is bare, open, and rough, nipped by winter and burnt by the sun, then you will see men who are slender, thin, well-jointed, well strung and sturdy. In such a nature as theirs you will see activity, cleverness, and wideawake spirit, and morals and passions which are self-willed and self-opinionated and partake of savagery rather than gentleness, and are keen and more intelligent than others with regard to arts and crafts, and are brave with regard to warfare. You will find too that all the other things produced in the soil follow on the nature of the soil.

Greek Heliocentric Theory

Aristarchus of Samos (c. 310–230 B.C.), known as "the Mathematician" to his contemporaries, is remembered both for his practical and his theoretical contributions to geography. He is credited with the invention of the *scaphē*. This bowl, with graduated circles and a stylus in the middle, was a much-improved version of the gnōmōn.

Aristarchus also occupies a place of particular importance in the history of geographical ideas for his elaboration of a heliocentric theory of the universe. The Pythagoreans alluded to such a theory, but its definitive formulation appears to be Aristarchus' work.

The following selection from Heath includes passages from Archimedes, Plutarch, Aetius, and Sextus Empirius referring to Aristarchus' theory, as well as Heath's own commentary.

19. Aristarchus of Samos: the first heliocentric theory

There is not the slightest doubt that Aristarchus was the first to put forward the heliocentric hypothesis. Ancient testimony is unanimous on the point, and the first witness is Archimedes, who was a younger contemporary of Aristarchus, so that there was no possibility of a mistake. Copernicus himself admitted that the theory was attributed to Aristarchus, though this does not seem to be generally known. Thus Schiaparelli quotes two passages from Copernicus's work in which he refers to the opinions of the ancients about the motion of the earth. One is in the dedicatory letter to Pope Paul III, where Copernicus mentions that he first found out from Cicero

From Sir Thomas Heath, *Aristarchus of Samos* (Oxford: Oxford University Press, 1913), pp. 301–302, 304–305. By permission of the Oxford University Press.

that one Nicetas (i.e. Hicetas) had attributed motion to the earth, and that he afterwards read in Plutarch that certain others held that opinion; he then quotes the *Placita,* according to which "Philolaus the Pythagorean asserted that the earth moved round the fire in an oblique circle, in the same way as the sun and moon." The other passage is in Book I, c. 5, where, after an allusion to the views of Heraclides, Ecphantus, and Nicetas (Hicetas), who made the earth rotate about its own axis at the centre of the universe, he goes on to say that it would not be very surprising if any one should attribute to the earth another motion besides rotation, namely revolution in an orbit in space; 'atque etiam (terram) pluribus motibus vagantem et unam ex astris Philolaus Pythagoricus sensisse fertur, Mathematicus non vulgaris.' Here, however, there is no question of the earth revolving round the *sun,* and there is no mention of Aristarchus. But it is a curious fact that Copernicus did mention the theory of Aristarchus in a passage which he afterwards suppressed: "Credibile est hisce similibusque causis Philolaum mobilitatem terrae sensisse, quod etiam nonnulli Aristarchum Samium ferunt in eadem fuisse sententia."

I will now quote the whole passage of Archimedes in which the allusion to Aristarchus's heliocentric hypothesis occurs, in order to show the whole context [*Arenarius,* I., 4–7].

"You are aware ["you" being King Gelon] that 'universe' is the name given by most astronomers to the sphere, the centre of which is the centre of the earth, while its radius is equal to the straight line between the centre of the sun and the centre of the earth. This is the common account (τὰ γραφόμεγα), as you have heard from astronomers. But Aristarchus brought out *a book consisting of certain hypotheses,* wherein it appears, as a consequence of the assumptions made, that the universe is many times greater than the 'universe' just mentioned. His hypotheses are that *the fixed stars and the sun remain unmoved, that the earth revolves about the sun in the circumference of a circle, the sun lying in the middle of the orbit,* and that the sphere of the fixed stars, situated about the same centre as the sun, is so great that the circle in which he supposes the earth to revolve bears such a proportion to the distance of the fixed stars as the centre of the sphere bears to its surface. Now it is easy to see that this is impossible; for, since the centre of the sphere has no magnitude, we cannot conceive it to bear any ratio whatever to the surface of the sphere. We must, however, take Aristarchus to mean this: since we conceive the earth to be, as it were, the centre of the universe, the ratio which the earth bears to what we describe as the universe is equal to the ratio which the sphere containing the circle in which he supposes the earth to revolve bears to the sphere of the fixed stars. For he adapts the proofs of the phenomena to a hypothesis of this kind, and in particular he appears to suppose the size of the sphere in which he makes the earth move to be equal to what we call the 'universe'."

We shall come back to the latter part of this passage; at present we are concerned only with the italicized words. The heliocentric hypothesis is stated in language which leaves no room for dispute as to its meaning. The sun, like the fixed stars, remains unmoved and forms the centre of a circular orbit in which the earth revolves round it; the sphere of the fixed stars has its centre at the centre of the sun.

Our next evidence is a passage of Plutarch:

"Only do not, my good fellow, enter an action against me for impiety in the style of Cleanthes, who thought it was the duty of Greeks to indict Aristarchus of Samos on the charge of impiety for putting in motion the Hearth of the Universe, this being the effect of his attempt to save the phenomena by supposing the heaven to remain at rest and the earth to revolve in an oblique circle, while it rotates, at the same time, about its own axis." [*De facie in orbe lunae* ch. VI. 922F-923A.]

Here we have the additional detail that Aristarchus followed Heraclides in attributing to the earth the daily rotation about its axis; Archimedes does not state this in so many words, but it is clearly involved by his remark that Aristarchus supposed that the fixed stars as well as the sun remain unmoved in space. When Plutarch makes Cleanthes say that Aristarchus ought to be indicted for the impiety of 'putting the Hearth of the Universe in motion', he is probably quoting the exact words used by Cleanthes, who doubtless had in mind the passage in Plato's *Phaedrus* where "Hestia abides alone in the House of the Gods." A similar expression is quoted by Theon of Smyrna from Dercyllides, who "says that we must suppose the earth, the Hearth of the House of the Gods according to Plato, to remain fixed, and the planets with the whole embracing heaven to move, and rejects with abhorrence the view of those who have brought to rest the things which move and set in motion the things which by their nature and position are unmoved, such a supposition being contrary to the hypotheses of mathematics"; the allusion here is equally to Aristarchus, though his name is not mentioned. A tract 'Against Aristarchus' is mentioned by Diogenes Laertius among Cleanthes' works; and it was evidently published during Aristarchus's lifetime (Cleanthes died about 232 B.C.).

Other passages bearing on our present subject are the following.

"Aristarchus sets the sun among the fixed stars and holds that the earth moves round the sun's circle (i.e. the ecliptic) and is put in shadow according to its (i.e. the earth's) inclinations." [*Aët.* II. 24.8]

One of the two versions of this passage has "*the disc* is put in shadow," and it would appear, as Schiaparelli says, "that the words 'the disc' were interpolated by some person who thought that the passage was an explanation of solar eclipses." It is indeed placed under the heading "Concerning the eclipse of the sun"; but this is evidently wrong, for we clearly have here in the concisest form an explanation of the phenomena of the seasons according to the system of Copernicus.

"Yet those who did away with the motion of the universe and were of opinion that it is the earth which moves, as Aristarchus the mathematician

held, are not on that account debarred from having a conception of time."
[Sextus Empiricus, *Adv. Math.* X. 174]

"Did Plato put the earth in motion, as he did the sun, the moon, and the five planets, which he called the instruments of time on account of their turnings, and was it necessary to conceive that the earth "which is globed about the axis stretched from pole to pole through the whole universe" was not represented as being held together and at rest, but as turning and revolving (στρεφουένην κοι ἀνειλουμένην), as Aristarchus and Seleucus afterwards maintained that it did, the former stating this as only a hypothesis (ὑποτιθέμενος μόνον), the latter as a definite opinion (κοι ἀποφαινόμενος)?" [Plutarch, *Plat. quaest,* VIII. 1, 1006C]

"Seleucus the mathematician, who had written in opposition to the views of Crates, and who himself too affirmed the earth's motion, says that the revolution (περιοτροφή) of the moon resists the rotation [and the motion] of the earth, and, the air between the two bodies being diverted and falling upon the Atlantic ocean, the sea is correspondingly agitated into waves." [*Aët.* III. 17.9]

Greek Travelers' Reports

Navigation, trade and travel were an essential part of the life of the Greek city-states. From an early date Greek merchants and seamen traveled not only throughout the Mediterranean but beyond it, westward to the Atlantic coasts of Europe, eastward to Arabia, Egypt, and India. The reports of these travels—describing business ventures, military undertakings, and diplomatic missions—were preserved in both written and graphic form, as is shown in Herodotus' description of the map prepared by order of Aristagoras, tyrant of the city of Miletus.

20. Herodotus describes the Royal Road of Persia, the Caspian Sea, Egypt, Libya, and the land of the Scythians

Herodotus (484–424 B.C.) is regarded as the father of history, yet his *Histories* contain much that is contemporary information not only about lands he himself had visited but also about places unknown to him. For the latter he uses information gathered from a variety of sources such as travelers' tales and diplomatic references.

The first of the following excerpts describes the unsuccessful diplomatic maneuver of Aristagoras, ruler of Miletus, to embroil Sparta in a war against the Persians. Aristagoras used a map to show the position of the two countries and to describe the Persians' royal road

The selection on the Nile is from E. H. Warmington, *Greek Geography*, The Library of Greek Thought Series (New York, 1934), pp. 40–42. Published in the United States by E. P. Dutton, and reprinted by their permission. The other selections are from *The Histories of Herodotus* trans. Henry Carey (New York: Appleton, 1904), pp. 290–293, 79, 86–89, 224–225, 221–222.

from the Aegean to their capital, Susa. The other excerpts from Herodotus, typical of his style, depict the Caspian, Egypt, Libya, and Scythia. The last excerpt, a statement of known facts, provides an illustrative contrast to the more analytical approach of Hippocrates' writings.

About Aristagoras and his map

Aristagoras, then, tyrant of Miletus, arrived at Sparta when Cleomenes held the government; and he went to confer with him, as the Lacedæmonians say, having a brazen tablet, on which was engraved the circumference of the whole earth, and the whole sea, and all rivers. And Aristagoras, having come to a conference, addressed him as follows: "Wonder not, Cleomenes, at my eagerness in coming here, for the circumstances that urge are such as I will describe. That the children of Ionians should be slaves instead of free is a great disgrace and sorrow to us, and above all others to you, inasmuch as you are at the head of Greece. Now, therefore, I adjure you, by the Grecian gods, rescue the Ionians, who are of your own blood, from servitude. It is easy for you to effect this, for the barbarians are not valiant; whereas you, in matters relating to war, have attained to the utmost height of glory: their mode of fighting is this, with bows and a short spear; and they engage in battle, wearing loose trousers and turbans on their heads, so they are easy to be overcome. Besides, there are treasures belonging to those who inhabit that continent, such as are not possessed by all other nations together; beginning from gold, there are silver, brass, variegated garments, beasts of burden, and slaves; all these you may have if you will. They live adjoining one another, as I will show you. Next these Ionians are the Lydians, who inhabit a fertile country, and abound in silver." As he said this he showed the circumference of the earth, which he brought with him, engraved on a tablet. "Next the Lydians," proceeded Aristagoras, "are these Phrygians to the eastward, who are the richest in cattle and in corn of all with whom I am acquainted. Next to the Phrygians are the Cappadocians, whom we call Syrians; and bordering on them, the Cilicians, extending to this sea in which the island of Cyprus is situated; they pay an annual tribute of five hundred talents to the king. Next to the Cilicians are these Armenians, who also abound in cattle; and next the Armenians are the Matienians, who occupy this country; and next them this territory of Cissia, in which Susa is situated on this river Choaspes, here the great king resides, and there are his treasures of wealth. If you take this city, you may boldly contend with Jupiter in wealth. But now you must carry on war for a country of small extent, and not very fertile, and of narrow limits, with the Messenians, who are your equals in valour, and with the Arcadians and Argives, who have nothing akin to gold and silver, the desire of which induces men to hazard their lives in battle. But when an opportunity is offered to conquer all Asia with ease, will you prefer anything else?" Aristagoras spoke thus, and Cleomenes answered him as follows, "Milesian friend, I defer to give you an answer until the third day." On that day they got so far. When the day appointed for the answer was come, and they had met at the appointed place, Cleomenes asked Aristagoras how many days' journey it was from the sea of the Ionians to the king. But Aristagoras, though he was cunning in other things, and had deceived him with much address, made a slip in this; for he should not have told the real fact if he wished to draw the Spartans into Asia; whereas he told him plainly that it was a three months' journey up there. But he, cutting short the rest of the description which Aristagoras was proceeding to give of the journey, said: "Milesian friend, depart from Sparta before sunset; for you speak no agreeable language to the Lacedæmonians in wishing to lead them a three months' journey from the sea." Cleomenes having spoken thus, went home. But Aristagoras, taking an olive branch in his hand, went to the house of Cleomenes, and having entered in, as a suppliant, besought Cleomenes to listen to him, having first sent away his little child; for his daughter, whose name was Gorgo, stood by him; she happened to be his only child, and was about eight or nine years of age. But Cleomenes bade him say what he would, and not refrain for the sake of the child. Thereupon Aristagoras began promising ten talents if he would do as he desired; and when Cleomenes refused, Aristagoras went on increasing in his offers, until he promised fifty talents; then the girl cried out, "Father, this stranger will corrupt you unless you quickly depart." Cleomenes, pleased with the advice of the child, retired to another apartment; and Aristagoras left Sparta altogether, nor could he get an opportunity to give further particulars of the route to the king's residence.

With respect to this road, the case is as follows: There are royal stations all along, and excellent inns, and the whole road is through an inhabited and safe country. There are twenty stations extending through Lydia and Phrygia, and the distance is ninety-four parasangs and a half. After Phrygia, the river Halys is met with, at which there are gates, through which it is absolutely necessary to pass, and thus to cross the river: there is also a considerable fort on it. When you cross over into Cappadocia, and traverse that country to the borders of Cilicia, there are eight-and-twenty stations, and one hundred and four parasangs; and on the borders of these people you go through two gates, and pass by two forts. When you have gone through these and made the journey through Cilicia, there are three stations and fifteen parasangs and a half. The boundary of Cilicia and Armenia is a river that is crossed in boats: it is called the Euphrates. In Armenia there are fifteen stations for resting places, and fifty-six parasangs and a half; there is also a fort in the stations. Four rivers that are crossed in boats flow through this country, which it is absolutely necessary to ferry over. First, the Tigris; then, the second and third have the same name, though they are not the same river, nor flow from the same source. For the first mentioned of these flows from the Armenians, and the latter from the Matienians. The fourth river is called the Gyndes, which Cyrus once distributed into three hundred and sixty channels. As you enter from Armenia into the country of Matiene, there are four stations; and from thence as you proceed to the Cissian territory there are eleven stations, and forty-two parasangs and a half, to the river Choaspes, which also must be crossed in boats: on this Susa is built. All these stations amount to one hundred and eleven: accordingly, the resting places at the stations are so many as you go up from Sardis to Susa. Now if the royal road has been correctly measured in parasangs, and if the parasang is equal to thirty stades, as indeed it is, from Sardis to the royal palace, called Memnonia, is a distance of thirteen thousand five hundred stades, the parasangs being four hundred and fifty; and by those who travel one hundred and fifty stades every day, just ninety days are spent on the journey. Thus Aristagoras the Milesian spoke correctly when he told Cleomenes the Lacedæmonian that it was a three months' journey up to the king's residence. But if any one should require a more accurate account than this, I will also point this out to him, for it is necessary to reckon with the above the journey from Ephesus to Sardis: I therefore say that the whole number of stades from the Grecian sea to Susa (for such is the name of the Memnonian city) amounts to fourteen thousand and forty; for from Ephesus to Sardis is a distance of five hundred and forty stades. And thus the three months' journey is lengthened by three days.

The Caspian Sea

The Caspian is a sea by itself, having no communication with any other sea; for the whole of that which the Grecians navigate, and that beyond the Pillars, called the Atlantic, and the Red Sea, are all one. But the Caspian is a separate sea of itself, being in length a fifteen days' voyage for a rowing-boat; and in breadth, where it is widest, an eight days' voyage. On the western shore of this sea stretches the Caucasus, which is in extent the largest, and in height the loftiest of all mountains; it contains within itself many and various nations of men, who for the most part live upon the produce of wild fruit trees. In this country, it is said, there are trees which produce leaves of such a nature that by rubbing them and mixing them with water the people paint figures on their garments; these figures, they say, do not wash out, but grow old with the wool, as if they had been woven in from the first. It is said that sexual intercourse among these people takes place openly, as with cattle. The Caucasus, then, bounds the western side of this sea, which is called the Caspian; and on the east, toward the rising sun, succeeds a plain in extent unbounded in the prospect.

On Egypt

But as concerns human affairs, they agree with one another in the following
account: That the Egyptians were the first to discover the year, which they
divided into twelve parts; and they say that they made this discovery from
the stars: and so far, I think, they act more wisely than the Grecians, in
that the Grecians insert an intercalary month every third year, on account
of the seasons; whereas the Egyptians, reckoning twelve months of thirty
days each, add five days each year above that number, and so with them
the circle of the seasons comes round to the same point. They say also that
the Egyptians were the first who introduced the names of the twelve gods,
and that the Greeks borrowed those names from them; that they were the
first to assign altars, images, and temples to the gods, and to carve the figures
of animals on stone; and most of these things they proved were so in fact.
They added, that Menes was the first mortal who reigned over Egypt, and
that in his time all Egypt, except the district of Thebes, was a morass, and
that no part of the land that now exists below Lake Myris was then above
water: to this place from the sea is a seven days' passage up the river. And
they seemed to me to give a good account of this region. For it is evident
to a man of common understanding, who has not heard it before, but sees
it, that the part of Egypt which the Greeks frequent with their shipping
is land acquired by the Egyptians, and a gift from the river; and the parts
above this lake, during a three days' passage, of which, however, they said
nothing, are of the same description. For the nature of the soil of Egypt is
of this kind: when you are first sailing to it, and are at the distance of a
day's sail from land, if you cast the lead you will bring up mud, and will
find yourself in eleven fathoms water: this so far shows that there is an
alluvial deposit.

The length of Egypt along the sea-coast is sixty schœni, according as we
reckon it to extend from the Plinthinetic Bay to Lake Serbonis, near which
Mount Casius stretches: from this point then the length is sixty schœni.
Now, all men who are short of land measure their territory by fathoms;
but those who are less short of land, by stades; and those who have much,
by parasangs; and such as have a very great extent, by schœni. Now, a
parasang is equal to thirty stades, and each schœnus, which is an Egyptian
measure, is equal to sixty stades. So the whole coast of Egypt is three
thousand six hundred stades in length. From thence, as far as Heliopolis,
inland, Egypt is wide, being all flat, without water, and a swamp. The
distance to Heliopolis, as one goes up from the sea, is about equal in length
to the road from Athens—that is to say, from the altar of the twelve gods
to Pisa and the Temple of Olympian Jupiter. For whoever will compare
these roads will find, by computation, that the difference between them is
but little, not exceeding fifteen stades; for the road from Athens to Pisa

is only fifteen stades short of one thousand five hundred stades, but the road from the sea to Heliopolis amounts to just that number. From Heliopolis upward Egypt is narrow, for on one side the mountain of Arabia extends from north to south and southwest, stretching up continuously to that which is called the Red Sea. In this mountain are the quarries whence the stone was cut for the pyramids at Memphis; here the mountain, deviating, turns to the parts above mentioned. But where its length is the greatest, I have heard that it is a two months' journey from east to west; and that eastward its confines produce frankincense. On that side of Egypt which borders upon Libya extends another rocky mountain, and covered with sand, on which the pyramids stand; and this stretches in the same direction as that part of the Arabian mountain that runs southward. So that from Heliopolis, the territory which belongs to Egypt is not very extensive; but for four days' sail up the river it is very narrow. Between the mountains before mentioned the land is level, and in the narrowest part appeared to me to be not more than two hundred stades in breadth, from the Arabian mountain to that called the Libyan; but above this Egypt again becomes wide. Such, then, is the character of this country. From Heliopolis to Thebes is a voyage up of nine days; the length of this journey is in stades four thousand eight hundred and sixty, which amounts to eighty-one schœni. Now, if we compute these stades together, the coast of Egypt, as I before explained, contains in length three thousand and six hundred stades: how far it is from the sea inland as far as Thebes I will next show, namely, six thousand one hundred and twenty stades; and from Thebes to the city called Elephantine, one thousand eight hundred stades.

The greater part of all this country, as the priests informed me, and as appeared to me also to be the case, has been acquired by the Egyptians. For the space between the above-mentioned mountains, that are situated beyond the city of Memphis, seem to me to have been formerly a bay of the sea; as is the case also with the parts about Ilium, Teuthrania, Ephesus, and the plain of the Mæander, if I may be permitted to compare small things with great; for of the rivers that have thrown up the soil that forms these countries, not one can justly be brought into comparison, as to size, with any one of the five mouths of the Nile. But there are other rivers not equal in size to the Nile, which have wrought great works; of these I could mention the names, and among them one of the most remarkable is the Achelous, which, flowing through Acarnania, and falling into the sea, has already converted one half of the Echinades Islands into continent. There is also in the Arabian territory, not far from Egypt, branching from the Red Sea, a bay of the sea, of the length and width I shall here describe: the length of the voyage, beginning from the innermost part of this bay to the broad sea, occupies forty days for a vessel with oars; and the width, where the bay is widest, half a day's passage: and in it an ebb and flow takes place daily; and I am of opinion that Egypt was formerly a similar bay; this stretching from the Northern Sea toward Ethiopia; and the Arabian

Bay, which I am describing, from the south toward Syria; and that they almost perforated their recesses so as to meet each other, overlapping to some small extent. Now, if the Nile were to turn its stream into this Arabian gulf, what could hinder it from being filled with soil by the river within twenty thousand years? For my part, I think it would be filled within ten thousand. How, then, in the time that has elapsed before I was born, might not even a much greater bay than this have been filled up by such a great and powerful river? I therefore both give credit to those who relate these things concerning Egypt, and am myself persuaded of their truth when I see that Egypt projects beyond the adjoining land; that shells are found on the mountains; that a saline humour forms on the surface so as even to corrode the pyramids; and that this mountain which is above Memphis is the only one in Egypt that abounds in sand: add to which, that Egypt, in its soil, is neither like Arabia or its confines, nor Libya, nor Syria (Syrians occupy the sea-coast of Arabia), but is black and crumbling, as if it were mud and alluvial deposit brought down by the river from Ethiopia; whereas we know that the earth of Libya is reddish, and somewhat more sandy; and that of Arabia and Syria is more clayey and flinty.

On the Nile

The Nile when it is in flood spreads over not only the Delta but also the region which is called Libyan, and the Arabian besides, in some places for a distance of two days' journey on either side, and sometimes even more than this, and sometimes less. But on the nature of the river I was not able to get any information either from the priests or from any other person; and I was eager to learn from them through what cause the Nile comes down in flood for one hundred days beginning at the summer solstice, and when it has approached that number of days, goes back again with failing stream, so that during the whole winter it is constantly low until the return of the summer solstice . . . And I often inquired why it is that alone of all rivers it does not afford breezes blowing from it. Nevertheless certain Greeks desiring to become conspicuous for cleverness have given explanations of this water in three different ways. Two of these ways I do not think worthy of mention beyond my desire simply to indicate their nature. The one explanation says that it is the etesian winds which are the cause of the river's flooding; they hinder the Nile from flowing out into the sea. But surely there is many a time when the etesians do not blow, and yet the Nile shows the same activity; and besides, if the etesians were the cause, it would needs follow that all other rivers also which flow against the etesians would be similarly affected with the Nile and with the same result; and all the more so in proportion to the fact that being smaller in size, they produce weaker currents; and there are many rivers in Syria, and many too in Libya which are affected in nothing like the same way as the Nile. The second explanation is based less on knowledge than the first mentioned, but one might really say it is more wonderful; it maintains that the river effects all this by flowing from the Ocean, and that the Ocean 'flows' round all the earth. Then the third explanation, though it appears most reasonable, is the most false; for this also is no explanation at all, alleging as it does that the Nile flows from the melting snow—the Nile which flows from Libya through the midst of the Ethiopians and only then pours into Egypt! How could it possibly flow from snow, since it flows from the hottest regions to colder? On this point there are many arguments—at any rate to the man who is capable of reasoning about such things—showing that it is not even reasonable that it should flow from snow. The first and most cogent piece of evidence is provided by the winds in that they blow hot from the regions mentioned; the second is that the country continues constantly without rain and ice, whereas on a fall of snow, it must of utter necessity rain within five days, so that if it ever snowed in those regions there would be rain in them also; and the third is the fact that the people are black through the burning heat; kites and swallows are resident throughout the year and do not leave; and cranes, fleeing from the winter which sets in in Scythia, regularly visit these localities for wintering: if, then, ever so little snow fell on this country through which the Nile flows, and from which it derives its source, none of these things would happen.

On Libya

Libya commences from Egypt. Now in Egypt this tract is narrow; for from this sea to the Red Sea are a hundred thousand orgyæ, which make one thousand stades. But from this narrow neck the tract which is called Libya becomes very wide. I wonder therefore at those who have described the limits of and divided Libya, Asia, and Europe, for the difference between them is not trifling; for in length Europe extends along both of them, but with respect to width, it is evidently not to be compared. Libya shows itself to be surrounded by water, except so much of it as borders upon Asia. Neco, King of Egypt, was the first whom we know of that proved this; he, when he had ceased digging the canal leading from the Nile to the Arabian Gulf, sent certain Phœnicians in ships, with orders to sail back through the Pillars of Hercules into the northern sea, and so to return to Egypt. The Phœnicians, accordingly, setting out from the Red Sea, navigated the southern sea; when autumn came, they went ashore, and sowed the land, by whatever part of Libya they happened to be sailing, and waited for harvest; then having reaped the corn, they put to sea again. When two years had thus passed, in the third, having doubled the Pillars of Hercules, they arrived in Egypt, and related what to me does not seem credible, but may to others, that as they sailed round Libya they had the sun on their right hand. Thus was Libya first known.

On Scythia

All this country which I have been speaking of is subject to such a severe winter that for eight months the frost is so intolerable that if you pour water on the ground you will not make mud, but if you light a fire you will make mud. Even the sea freezes, and the whole Cimmerian bosphorus; and the Scythians who live within the trench, lead their armies and drive their chariots over the ice to the Sindians, on the other side. Thus winter continues eight months, and during the other four it is cold there. And this winter is different in character from the winters in all other countries; for in this no rain worth mentioning falls in the usual season, but during the summer it never leaves off raining. At the time when there is thunder elsewhere there is none there, but in summer it is violent: if there should be thunder in winter, it is accounted a prodigy to be wondered at. So should there be an earthquake, whether in summer or winter, in Scythia it is accounted a prodigy. Their horses endure this cold, but their asses and mules can not endure it at all; but in other places horses that stand exposed to frost become frost-bitten in the cold, waste away; but asses and mules endure it. On this account also the race of beeves appears to me to be defective there, and not to have horns; and the following verse of Homer, in his "Odyssey," confirms my opinion. "And Libya, where the lambs soon put forth their horns"; rightly observing that in warm climates horns shoot out quickly; but in very severe cold, the cattle either do not produce them at all, or if they do produce them they do so with difficulty. Here, then, such are the effects of the cold.

21. Xenophon on western Asia

Xenophon (c. 445–c. 355 B.C.) was a philosopher, historian, and general of Athens who studied under Socrates. During an internal struggle for the throne of Persia he served as one of the leaders of a Greek mercenary force. After their defeat in 400 B.C., the ten thousand Greeks were led by Xenophon in a long march called the Anabasis from Mesopotamia across Asia Minor to the shores of the Black Sea.

Thence he marched through Arabia, keeping the Euphrates on the right, five stages of desert, 35 leagues. In this part the country was one great flat plain like a sea, but covered with absinth; whatever there was besides of

---

From W. H. D. Rouse, *The March Up Country,* a translation of Xenophon's *Anabasis,* (Ann Arbor: The University of Michigan Press, 1964), pp. 15–16, 42, 96–97, 106–108, by permission of the publisher.

wood or reeds, all were sweet-smelling like spices. There was no tree, but all sorts of animals, troops of wild asses, plenty of ostriches, bustards also and gazelles; and the horsemen often chased these. The asses, if they were chased, ran ahead and stood still, for they ran much faster than horses; again when the horses came near they ran as before, and they could not be caught unless the horsemen placed themselves at intervals and hunted in relays. The meat of those caught was like venison but more tender. No-one caught an ostrich; horsemen in chase soon gave up, for she got far away before them, running on her feet and flying too, using her wings like a sail. Bustards, however, can be caught if you put them up smartly, for they fly in short spurts like a partridge and quickly tire. Their flesh was very good . . .

At last they reached villages from which the guides directed them to take the provisions. There was plenty of corn and date-wine and a sour drink made by boiling dates. The palm-nuts or date-nuts which we see in our country are such as were kept for servants; but those reserved for the masters were select, wonderfully fine and large, just like amber in colour: and some they dry and store for delicacies. These were delicious as a relish for wine, but they give headache. That was the first time the men tasted the brain or top-cabbage of the palm, and most of them were surprised at the shape and peculiar flavour. This also was very apt to give headache. When the brain was taken from the palm, the whole tree withered.

Next day they decided to march as quickly as possible, before the enemy could collect and occupy the narrows. They packed up and marched at once, through deep snow, having several guides; and the same day they got over the height where Tiribazos meant to attack them, and encamped there. From that place they marched three desert stages, 15 leagues, and reached the River Euphrates, which they crossed waist-deep. The source was said to be not far away. From there they marched through deep snow over a plain, three stages, 15 leagues. The third was difficult, and a north wind blew in their faces, parching everything everywhere and freezing the men. There one of the seers advised them to sacrifice to the wind, and sacrifice they did. The answer to their prayers was manifest at once to all, when the violence of the wind abated. But the snow was a fathom deep, so that many animals and slaves were lost, and soldiers too, about thirty. They kept fire burning all night, and so got through; there was plenty of wood in that stage, but those who came up late found none. Those who had been there long and lit the fire would not let the late ones come near the fire unless they gave them some of their wheat or anything else they had eatable. Here, then, they all shared together what they had. Where the fire burned great holes were made in the snow right down to the soil, and so it was possible to measure the depth of the snow . . .

After this the Hellenes reached the Harpasos River, width 400 feet. From there they marched through the Scythenians, four stages, 20 leagues, over a plain to some villages, in which they stayed three days and got supplies. From these they passed on four stages, 20 leagues, to an inhabited city, large and prosperous, which was called Gymnias.

Here the governor of the place sent the Hellenes a guide to lead them through a country which was hostile to himself. This man, when he came, told them that in five days he would take them to a place where they could have sight of the sea; they might kill him if he didn't. As soon as he passed the boundaries he advised them to burn and destroy the country; so it was clear that that was why he came, not friendliness to the Hellenes.

They reached the mountain in the fifth day; its name was Theches. When the first men reached the summit and caught sight of the sea there was loud shouting. Xenophon and the rearguard, hearing this, thought that more enemies were attacking in front; for some were following behind them from the burning countryside, and their own rearguards had killed a few men and captured others, and taken wicker shields, covered with raw hairy oxhides, about twenty. But when the shouts grew louder and nearer, as each group came up it went pelting along to the shouting men in front, and the shouting was louder and louder as the crowds increased. Xenophon thought it must be something very important; he mounted his horse, and took Lycios with his horsemen, and galloped to bring help. Soon they heard the soldiers shouting "Sea! sea!" and passing the word along.

Then the rearguard also broke into a run, and the horses and baggage animals galloped too. When they all reached the summit then they embraced each other, captains and officers and all, with tears running down their cheeks. And suddenly—whoever sent the word round—the soldiers brought stones and made a huge pile. Upon it they threw heaps of raw hides and sticks and the captured shields, and the guide cut them up with his own hands and told them to do the same.

After this the Hellenes let the guide go; they presented him publicly with a horse and a silver bowl and Persian dress and ten darics; he asked especially for their rings, and the soldiers gave him great numbers of them. He pointed out a village where they could encamp, and the road to the Macronian country, and in the evening he departed to travel through the night.

22. An early description of southernmost Persia

During Alexander the Great's campaign in India, the Macedonian fleet, commanded by Nearkhos, sailed from the mouth of the Indus to the head of the Persian Gulf. The description of that voyage, which

From a translation of Arrian's account of the voyage of Nearkhos in J. W. Mc-Crindle, *The Commerce and Navigation of the Erythrean Sea* (Calcutta, 1879), pp. 183–184, 195.

took place in 325 B.C., includes detailed reports on the people who dwelt along the southern coast of Persia and Baluchistan. The passage quoted here is from Arrian's *Indica*. Arrian, or Flavius Arrianus (c. 96–180 A.D.), was a historian and philosopher; his works are considered the best source for information about the campaigns of Alexander.

From this place they bore away with a fresh breeze, and having made good a course of 500 stadia anchored near a winter torrent called the Tomeros, which at its mouth expanded into an estuary. The natives lived on the marshy ground near the shore in cabins close and suffocating. Great was their astonishment when they descried the fleet approaching, but they were not without courage, and collecting in arms on the shore, drew up in line to attack the strangers when landing. They carried thick spears about 6 cubits long, not headed with iron, but what was as good, hardened at the point by fire. Their number was about 600, and when Nearkhos saw that they stood their ground prepared to fight, he ordered his vessels to advance, and then to anchor just within bowshot of the shore, for he had noticed that the thick spears of the barbarians were adapted only for close fight, and were by no means formidable as missiles. He then issued his directions: those men that were lightest equipped, and the most active and best at swimming were to swim to shore at a given signal: when any one had swum so far that he could stand in the water he was to wait for his next neighbour, and not advance against the barbarians until a file could be formed of three men deep: that done, they were to rush forward shouting the war-cry. The men selected for this service at once plunged into the sea, and swimming rapidly touched ground, still keeping due order, when forming in file, they rushed to the charge, shouting the war-cry, which was repeated from the ships, whence all the while arrows and missiles from engines were launched against the enemy. Then the barbarians terrified by the glittering arms and the rapidity of the landing, and wounded by the arrows and other missiles, against which they had no protection, being all but entirely naked, fled at once without making any attempt at resistance. Some perished in the ensuing flight, others were taken prisoners, and some escaped to the mountains. Those they captured had shaggy hair, not only on their head but all over their body; their nails resembled the claws of wild beasts, and were used, it would seem, instead of iron for dividing fish and splitting the softer kinds of wood. Things of a hard consistency they cut with sharp stones, for iron they had none. As clothing they wore the skins of wild beasts, and occasionally also the thick skins of the large sorts of fish . . .

In navigating the Ikhthyophagi coast the distance traversed was not much short of 10,000 stadia. The people, as their name imports, live upon fish. Few of them, however, are fishermen, and what fish they obtain they owe mostly to the tide at whose reflux they catch them with nets made for this purpose. These nets are generally about 2 stadia long, and are composed of the bark (or fibres) of the palm, which they twine into cord in the same way as the fibres of flax are twined. When the sea recedes, hardly any fish are found among the dry sands, but they abound in the depressions of the surface where the water still remains. The fish are for the most part small, though some are caught of a considerable size, these being taken in the nets. The more delicate kinds they eat raw as soon as they are taken out of the water. The large and coarser kinds they dry in the sun, and when properly dried grind into a sort of meal from which they make bread. This meal is sometimes also used to bake cakes with. The cattle as well as their masters fare on dried fish, for the country has no pastures, and hardly even a blade of grass. In most parts crabs, oysters and mussels add to the means of subsistence. Natural salt is found in the country.

23. Pytheas of Marseille on northern Europe

Pytheas, a citizen of the Greek colony located at the present site of Marseille, was charged by his community to make a firsthand reconnaissance of the routes used for the transport of tin from the mines in Cornwall to the Mediterranean. Pytheas was a keen observer not only of heavenly phenomena but also of the customs of little-known distant people, and his description of Britain and the countries of northern Europe is among the earliest we possess.

Elevation of the sun and latitude

By the Borysthenes River, and in the land of the Kelts, the sun shines all night during the summer, its light moves from west to east. In those regions, the elevation of the sun as shown on the gnomon represents nine ells at the time of the winter solstice; in the land of the Kelts, who live 6300 stadia (1185 kilometers) distance from Marseille, this phenomenon is even more pronounced. In the land of the Kelts the height of the sun at the winter solstice is six ells, at a distance of 9100 stadia (1680 kilometers) its height is only four ells, and at an even greater distance only three ells. In

---

From Pytheas of Marseille, "About the Ocean," from the German translation by D. Stichtenoth in Josef Schmithüsen, *Geschichte der Geographischen Wissenschaft* (Mannheim: Bibliographisches Institut, 1970), pp. 29–32, by permission of the author. Editor's translation.

those regions (where the winter height of the sun is three ells) daylight on the day of the equinox lasts nineteen hours; where the height of the sun is four ells, eighteen hours; but in Marseille only fifteen and a quarter hours. The gnomon there shows a ratio of 120 : 41 4/5. It was found, on the basis of these observations, that there isn't any star at the celestial pole, only an empty space, but next to it are three stars that form almost a quadrangle with the exact location of the celestial pole.

The triangle called Britain

Britain is the name given to the isles of the North, the most important among these is called Albion. I have taken its measure, pacing across it, in the manner of surveyors. Britain, like Sicily, has a triangular shape. Cape Kantion lies one hundred stadia from the continent, the so-called Cape Belerion at a distance of four days' sailing. This last-named cape extends into the Ocean, it is called Orkan. The shortest side of the triangle that is Britain, a distance of 7500 stadia (1390 kilometers) runs parallel to Europe; the second side, from the narrows (separating it from Europe) to its point is 15,000 stadia long (2800 kilometers), the third side, 20,000 stadia long (3700 kilometers). Outside this triangle lie Ierne, Caledonia, and Thule.

Thule, where the sun sleeps

The northernmost of the isles we call Britain is Thule; it is located near the latitude of the "Curdled Sea," six days' sail north of Albion. Thule is on the Arctic Circle, where that line coincides with that running through the celestial pole and the constellation of the Bear. The barbarians there showed us the district where the sun sleeps; she was supposed to be there permanently, as it were. As a matter of fact, we found that in this region nights become very short, only two hours in some, three hours in other places: the sun thus rises but a short time following its setting. In this region days are continuous in the summer, nights in the winter, and there must be regions where daylight lasts half a year, and night lasts half a year.

The journey from Caledonia to Thule

Sailing from Cape Caledonia towards Thule, one reaches after two days the five islands called Ebudes; people there do not cultivate any crops and live exclusively on fish and meat. The second stage of the journey takes one to the Orcades, which lie seven days and nights' sail from the Ebudes. These islands have no inhabitants at all, do not have woods, only sedges are found in abundance, the rest of it is covered with barren sand and rocks. Thule is five days and nights' sail from the Orcades. It is a large island, and bears much fruit. The people there depend on plants in the spring, that supports their cattle, later on on milk; they collect the produce of the fields for the winter. They share women in common, no one lives in permanent wedlock.

Plants and animals of the northern temperate zone

In those regions near the frigid zone, cultivated plants and domesticated animals become rare, or else they cannot be found at all. Millet, vegetables, fruit and roots provide food. In those places where they do cultivate plants and collect honey, they use these to brew their drink. Grain is trashed in sheds, because of the lack of sunshine and heavy rains; these are large structures where they bring the ears of grain.

Tides and the marine lung

North of Britain tides rise eight [eighty?] ells. On the other hand, there are no tidal movements beyond Cape Hiberon, this makes a voyage to the land of the Kelts easier than in the opposite direction. High and low tides are a result of the waxing and waning of the moon. I myself have observed a remarkable mixture, similar to a marine lung. In that region there is no land, nor sea, nor air, but a mixture similar to the marine lung and consisting of all of these ingredients; land, sea and all objects are suspended in it, and it is supposed to be a segment of the universe that cannot be traversed either on foot or by a vessel. An exact opposite of this phenomenon is supposed to exist in the Aeolian Islands of Lipari and Stromboli in the Mediterranean, the abode of Hephaistos, where the sound of fire and the din of hammers can be heard. People are supposed to be able to bring pig iron there, and then return the next day to pick up a sword or whatever they desire, and leave payment for these.

Scandinavia and the island of tin

Near Thule are the Scandinavian islands of Dumna, Vergos and Berrike. Dumna lies opposite Cape Orkan; Berrike is the largest of the three, and the usual starting place for a journey to Thule. At a distance of six days' sail from Albion, on the island of Viktis, tin is mined. The furthest island, Vexisame, lies three days' sail from Cape Kantion. The Ostideans sail to the island of tin from the Oestrymnian Sea and the Oestrymnian Islands. The Oestrymnian Sea begins south of a land noted for its rocky mountain. In two days' sail thence, one can reach the sacred Island of Amber. The Ostideans are clever seamen. They do not use wooden boats, but boats woven from branches, covered with skins sewn together and with thick leather.

The Sacred Isle of Abalus, Isle of Amber

The sacred isle of amber, Abalus, the "Isle of Kings," or Basilea, lies in front of the mouth of the amber river, Eridanos. This stream has seven mouths, which end in a coastal lagoon near the "Silver Mountain," the rock of Argyrios. The river is also known as Rhodanos, and according to some people as the Tartessian stream. The river Panisos also drains to the coastal lagoon. One of the mouths [of the amber river] drains to the ocean, one into the Ionian Sea, and one into the Sea of Sardinia. The hinterland is occupied by Kelts and Lygians. It is a days' sail from the lagoon to the sacred amber isle of Abalus. The Lygian Ostideans burn amber on their altars, or sell it to the neighboring Hibernians. The length of the lagoon, that is, the length of the amber coast, is 6000 stadia (1100 kilometers). I have observed this myself, sailing back from Thule along the entire coast from Gadir to the Tanais. The mouth of the Tanais appears to be on the same meridian as Alexandria and the Nile delta, and also Thule.

The Silver Mountain, Gadir, and Erytheia

The rock Argyrion is called "Silver Mountain," because it has a metallic shine on all sides, particularly so in the rays of the sun. Two days' journey upstream lies the urban area of Gadir, also called Tartessos. In the language of the natives, "Gadir" means a garden. At one time it was a large and respected settlement, now it is poor, small, abandoned, a heap of ruins. The inhabitants were driven to the interior, to impassable districts, by the Kelts. The important sanctuary on the island of Erytheia used to be ruled by Gadir and the Lygians. It is five stadia (900 meters) distance from the mainland, and lies to the west of the castle of the Elder on the Isle of Amber. The sanctuary is dedicated to Aphrodite, born from sea foam; there is a temple of Aphrodite, a secret cave, and an oracle. The amber river is called Eridanos, "River of Erytheia." This river has its source in the land of the Tylangians, and flows through the land of the Dalitanians. It is navigable in its entire length. It empties in the lagoon of Albion, where it receives the waters of the Panisos, and then reaches the ocean through three mouths in the east, and twice two mouths located between islands in the west. This river separates Europe from the land of the Lygians.

The northern column

Two of these islands lie thirty stadia (5.6 kilometers) apart. There are columns on them dedicated to Herakles, temples, and altars. According to the natives, foreign seamen came there, made their sacrifices, and left the island right away, since it is considered a sacrilege to tarry there. The waters surrounding these islands are so shallow that ships must be partially unloaded to reach them. There is also a long chain of cliffs and sandy shore, protecting the land and waters lying behind it, called "the way of Herakles."

24. Megasthenes describes India

Seleucus I Nicator, ruler of Syria, sent Megasthenes as his envoy to the Indian ruler, Chandragupta, in about 300 B.C. Megasthenes' report on the Indian subcontinent was preserved, in fragments, by Greek writers of the Hellenistic Age.

Diodorus

India is in shape four-sided; the side which inclines towards the east and the side inclining towards the south are encompassed by the Great Sea; on the northern side the Emodus range (*Himalayas*) is the barrier between it and Scythia, which is inhabited by Scythians called Sacae. The fourth side, turned towards the midday, is bounded by the river called Indus ... The size of India as a whole is, they say, 28,000 stades from east to west, 32,000 stades from north to south. Being so great in size it embraces the tropic of the summer solstice more than all the rest of the world, and at many places on the extreme headland (*Cape Comorin*), it is possible to see sundials casting no shadow, and to find the Bears invisible ... they say that shadows fall towards the south.

Arrian

Eratosthenes and Megasthenes, who stayed with Sibyrtius, the satrap of Arachosia, and states that he often attended the court of Sandrocottus (*Chandragupta*), a king of the Indians, hold that southern Asia being divided into four parts the greatest portion is occupied by India ... They say that the land of India is shut in by the great sea towards the dawn and the east wind southward; and on the north by the Caucasus range to its junction with the Taurus; while the river Indus is the dividing line on the west and the west-north-west wind as far as the great sea. Much of it is plain-land, and this they conjecture is the result of accumulated silt from the rivers.

From E. H. Warmington, *Greek Geography*, The Library of Greek Thought Series New York, 1934), pp. 163–164. Published in the United States by E. P. Dutton and reprinted by their permission.

Strabo

The whole length (*of India along the north*), where it is shortest, will be 16,000 stades (*according to Eratosthenes*) ... Megasthenes's statement in this case agrees with his. Patrocles, however, makes it 1,000 less ... It may be seen from these examples how far the accounts of others differ; for Ctesias says that India is not less in size than the rest of Asia, Onesicritus that it is one-third of the inhabited earth, Nearchus that the journey through the plain itself is one of four months, while Megasthenes and Deimachus give more moderate estimates; they fix the distance from the southern sea to the Caucasus (*Himalayan range*) at 20,000 stades.

Arrian

There are so many rivers in India as to exceed the number in all the rest of Asia. The biggest are the Ganges and the Indus ... both being bigger than the Nile ... and the Ister ... and would still be bigger if both the latter united their waters into one. In my opinion the Acesines also is bigger than the Nile and the Ister where, having taken in along its course the Hydaspes, the Hydraotes, and the Hyphasis, it casts its waters into the Indus, the result being a breadth here of 30 stades; and it may be that there are many other still greater rivers in India.

Geography in the Hellenistic Age

Classical geography reached its highest level during the late Hellenistic Age and the early years of the Roman Empire. The main centers of learning were in Alexandria, Greece, and Asia Minor; the principal contributors to geography all wrote in Greek. We know of some of them only through fragments or secondhand reports, as is the case with Eratosthenes and Hipparchus, but we are fortunate to possess substantial portions or even complete works of others, among them Polybius and Ptolemy.

25. Eratosthenes measures the earth

Eratosthenes (fl. c. 275–195 B.C.), a native of Cyrene in North Africa, was librarian of the great Museum in Alexandria. Some of his contemporaries referred to him as "Pentathlos," a compliment to his versatility; others called him "Beta," second in many fields, first in none. His *Geographika*, known only through references made to it by later writers, was the first work to bear that title. It has been described as the first scientific attempt to give geographical studies a mathematical basis. During his long tenure (235–195 B.C.) at the

The first selection is from Cleomedes, "On the Circular Movement of the Heavenly Bodies," trans. Ivor Thomas in *The Portable Greek Reader*, ed. W. H. Auden (New York: Viking Press, 1950) pp. 435–437. Copyright 1948 by the Viking Press, Inc. Reprinted by permission of The Viking Press and Chatto & Windus. The second selection is from H. F. Tozer, *A History of Ancient Geography* (Cambridge, 1897), pp. 170–172. The third selection is from George Sarton, *A History of Science* (Cambridge, Mass.: Harvard University Press, 1959), pp. 104–105.

Museum, Eratosthenes carried out the experiment that made his name immortal: the elegant measurement of the circumference of the earth.

Eratosthenes' method of investigating the size of the earth depends on a geometrical argument, and gives the impression of being more obscure. What he says will, however, become clear if the following assumptions are made. Let us suppose, in this case also, first that Syēnē and Alexandria lie under the same meridian circle; secondly, that the distance between the two cities is 5000 stades; and thirdly, that the rays sent down from different parts of the sun upon different parts of the earth are parallel; for the geometers proceed on this assumption. Fourthly, let us assume that, as is proved by the geometers, straight lines falling on parallel straight lines make the alternate angles equal, and fifthly, that the arcs subtended by equal angles are similar, that is, have the same proportion and the same ratio to their proper circles—this also being proved by the geometers. For whenever arcs of circles are subtended by equal angles, if any one of these is (say) one-tenth of its proper circle, all the remaining arcs will be tenth parts of their proper circles.

Anyone who has mastered these facts will have no difficulty in understanding the method of Eratosthenes, which is as follows. Syēnē and Alexandria, he asserts, are under the same meridian. Since meridian circles are great circles in the universe, the circles on the earth which lie under them are necessarily great circles also. Therefore, of whatever size this method shows the circle on the earth through Syēnē and Alexandria to be, this will be the size of the great circle on the earth. He then asserts, as is indeed the case, that Syēnē lies under the summer tropic. Therefore, whenever the sun, being in the Crab at the summer solstice, is exactly in the middle of the heavens, the pointers of the sundials necessarily throw no shadows, the sun being in the exact vertical line above them; and this is said to be true over a space 300 stades in diameter. But in Alexandria at the same hour the pointers of the sundials throw shadows, because this city lies farther to the north than Syēnē. As the two cities lie under the same meridian great circle, if we draw an arc from the extremity of the shadow of the pointer to the base of the pointer of the sundial in Alexandria, the arc will be a segment of a great circle in the bowl of the sundial, since the bowl lies under the great circle. If then we conceive straight lines produced in order from each of the pointers through the earth, they will meet at the centre of the earth. Now since the sundial at Syēnē is vertically under the sun, if we conceive a straight line drawn from the sun to the top of the pointer of the sundial, the line stretching from the sun to the centre of the earth will be one straight line. If now we conceive another straight line drawn upwards from the extremity of the shadow of the pointer of the sundial in Alexandria, through the top of the pointer to the sun, this straight line and the aforesaid straight line will be parallel, being straight lines drawn through from different parts of the sun to different parts of the earth. Now on these parallel straight lines there falls the straight line drawn from the centre of the earth to the pointer at Alexandria, so that it makes the alternate angles equal; one of these is formed at the centre of the earth by the intersection of the straight lines drawn from the sundials to the centre of the earth; the other is at the intersection of the top of the pointer in Alexandria and the straight line drawn from the extremity of its shadow to the sun through the point where it meets the pointer. Now this latter angle subtends the arc carried round from the extremity of the shadow of the pointer to its base, while the angle at the centre of the earth subtends the arc stretching from Syēnē to Alexandria. But the arcs are similar since they are subtended by equal angles. Whatever ratio, therefore, the arc in the bowl of the sundial has to its proper circle, the arc reaching from Syēnē to Alexandria has the same ratio. But the arc in the bowl is found to be the fiftieth part of its proper circle. Therefore the distance from Syēnē to Alexandria must necessarily be a fiftieth part of the great circle of the earth. And this distance is 5000 stades. Therefore the whole great circle is 250,000 stades. Such is the method of Eratosthenēs.

The method of investigation which Eratosthenēs pursued . . . was one which guaranteed more accurate results. The gnōmōn which he used as the instrument for his observations was an upright staff set in the midst of a scaphe or bowl, which was so arranged as to correspond to the celestial hemisphere, only inverted, and was marked with lines like a dial. By means of this he discovered that, at Alexandria at the summer solstice, the shadow of the gnōmōn at midday measured one-fiftieth part of the meridian—i.e. of a great circle of the heavens, as measured on the scaphe. He assumes at starting the following points—(1) that all the sun's rays fall to the earth parallel to one another; (2) that when one straight line falls on two parallel straight lines, the alternate angles are equal; (3) that, if arcs of different circles subtend equal angles at the centre, they bear the same proportion to the whole circumference of the circles of which they are parts; so that, if one of them, for instance, is a tenth part of its circle, the same will be the case with the others. He also takes it as proved that Syēnē and Alexandria are on the same meridian, and that the distance between them is 5,000 stadia. He then proceeds to argue as follows. In Syēnē (B), which was regarded by the ancients as lying under the tropic, at the summer solstice a ray of the sun (AB) when on the meridian, falling on the point of the gnōmōn, would coincide with the axis of the gnōmōn, so as to cast no shadow, and, as produced, would strike the earth's centre (E). In Alexandria at the same time a ray of the sun (CD), falling on the point of the gnōmōn (D), would form an angle with the axis of the gnōmōn (DG). Now, if the axis of the gnōmōn at Alexandria (DG) is produced

Northern Ray   Southern Ray

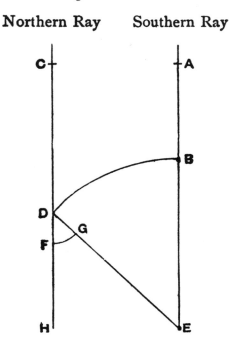

to the centre of the earth (E), the line thus drawn (DGE) would intersect the lines formed by these two rays (CDH, ABE)), and form alternate angles with them; one of these angles (BED) being at the centre of the earth (E), the other (EDH) at the point where the northern-most of the two rays meets the point of the gnōmōn at Alexandria (D). But, since alternate angles are equal, and the arcs of a circle which subtend equal angles bear the same proportion to the circles of which they are segments, the arc of the celestial sphere marked by the shadow of the gnomon on the scaphe at Alexandria (FG) must correspond to the arc of the great circle of the earth which lies between Syēnē and Alexandria (BD). Consequently, since it has been already shewn that the shadow of the gnōmōn at Alexandria measures one-fiftieth part of the great circle of the heavens, the arc between Syēnē and Alexandria must be one-fiftieth of the great circle of the earth; and as the distance between Syēnē and Alexandria is 5,000 stadia, the great circle of the earth must be 250,000 stadia, or 25,000 geographical miles.

This calculation involved two minor errors. One of these arose from the belief that the earth was a perfect sphere, instead of being flattened at the poles—a mistake which was unavoidable according to the knowledge of that time. The other was caused by Syēnē being regarded as lying directly under the tropic, whereas in reality it was 37 miles to the northward of it. This slight inaccuracy was caused by the imperfection of the methods of observation then employed, for the position of the tropic was calculated partly from the sun being seen from the bottom of a well at the summer solstice, and therefore being considered vertical, and partly from the gnōmōn casting no shadow at that time. A more considerable error proceeded from the distance between Syēnē and Alexandria being overestimated, for the number of stadia assumed for this by Eratosthenēs is in excess by more than one-fifth. Still, after all deductions have been made, the general accuracy of the result is very striking; for whereas the real circumference of the earth at the equator is 25,000 English miles, Eratosthenēs estimates the great circle of the meridian at 25,000 geographical miles, which is about one-seventh part in excess. By the ancients it was regarded as an extraordinary achievement of science, and immense importance was attached to it.

It is certain . . . that Eratosthenēs measured the Earth, and his measurement was astoundingly accurate.

His method consisted in measuring the distance between two places located on the same meridian. If the difference of latitude of these two places is known, it is easy to deduce the length of 1° or of the whole meridian. I do not say 360° because Eratosthenēs divided a great circle into 60 parts; Hipparchos was probably the first to divide it into 360°.

Eratosthenēs' was not the first estimate; according to Aristotle the circumference of the Earth amounted to 400,000 stadia; according to Archimēdēs, 300,000 stadia; according to Eratosthenēs, 252,000 stadia. It is said by Cleomēdēs that his result was $50 \times 5000 = 250,000$ stadia, but he made various measurements and accepted 252,000 as the final result. These measurements were not accurate in the modern sense; they were approximations and the final result was probably made more acceptable for non-experimental reasons ($252 = 2^2 \times 3^2 \times 7$).

In order to determine the latitude, Eratosthenēs used a gnōmōn or a sciothēron. In Syēnē, at the time of the summer solstice there was no shadow at all and he concluded that that place was located on the Tropic of Cancer; Syēnē and Alexandria were, he believed, on the same meridian, their difference of latitude was 7°12' (1/50 of a great circle) and the distance between them amounted to 5000 stadia. Thus the length of the circumference was 250,000 stadia, a result which he corrected eventually to 252,000. These assumptions were not quite correct. The differences in longitude and latitude of the two places are 3°4' (instead of 0°) and 7°7' (instead of 7°12'); the distance 5000 stadia is obviously an approximation in a round number. The distance was measured by a bēmatistēs (a surveyor trained to walk with equal steps and to count them). It is clear that Eratosthenēs was satisfied with approximations: the original figures, 1/50 of the circumference and 5000 stadia, are too good to be true.

It is said that he determined the position of the Tropic by means of a deep well; the sun at noon on the summer solstice would light up the well right down to the water and cast no shadow on the walls. That is not impossible, though a well would hardly be a better instrument than the sciothēron. The "well of Eratosthenēs" is not in Syēnē proper but in Elephantine, an island in the Nile (Jazirat Aswān), opposite Syēnē just below the First Cataract; this makes no difference. The well which can be seen to this day in Elephantine is probably the nilometer (miqyās) described by Strabōn.

If we admit the measurement 252,000 stadia our difficulties are not yet over, for how long was a stadium? There were differences between the various stadia in different times and places, and ancient geographers were hardly conscious of them. Perhaps the most acceptable solution of that insolvable puzzle is the one given by Pliny (XII, 53), according to whom one schoinos equals 40 stadia. On the other hand, according to Egyptologists the schoinos equals 12,000 cubits and the Egyptian cubit equals 0.525 meter. If so, the schoinos equals 6300 meters and Eratosthenes' circumference equals 6300 schoinoi or 39,690 kilometers. That result was almost unbelievably close to the real value (40, 120 km), the error being not much above 1 percent. On that basis, the Eratosthenian stadium equaled 157.5 m, shorter than the Olympic Stadium (185 m) and than the Ptolemaic or Royal one (210 m).

There were about 9.45 Eratosthenian stadia to a mile; according to another interpretation, his stadium was even smaller, 10 to a mile. The other stadia were larger (9, 8 1/3, 8, 7 1/2 to a mile). The smallest of these (9 to a mile) would give a circumference of 41,664 km (too large by less than 4 percent) and the others would increase the error. That does not matter very much, however. Eratosthenēs' achievement lies in his method; whichever the stadium used it would give a not implausible value of the Earth's size. That was a great mathematical achievement.

26. From the writings of Hipparchus

Hipparchus (c. 192–120 B.C.) has been called the greatest astronomer of antiquity. The following excerpts, taken from Strabo, represent Hipparchus' ideas and the esteem in which he was held by his successors, as well as the clarity of exposition for which he was famous.

First of all, it has been rightly asserted both by myself and by my predecessors, one of whom was Hipparchus, that Homer was the founder of the science of geography.

From D. R. Dicks, *The Geographical Fragments of Hipparchus* (London: Athlone Press of the University of London, 1960), pp. 57–58, 77–78, by permission of the publisher.

Now to seek to credit Homer with omniscience might be regarded as a characteristic of the over-zealous: Hipparchus says that to endow Homer with all knowledge and every art is as if a man should hang apples and pears, and other things it cannot bear, on an Attic harvest-wreath.

Moreover, in the Catalogue [of ships] he does not name the cities in their right order, for it is unnecessary; but he does name the nations in the right order. And the case is the same with regard to the far distant nations: "I roamed through Cyprus, Phoenicia and Egypt, and reached the Ethiopians and Sidonians and Erembians and Libya." Hipparchus also notes this . . .

At all events it is a fact that many men have spoken of the necessity for wide learning in relation to this subject [i.e. geography]. Hipparchus also rightly points out in his treatise against Eratosthenes that, while geographical knowledge is the concern of everyone whether layman or scholar, it is impossible to attain it without consideration of the heavens and of the observations of eclipses; thus one cannot determine whether Alexandria in Egypt is north or south of Babylon, or by how much, without investigation by means of the *climata*. Similarly one cannot decide accurately whether places are situated to a greater or less degree towards the east or west except by comparison of [the times of] eclipses of the sun and moon. This is what Hipparchus says, anyway . . .

Later authorities do not agree that the earth is of the size that Eratosthenes has stated, and do not approve of the measurement. Nevertheless, for his list of the celestial phenomena for each inhabited region, Hipparchus makes use of these distances along the meridian through Meroë, Alexandria and [the mouth of] the Borysthenes, and says that they differ little from the truth.

Moreover, Hipparchus' views are largely in agreement with all this. For he says that, on the assumption that the size of the earth is what Eratosthenes gives, it is then necessary to consider the inhabited part of the world separately; for it will not make much difference with respect to the celestial phenomena for each inhabited region, whether the measurement of Eratosthenes is accepted as it stands, or that given by the later authorities.

For very often—in fact in most cases—places lie on a line which is not straight—a straight line cannot occur [on a spherical surface]; but it is possible to speak [of them as lying] on a segment of a circle, and [to determine] what ratio the distance between them bears to the great circle drawn on the earth. Then, as was shown by Hipparchus and Ptolemy himself, if we take the stars directly overhead and measure the distances in degrees between them, we shall find what ratio such a distance bears to the greatest circle, and this ratio will be the same on the earth also, since both the circle of the heavens and that drawn on the earth will comprise similar circumferences. For, let AB be a circle of the heavens, GD one on the earth, E and Z the given places, and η and θ the points directly overhead whose positions we shall find if we produce the line of the circle.

For when we have found by means of the 'meteoroscope' the distance in degrees of the stars from each other, we shall have also the number of stades equivalent to this distance. If we stand in the given places, and observe the star immediately overhead by means of the instrument, we shall find that on the earth also the distance between them is the same, according to the number of stades underlying each celestial degree in that region. Thus it is unnecessary to relate this ratio to the perimeter of the whole earth; and this holds good even if the given distance is not straight and direct.

27. Posidonius on the size of the earth and on zones

Posidonius (c. 135–c. 51 B.C.), a latter-day successor to the Alexandrine scholars who studied heaven and earth, is important because of the manner in which he summarized the findings of his predecessors. In his treatise on the ocean, having first accepted a figure for the circumference of the earth closely approximating that of Eratosthenes, he later reduced it substantially and, in fact, suggested that a voyage from western Europe to the Indies would not be a particularly difficult undertaking. This mistaken idea of the size of the earth haunted generations of geographers and, together with certain concepts of Ptolemy, was still a part of the intellectual background of the Age of Discovery.

The fragments quoted here about Posidonius are from the writings of Cleomedes (*De Motu Circulari*) and Strabo (*geography*).

On the circumference of the earth

Cleomedes
The great circle of the earth (says Posidonius) is found to measure 240,000 stades, if we assume that the distance from Rhodes to Alexandria is 5,000 stades; but if it is not, the measurement is in proportion to the distance.

Strabo
Posidonius judges the amount (of the circumference) to be round about 180,000 stades ... He suspects that the length of the inhabited earth is about 70,000 stades, and that it forms one-half of the whole circle on which it is taken, so that, according to him, a man sailing from the west on a straight voyage would, within so many thousand stades, come to the Indians.

The first two excerpts are from E. H. Warmington, *Greek Geography*, The Library of Greek Thought Series (New York, 1934), pp. 247–248. Published in the United States by E. P. Dutton, and reprinted by their permission. The third is from Cohen and Drabkin, *A Source Book in Greek Science*, (Cambridge, Mass.: Harvard University Press, 1948), pp. 156–158.

On zones

Strabo

Now let us see what Poseidonius has to say in his treatise on Oceanus. For in it he seems to deal mainly with geography, treating it partly from the point of view of geography properly so called, and partly from a more mathematical point of view. And so it will not be out of place for me to pass judgment upon a few of Poseidonius' statements, some of them now, and others in my discussion of the individual countries, as occasion offers, always observing a kind of standard. Now it is one of the things proper to geography to take as an hypothesis that the earth as a whole is spheroidal —just as we do in the case of the universe—and accept all the conclusions that follow this hypothesis, one of which is that the earth has five zones.

Poseidonius, then, says that Parmenides was the originator of the division into five zones, but that Parmenides represents the torrid zone as almost double its real breadth, inasmuch as it falls beyond both the tropics and extends into the two temperate zones, while Aristotle calls "torrid" the region between the tropics, and "temperate" the regions between the tropics and the "arctic circles." But Poseidonius censures both systems, and with justice, for by "torrid," he says, is meant only the region that is uninhabitable on account of heat; and, of the zone between the tropics, more than half is uninhabitable if we may base a conjecture upon the Ethiopians who live south of Egypt—if it be true, first, that each division of the torrid zone made by the equator is half the whole breadth of that zone and, secondly, that, of this half, the part that reaches to Meroë from Syēnē (which is a point on the boundary line of the summer tropic) is five thousand stadia in breadth, and the part from Meroë to the parallel of the Cinnamon-producing Country, on which parallel the torrid zone begins, is three thousand stadia in breadth. Now the whole of these two parts can be measured, for they are traversed both by water and by land; but the rest of the distance, up to the equator, is shown by calculation based upon the measurement which Eratosthenes made of the earth to be eight thousand eight hundred stadia. Accordingly, as is the ratio of the sixteen thousand eight hundred stadia to the eight thousand eight hundred stadia, so would be the ratio of the distance between the two tropics to the breadth of the torrid zone. And if, of the more recent measurements of the earth, the one which makes the earth smallest in circumference be introduced— I mean that of Poseidonius, who estimates its circumference at about one hundred and eighty thousand stadia—this measurement, I say, renders the breadth of the torrid zone somewhere about half the space between the tropics, or slightly more than half, but in no wise equal to, or the same as, that space. And again, Poseidonius asks how one could determine the limits of the temperate zones, which are non-variable, by means of the "arctic circles," which are neither visible among all men nor the same everywhere. Now the fact that the "arctic circles" are not visible to all could be of no aid to his refutation of Aristotle, because the "arctic circles" must be visible to all who live in the temperate zone, with reference to whom alone the term "temperate" is in fact used. But this point that the "Arctic circles" are not everywhere visible in the same way, but are subject to variations, has been well taken.

When Poseidonius himself divides the earth into the zones, he says that five of them are useful with reference to the celestial phenomena; of these five, two—those that lie beneath the poles and extend to the regions that have the tropics as arctic circles—are "periscian"; and the two that come next and extend to the people who live beneath the tropics are "heteroscian"; and the zone between the tropics, "amphiscian." But for purposes of human interest there are, in addition to these five zones, two other narrow ones that lie beneath the tropics and are divided into two parts by the tropics; these have the sun directly overhead for about half a month each year. These two zones, he says, have a certain peculiarity, in that they are parched in the literal sense of the word, are sandy, and produce nothing except silphium and some pungent fruits that are withered by the heat; for those regions have in their neighbourhood no mountains against which the clouds may break and produce rain, nor indeed are they coursed by rivers; and for this reason they produce creatures with woolly hair, crumpled horns, protruding lips, and flat noses (for their extremities are contorted by the heat); and the "fish-eaters" also live in these zones. Poseidonius says it is clear that these things are peculiar to those zones from the fact that the people who live farther south than they do have a more temperate atmosphere, and also a more fruitful, and a better-watered, country.

28. Polybius describes the Black Sea and Italy

Polybius (205?–?125 B.C.) was a Greek historian and statesman who traveled widely throughout the Mediterranean region and even as far as the Black Sea. He was a close friend of several distinguished Romans, including Scipio the conqueror of Carthage, and spent many years writing a history of his time. His writings include descriptions of the areas he visited.

The Black Sea

Since many great rivers from Asia issue into the basins here mentioned, and still more greater ones from Europe, it comes about that Lake Maeotis

---

From E. H. Warmington, *Greek Geography,* The Library of Greek Thought Series (New York, 1934), pp. 52–53, 193–195. Published in the United States by E. P. Dutton and reprinted by their permission.

is filled full by those and flows into the Pontus through the mouth; likewise the Pontus flows into the Propontis . . .

A second cause [sc. of the flow" from Maeotis and Pontus] is that since in times of intense rainfall the rivers carry into these basins quantities of silt of all kinds, the water, under pressure from the accumulation of silt-banks, is displaced and mounts up . . . I say that the silting of the Pontus has been going on from of old and still goes on, and also that in course of time both it and Lake Maeotis will be completely silted up, provided of course that the present local conditions continue as they are and that the causes of the silting up go on acting without interruption . . . What I say will not take place at some remote time, but soon. It can, I assure you, be seen happening now; for it is a fact that Maeotis is already silted up, for the greater part of it varies in depth between five and seven fathoms, so that men cannot any longer navigate it in large ships without a pilot. Again, while it was, as the ancient writers agree, at first a sea confluent with the Pontus, it is now a fresh-water lake . . . something very like this will happen, and is now happening, with regard to the Pontus . . .

For, since the Ister is casting its waters from Europe by several mouths into the Pontus, it comes about that, opposite to this river, a ribbon of land about 1,000 stades long is accumulating at a distance of a day's course from the land, formed by the mud being discharged by the mouths . . . As far as the currents of the rivers prevail through the strength of flow and force a way through the sea, the earth and all that is conveyed by their currents must of necessity be thrust forward . . . But whenever the streams lose their power through the depth and volume of the sea, then at once the silt is naturally borne downwards, settles and comes to rest. Hence it is that the deposits of large and violent rivers accumulate at a distance, and the parts along the land are deep, while in the case of lesser and gently flowing rivers the banks accumulate right at the mouths. This becomes especially clear at times of heavy rains; for then even ordinary streams, when they over-power the waves at their mouths, thrust the silt forward into the sea for a distance proportionate to the force of their respective currents . . . Inas-much as Maeotis is now sweeter than the Pontic Sea, so the Pontic is found to be appreciably different from our own sea. From which it is clear that when that time has passed which stands to the time required to fill up Maeotis in the same proportion as the cubic capacity of the larger basin to that of the other, then the Pontus also will become a shallow swamp and sweet and resembling a lake, much like Lake Maeotis. Indeed we must suppose that this will happen still sooner, inasmuch as the streams of the rivers which fall into this Pontus are greater and more numerous.

Italy

Italy as a whole is in shape a triangle of which one side towards the east is bounded by the Ionian Strait, and then continuously by the Adriatic

Gulf, while the other side, which is turned towards the south and west, is bounded by the Sicilian and Tyrrhenian Sea.

These two sides meeting each other form the apex of the triangle at the jutting headland of Italy towards the south, called Cocynthus, and separating the Ionian Strait from the Sicilian Sea. The remaining side of the triangle stretching along the north and inland is bounded without a break by the mountain-side of the Alps, which begins at Massalia and the regions above the Sardinian Sea and extends continuously as far as the recess of the Adriatic, except a short distance by which the range fails to touch that sea by stopping short. Along the aforesaid chain, which we should conceive as the base of the triangle, on its southern side, lies the last low plain of all Italy towards the north . . . surpassing in fertility and size all other plains in Europe so far as we have investigated. The shape as a whole of the lines which bound this plain also is triangular. The apex of this figure is formed by the junction of the mountains called Apennine, and the Alps not far from the Sardinian Sea above Massalia. Of its sides . . . the Alps themselves are found to stretch along the northern for about 2,200 stades, and the Apennine range the southern side for about 3,600. The position of the base of the whole figure is held by the sea-coast of the Adriatic Gulf, its size from the city Sena to the recess being above 2,500 stades, so that the measurement of the whole circumference of the plain mentioned falls not far short of 10,000 stades . . . Of the regions provided with hills and good soil on either slope of the Alps (that which bears towards the Rhodanus, and that which bears towards the plain mentioned), the regions turned towards the Rhodanus and the north are inhabited by Gauls called Transalpine, and the regions towards the plain by the Taurisci and Agones, and several other barbarous nations . . . The summits (*sc. of the Alps*) are utterly uninhabitable because of their ruggedness and the quantity of snow which remains on them always. The Apennine range from its beginning above Massalia and its meeting with the Alps is inhabited by Ligystini (*Ligurians*); they possess both sides of the range, one of which inclines towards the Tyrrhenian Sea and the other towards the plain; they hold the sea-coast as far as the city Pisa, which is the first city of western Tyrrhenia (*Etruria*), and the inland as far as the territory of Arretium. Next in order are the Tyrrheni; adjoining these on either slope dwell the Umbrians. After this, the Apennines, keeping a distance of about 500 stades from the Adriatic Sea, leave the plain behind and, bearing to the right, and extending through the middle of the rest of Italy, stretch down to the Sicilian Sea. The remaining flat part of this side (*sc. of the triangle*) goes down to the sea and the city Sena. The river Padus, celebrated by poets as Eridanus, has its sources in the Alps, rather towards the apex of the figure, mentioned above, and is borne down to the plain, sending its stream towards the south. When it reaches these flat regions, its stream takes a turn to the east, and is borne through the plain, and issues by two mouths into the waters of the Adriatic. It cuts off a large part of the plainland towards the Alps, and the recess of the Adriatic. It brings a volume of water larger than any other rivers in Italy because all the streams which bear down from the Alps and the Apennine mountains fall into it on all sides. It flows at its highest and grandest about the rising of the Dog-star (*in the middle of July*), being then increased by the quantity of melting snow on the mountains mentioned above. From the sea at the mouth called Olana it is navigable upstream for about 2,000 stades; while at first it flows from its sources with a single stream, it is split into two parts at the place called Trigaboli; one of the mouths is named Padua, and the other Olana. At the latter lies a harbour, which provides for navigators who anchor in it greater safety than any other in the Adriatic. Among the natives the river is called Bodecus.

29. Strabo: the summing up of Greek geography

Strabo (c. 63 B.C.–c. 25 A.D.), a native of Amasia in Asia Minor, was Greek by language and education but was also a great admirer of things Roman. Of his writings only one remains—the *Geographica*. It is one of the most valuable works of classical times because it reflects the author's knowledge of a wide variety of Greek sources that are now lost. In essence, Strabo's *Geographica* is a detailed survey of the *oikumene* (the habitable world, according to the Greeks) liberally sprinkled with references to the views of his distinguished predecessors.

On geography in general

The science of Geography, which I now propose to investigate, is, I think, quite as much as any other science, a concern of the philosopher; and the correctness of my view is clear for many reasons. In the first place, those who in earliest times ventured to treat the subject were, in their way, philosophers—Homer, Anaximander of Miletus, and Anaximander's fellow-citizen Hecataeus—just as Eratosthenes has already said; philosophers,

---

The first, third, fourth, and fifth selections are from *The Geography of Strabo,* trans. H. L. Jones, Loeb Classical Library (London: William Heinemann, 1917 and 1924), I, 3, 23–29, 41–43; I, 179–187; III, 87-89; I, 419–433. The ninth and tenth selections are from the *Geography,* trans. H. C. Hamilton and W. Falconer (London, 1889), pp. 182–183, 220–225. The other selections are from E. H. Warmington, *Greek Geography,* The Library of Greek Thought Series (New York, 1934), pp. 250, 252–253; 179–180; 193; 213–214; 182–184. Published in the United States by E. P. Dutton and reprinted by their permission.

too, were Democritus, Eudoxus, Dicaearchus, Ephorus, with several others of their times; and further, their successors—Eratosthenes, Polybius, and Poseidonius—were philosophers. In the second place, wide learning, which alone makes it possible to undertake a work on geography, is possessed solely by the man who has investigated things both human and divine— knowledge of which, they say, constitutes philosophy. And so, too, the utility of geography—and its utility is manifold, not only as regards the activities of statesmen and commanders but also regards knowledge both of the heavens and of things on land and sea, animals, plants, fruits, and everything else to be seen in various regions—the utility of geography, I say, presupposes in the geographer the same philosopher, the man who busies himself with the investigation of the art of life, that is, of happiness . . .

For the moment what I have already said is sufficient, I hope, to show that Homer was the first geographer. And, as every one knows, the successors of Homer in geography were also notable men and familiar with philosophy. Eratosthenes declares that the first two successors of Homer were Anaximander, a pupil and fellow-citizen of Thales, and Hecataeus of Miletus; that Anaximander was the first to publish a geographical map, and that Hecataeus left behind him a work on geography, a work believed to be his by reason of its similarity to his other writings.

Assuredly, however, there is need of encyclopaedic learning for the study of geography, as many men have already stated; and Hipparchus, too, in his treatise *Against Eratosthenes,* correctly shows that it is impossible for any man, whether layman or scholar, to attain to the requisite knowledge of geography without the determination of the heavenly bodies and of the eclipses which have been observed; for instance, it is impossible to determine whether Alexandria in Egypt is north or south of Babylon, or how much north or south of Babylon it is, without investigation through the means of the "climata." In like manner, we cannot accurately fix points that lie at varying distances from us, whether to the east or the west, except by a comparison of the eclipses of the sun and the moon. That, then, is what Hipparchus says on the subject.

All those who undertake to describe the distinguishing features of countries devote special attention to astronomy and geometry, in explaining matters of shape, of size, of distances between points, and of "climata," as well as matters of heat and cold, and, in general, the peculiarities of the atmosphere. Indeed, an architect in constructing a house, or an engineer in founding a city, would make provision for all these conditions; and all the more would they be considered by the man whose purview embraced the whole inhabited world; for they concern him more than anyone else. Within the area of small countries it involves no very great discrepancy if a given place be situated more towards the north, or more towards the south; but when the area is that of the whole round of the inhabited world, the north extends to the remote confines of Scythia and Celtica, and the south to the remote confines of Ethiopia, and the difference between these two extremes is very great. The same thing holds true also as regards a man's living in India or Iberia; the one country is in the far east, and the other is in the far west; indeed, they are, in a sense, the antipodes of each other, as we know.

Everything of this kind, since it is caused by the movement of the sun and the other stars as well as by their tendency towards the centre, compels us to look to the vault of heaven, and to observe the phenomena of the heavenly bodies peculiar to our individual positions; and in these phenomena we see very great variation in the position of inhabited places. So, if one is about to treat of the differences between countries, how can he discuss his subject correctly and adequately if he has paid no attention, even superficially, to any of these matters? For even if it be impossible in a treatise of this nature, because of its having a greater bearing on affairs of state, to make everything scientifically accurate, it will naturally be appropriate to do so, at least in so far as the man in public life is able to follow the thought.

Moreover, the man who has once thus lifted his thoughts to the heavens will surely not hold aloof from the earth as a whole; for it is obviously absurd, if a man who desired to give a clear exposition of the inhabited world had ventured to lay hold of the celestial bodies and to use them for the purposes of instruction, and yet had paid no attention to the earth as a whole, of which the inhabited world is but a part—neither as to its size, nor its character, nor its position in the universe, nor even whether the world is inhabited only in the one part in which we live, or in a number of parts, and if so, how many such parts there are; and likewise how large the uninhabited part is, what its nature is, and why it is uninhabited. It seems, then, that the special branch of geography represents a union of meteorology and geometry, since it unites terrestrial and celestial phenomena as being very closely related, and in no sense separated from each other as heaven is high above the earth . . .

Most of all, it seems to me, we need, as I have said, geometry and astronomy for a subject like geography; and the need of them is real indeed; for without such methods as they offer it is not possible accurately to determine our geometrical figures, "climata", dimensions, and the other cognate things; but just as these sciences prove for us in other treatises all that has to do with the measurement of the earth as a whole and as I must in this treatise take for granted and accept the propositions proved there, so I must take for granted that the universe is spheroidal, and also that the earth's surface is spheroidal, and, what is more I must take for granted the law that is prior to these two principles, namely that the bodies tend toward the centre; and I need only indicate, in a brief and summary way, whether a proposition comes—if it really does—within the range of sense-perception or of intuitive knowledge. Take, for example, the proposition that the earth is spheroidal: whereas the suggestion of this proposition comes to us mediately from the law that bodies tend toward the centre and that each body inclines toward its own centre of gravity, the suggestion comes immediately from the phenomena observed at sea and in the heavens; for our sense-perception and also our intuition can bear testimony in the latter case. For instance, it is obviously the curvature of the sea that prevents sailors from seeing distant lights that are placed on a level with their eyes. At any rate, if the lights are elevated above the level of the eyes, they become visible, even though they be at a greater distance from the eyes; and similarly if the eyes themselves are elevated, they see what was before invisible; This fact is noted by Homer, also, for such is the meaning of the words: "With a quick glance ahead, being upborne on a great wave, he saw the land very near." So, also, when sailors are approaching land, the different parts of the shore become revealed progressively, more and more, and what at first appeared to be low-lying land grows gradually higher and higher. Again, the revolution of the heavenly bodies is evident on many grounds, but it is particularly evident from the phenomena of the sun-dial; and from these phenomena our intuitive judgment itself suggests that no such revolution could take place if the earth were rooted to an infinite depth.

On the *oikumene*

The inhabited earth is shaped like a soldier's cloak, and is an island in the Atlantic and is less than one-half of the quadrilateral *(in which it lies)* . . .

Its greatest breadth is marked out by the line along the Nile, beginning at the parallel through the Cinnamon-bearing country . . . and reaching to the parallel through Ierne, while its length is marked by the line drawn at right angles to this from the west through the Pillars and the Sicilian Strait to Rhodes and the Gulf of Issus, then running along the Taurus, which girdles Asia, and finishing at the eastern sea between the Indians and the Scythians above Bactriane. We must imagine a parallelogram within which the cloak-shaped figure is drawn so that the greatest length and breadth of the one correspond with the greatest length and breadth of the other . . . the extremities at either end of the length *(of the inhabited earth)* tail off to a point, and take away some of the breadth, being washed by the sea. This is clear from those who have sailed round the eastern and western parts in both directions *(sc. north and south)*; they declare that the island called Taprobane is far to the south of India, but inhabited nevertheless and rising in the same latitude as the island of the Egyptians and the land which bears the cinnamon; for the temperature of their atmospheres is very similar. Again, the regions round the mouth of the Hyrcanian *(Caspian)* Sea are farther north than farthest Scythia is beyond the Indians; more so still are the regions of Ierne. Similar statements are made about the land outside the Pillars; it is said that the westernmost point of the inhabited earth is the promontory (*Cape St. Vincent*) of Iberia which they call Sacred. It lies roughly on the line through Gadeira, the Pillars and the Sicilian Strait, and Rhodes ... As you sail towards the southern parts you find Libya. The westernmost regions of this project a little beyond Gadeira, and then having formed a narrow headland recede south-eastwards; they broaden little by little until they touch upon the western Ethiopians; these are the farthest of the peoples who live beneath the regions of Carthage, and touch upon the parallel through the Cinnamon-bearing country. In the opposite direction as you sail from the Sacred Promontory as far as the people called Artabri (*round Cape Finisterre*) your voyage goes towards the north, and you have Lusitania on the right. Then all the rest of the voyage makes an obtuse angle eastwards as far as the extremities of Pyrene, which end at the Ocean. Opposite to these and towards the north lie the western parts of Brettanice, and likewise opposite the Artabri and towards the north lie the islands called Tin Islands; they are in the open sea, and situated roughly in the same clime as Brettanice ...

On changes on the earth's surface

Now after Eratosthenes has himself told what great advances in the knowledge of the inhabited world had been made not only by those who came after Alexander but by those of Alexander's own times, he passes to his discussion of the shape of the world, not indeed of the inhabited world— which would have been more appropriate to his discussion of that subject —but of the earth as a whole; of course, one must discuss that point too, but not out of its proper place. And so, after he has stated that the earth as a whole is spheroidal—not spheroidal indeed as though turned by a sphere-lathe, but that it has certain irregularities of surface—he proceeds to enumerate the large number of its successive changes in shape—changes which take place as the result of the action of water, fire, earthquakes, volcanic eruptions, and other similar agencies; and here too he does not preserve the proper order. For the spheroidal shape that characterises the earth as a whole results from the constitution of the universe, but such changes as Eratosthenes mentions do not in any particular alter the earth as a whole (changes so insignificant are lost in great bodies), though they do produce conditions in the inhabited world that are different at one time from what they are at another, and the immediate causes which produce them are different at different times.

Eratosthenes says further that this question in particular has presented a problem: how does it come about that large quantities of mussel-shells, oyster-shells, scallop-shells, and also salt-marshes are found in many places in the interior at a distance of two thousand or three thousand stadia from the sea—for instance (to quote Eratosthenes) in the neighbourhood of the temple of Ammon and along the road, three thousand stadia in length, that leads to it? At that place, he says, there is a large deposit of oyster-shells, and many beds of salt are still to be found there, and jets of salt-water rise to some height; besides that, they show pieces of wreckage from sea-faring ships which the natives said had been cast up through a certain chasm, and on small columns dolphins are dedicated that bear the inscription: "Of Sacred Ambassadors of Cyrene." Then he goes on to praise the opinion of Strato, the physicist, and also that of Xanthus of Lydia. In the first place he praises the opinion of Xanthus, who says that in the reign of Artaxerxes there was so great a drought that the rivers, lakes, and wells dried up; that far from the sea, in Armenia, Matiene, and Lower Phrygia, he himself had often seen, in many places, stones in the shape of a bivalve, shells of the pecten order, impressions of scallop-shells, and a salt-marsh, and therefore was persuaded that these plains were once sea. Then Eratosthenes praises the opinion of Strato, who goes still further into the question of causes, because Strato says he believes the Euxine Sea formerly did not have its outlet at Byzantium, but the rivers which empty into the Euxine forced and opened a passage, and then the water was discharged into the Propontis and the Hellespont. The same thing, Strato says, happened in the Mediterranean basin also; for in this case the passage at the Pillars was broken through when the sea had been filled by the rivers, and at the time of the outrush of the water the places that had hitherto been covered with shoal-waters were left dry. Strato proposes as a cause of this, first, that the beds of the Atlantic and the Mediterranean are on different levels, and, secondly, that at the Pillars even at the present day a submarine ridge stretches across from Europe to Libya, indicating that the Mediterranean and the Atlantic could not have been one and the same formerly. The seas of the Pontus region, Strato continues, are very shallow, whereas the Cretan, the Sicilian, and the Sardinian Seas are very deep; for since the rivers that flow from the north and east are very numerous and very large, the seas there are being filled with mud, while the others remain deep; and herein also is the reason why the Pontus is sweetest, and why its outflow takes place in the direction of the inclination of its bed. Strato further says it is his opinion that the whole Euxine Sea will be silted up at some future period, if such inpourings continue; for even now the regions on the left side of the Pontus are already covered with shoal waters; for instance, Salmydessus, and the land at the mouth of the Ister, which sailors call "the Breasts," and the desert of Scythia; perhaps too the temple of Ammon was formerly on the sea, but is now situated in the interior because there has been an outpouring of the sea. Strato conjectures that the oracle of Ammon with good reason became so distinguished and so well-known as it is if it was situated on the sea, and that its present position so very far from the sea gives no reasonable explanation of its present distinction and fame; and that in ancient times Egypt was covered by the sea as far as the bogs about

Pelusium, Mt. Casius, and Lake Sirbonis; at all events, even to-day when the salt-lands in Egypt are dug up, the excavations are found to contain sand and fossil-shells, as though the country had been submerged beneath the sea and the whole region round Mt. Casius and the so-called Gerrha had once been covered with shoal water so that it connected with the Gulf of the Red Sea; and when the sea retired, these regions were left bare, except that the Lake Sirbonis remained; then the lake also broke through to the sea, and thus became a bog. In the same way, Strato adds, the beaches of the so-called Lake Moeris more nearly resemble sea-beaches than river-banks. Now one may admit that a great part of the continents was once covered by water for certain periods and was then left bare again; and in the same way one may admit also that the whole surface of the earth now submerged is uneven, at the bottom of the sea, just as we might admit, of course, that the part of the earth above water, on which we live, is subject to all the changes mentioned by Eratosthenes himself.

On volcanoes

Near Centoripa is the town of Aetna, which was mentioned a little above, whose people entertain and conduct those who ascend the mountain; for the mountain-summit begins here. The upper districts are bare and ash-like and full of snow during the winter, whereas the lower are divided up by forests and plantations of every sort. The topmost parts of the mountain appear to undergo many changes because of the way the fire distributes itself, for at one time the fire concentrates in one crater, but at another time divides, while at one time the mountain sends forth lava, at another, flames and fiery smoke, and at still other times it also emits red-hot masses; and the inevitable result of these disturbances is that not only the underground passages, but also the orifices, sometimes rather numerous, which appear on the surface of the mountain all round, undergo changes at the same time. Be this as it may, those who recently made the ascent gave me the following account: They found at the top a level plain, about twenty stadia in circuit, enclosed by a rim of ashes the height of a house-wall, so that any who wished to proceed into the plain had to leap down from the wall; they saw in the centre of the plain a mound of the colour of ashes, in this respect being like the surface of the plain as seen from above, and above the mound a perpendicular cloud rising straight up to a height of about two hundred feet, motionless (for it was a windless day) and resembling smoke; and two of the men had the hardihood to proceed into the plain, but because the sand they were walking on got hotter and deeper, they turned back, and so were unable to tell those who were observing from a distance anything more than what was already apparent. But they believed, from such a view as they had, that many of the current stories are mythical, and particularly those which some tell about Empedocles, that he leaped down into the crater and left behind, as a trace of the fate he suffered, one of the brazen sandals which he wore; for it was found, they say, a short distance outside the rim of the crater, as though it had been thrown up by the force of the fire. Indeed, the place is neither to be approached nor to be seen, according to my informants; and further, they surmised that nothing could be thrown down into it either, owing to the contrary blasts of the winds arising from the depths, and also owing to the heat, which, it is reasonable to suppose, meets one long before one comes near the mouth of the crater; but even if something should be thrown down into it, it would be destroyed before it could be thrown up in anything like the shape it had when first received; and although it is not unreasonable to assume that at times the blasts of the fire die down when at times the fuel is deficient, yet surely this would not last long enough to make possible the approach of man against so great a force. Aetna dominates more especially the seaboard in the region of the Strait and the territory of Catana, but also that in the region of the Tyrrhenian Sea and the Liparaean Islands. Now although by night a brilliant light shines from the summit, by day it is covered with smoke and haze . . .

Phenomena akin both to these and to those in Sicily are to be seen about the Liparaean Islands and Lipara itself. The islands are seven in number, but the largest is Lipara (a colony of the Cnidians), which, Thermessa excepted, lies nearest to Sicily. It was formerly called Meligunis; and it not only commanded a fleet, but for a long time resisted the incursions of the Tyrrheni, for it held in obedience all the Liparaean Islands, as they are now called, though by some they are called the Islands of Aeolus. Furthermore, it often adorned the temple of Apollo at Delphi with dedications from the first fruits of victory. It has also a fruitful soil, and a mine of styptic earth that brings in revenues, and hot springs, and fire-blasts. Between Lipara and Sicily is Thermessa, which is now called Hiera of Hephaestus; the whole island is rocky, desert, and fiery, and it has three fire blasts, rising from three openings which one might call craters. From the largest the flames carry up also red-hot masses, which have already choked up a considerable part of the Strait. From observation it has been believed that the flames, both here and on Aetna, are stimulated along with the winds and that when the winds cease the flames cease too. And this is not unreasonable, for the winds are begotten by the evaporations of the sea and after they have taken their beginning are fed thereby; and therefore it is not permissible for any who have any sort of insight into such matters to marvel if the fire too is kindled by a cognate fuel or disturbance. According to Polybius, one of the three craters has partially fallen in, whereas the others remain whole; and the largest has a circular rim five stadia in circuit, but it gradually contracts to a diameter of fifty feet; and the altitude of this crater above the level of the sea is a stadium, so that the crater is visible on windless days. But if all this is to be believed, perhaps one should also believe the mythical story about Empedocles. Now if the south wind is about to blow, Polybius continues, a cloud-like mist pours down all round the island, so that not even Sicily is visible in the distance; and when the north wind is about to blow, pure flames rise aloft from the aforesaid crater and louder rumblings are sent forth; but the west wind holds a middle position, so to speak, between the two; but though the two other craters are like the first in kind, they fall short in the violence of their spoutings; accordingly, both the difference in the rumblings, and the place whence the spoutings and the flames and the fiery smoke begin, signify beforehand the wind that is going to blow again three days afterward; at all events, certain of the men in Liparae, when the weather made sailing impossible, predicted, he says, the wind that was to blow, and they were not mistaken; from this fact, then, it is clear that that saying of the Poet which is regarded as most mythical of all was not idly spoken, but that he hinted at the truth when he called Aeolus "steward of the winds." However, I have already discussed these matters sufficiently. It is the close attention of the Poet to vivid description, one might call it, . . . for both are equally present in rhetorical composition and vivid description; at any rate pleasure is common to both.

On the mathematical and physical principles of geography; on zones; and on determining the extent of the *oikumene*

Since the taking in hand of my proposed task naturally follows the criticisms of my predecessors, let me make a second beginning by saying that the person who attempts to write an account of the countries of the earth must take many of the physical and mathematical principles as hypotheses and elaborate his whole treatise with reference to their intent and authority. For, as I have already said, no architect or engineer would be competent even to fix the site of a house or a city properly if he had no conception beforehand of "climata" and of the celestial phenomena, and of geometrical figures and magnitudes and heat and cold and other such things—much less a person who would fix positions for the whole of the inhabited world. For the mere drawing on one and the same plane surface of Iberia and India and the countries that lie between them and, in spite of its being a plane surface, the plotting of the sun's position as its settings, risings, and in meridian, as though these positions were fixed for all the people of the world—merely this exercise gives to the man who has previously conceived of the arrangement and movement of the celestial bodies and grasped the fact that the true surface of the earth is spherical but that it is depicted for the moment as a plane surface for the convenience of the eye—merely this exercise, I say, gives to that man instruction that is truly geographical, but to the man not thus qualified it does not. Indeed, the case is not the same with us when we are dealing with geography as it is when we are travelling over great plains (those of Babylonia, for example) or over the sea: then all that is in front of us and behind us and on either side of us is presented to our minds as a plane surface and offers no varying aspects with reference to the celestial bodies or the movements or the positions of the sun and the other stars relatively to us; but when we are dealing with geography the like parts must never present themselves to our minds in that way. The sailor on the open sea, or the man who travels through a level country, is guided by certain popular notions (and these notions impel not only the uneducated man but the man of affairs as well to act in the self-same way), because he is unfamiliar with the heavenly bodies and ignorant of the varying aspects of things with reference to them. For he sees the sun rise, pass the meridian, and set, but how it comes about he does not consider; for, indeed, such knowledge is not useful to him with reference to the task before him, any more than it is useful for him to know whether or not his body stands parallel to that of his neighbour. But perhaps he does consider these matters, and yet holds opinions opposed to the principles of mathematics—just as the natives of any given place do; for a man's place occasions such blunders. But the geographer does not write for the native of any particular place, nor yet does he write for the man of affairs of the kind who has paid no attention to the mathematical sciences properly so-called; nor, to be sure, does he write for the harvest-hand or the ditch-digger, but for the man who can be persuaded that the earth as a whole is such as the mathematicians represent it to be, and also all that relates to such an hypothesis. And the geographer urges upon his students that they first master those principles and then consider the subsequent problems; for, he declares, he will speak only of the results which follow from those principles; and hence his students will the more unerringly make the application of his teachings if they listen as mathematicians; but he refuses to teach geography to persons not thus qualified.

Now as for the matters which he regards as fundamental principles of his science, the geographer must rely upon the geometricians who have measured the earth as a whole; and in their turn the geometricians must rely upon the astronomers; and again the astronomers upon the physicists. Physics is a kind of *Arete;* and by *Aretai* they mean those sciences that postulate nothing but depend upon themselves, and contain within themselves their own principles as well as the proofs thereof. Now what we are taught by the physicists is as follows: The universe and the heavens are sphere-shaped. The tendency of the bodies that have weight is towards the centre. And, having taken its position about this centre, the earth is spherically concentric with the heavens, and it is motionless as is also the axis through it, which axis extends also through the centre of the heavens. The heavens revolve round both the earth and its axis from east to west; and along with the heavens revolve the fixed stars, with the same rapidity as the vault of the heavens. Now the fixed stars move along parallel circles, and the best known parallel circles are the equator, the two tropics, and the arctic circles; whereas the planets and the sun and the moon move along certain oblique circles whose positions lie in the zodiac. Now the astronomers first accept these principles, either in whole or in part, and then work out the subsequent problems, namely, the movements of the heavenly bodies, their revolutions, their eclipses, their sizes, their respective distances, and a host of other things. And, in the same way, the geometricians, in measuring the earth as a whole, adhere to the doctrines of the physicists and the astronomers, and, in their turn, the geographers adhere to those of the geometricians.

Thus we must take as an hypothesis that the heavens have five zones, and that the earth also has five zones, and that the terrestrial zones have the same names as the celestial zones (I have already stated the reasons for this division into zones). The limits of the zones can be defined by circles drawn on both sides of the equator and parallel to it, namely, by two circles which enclose the torrid zone, and by two others, following upon these, which form the two temperate zones next to the torrid zone and the two frigid zones next to the temperate zones. Beneath each of the celestial circles falls the corresponding terrestrial circle which bears the same name: and, in like manner, beneath the celestial zone, the terrestrial zone. Now they call "temperate" the zones that can be inhabited; the others they call uninhabitable, the one on account of the heat, and the other two on account of the cold. They proceed in the same manner with reference to the tropic and the arctic circles (that is, in countries that admit of arctic circles): they define their limits by giving the terrestrial circles the same names as the celestial—and thus they define all the terrestrial circles that fall beneath the several celestial circles. Since the celestial equator cuts the whole heavens in two, the earth also must of necessity be cut in two by the terrestrial equator. Of the two hemispheres— I refer to the two celestial as well as the two terrestrial hemispheres—one is called "the northern hemisphere" and the other "the southern hemisphere"; so also, since the torrid zone is cut in two by the same circle, the one part of it will be the northern and the other the southern. It is clear that, of the temperate zones also, the one will be northern and the other southern, each bearing the name of the hemisphere in which it lies. That hemisphere is called "northern hemisphere" which contains that temperate zone in which, as you look from the east to the west, the pole is on your right hand and the equator on your left, or in which, as you look towards the south, the west is on your right hand and the east on your left; and that hemisphere is called "southern hemisphere," in which the opposite is true; and hence it is clear that we are in one of the two hemispheres (that is, of course, in the northern), and that it is impossible for us to be in both. "Between them are great rivers; first, Oceanus", and then the torrid zone. But neither is there an Oceanus in the centre of our whole inhabited world, cleaving the whole of it, nor, to be sure, is there a torrid spot in it; nor yet, indeed, is there a portion of it to be found whose "climata" are opposite to the "climata" which I have given for the northern temperate zone.

By accepting these principles, then, and also by making use of the sun-dial and the other helps given him by the astronomer—by means of which are found, for the several inhabited localities, both the circles that are parallel to the equator and the circles that cut the former at right angles, the latter being drawn through the poles—the geometrician can measure the inhabited portion of the earth by visiting it and the rest of the earth by his calculation of the intervals. In this way he can find the distance from the equator to the pole, which is a fourth part of the earth's largest circle; and when he has this distance, he multiplies it by four; and this is the circumference of the earth. Accordingly, just as the man who measures the earth gets his principles from the astronomer and the astronomer his from the physicist, so, too, the geographer must in the same way first take his point of departure from the man who has measured the earth as a whole, having confidence in him and in those in whom he, in his turn, had confidence, and then explain, in the first instance, our inhabited world—its size, shape, and character, and its relations to the earth as a whole; for this is the peculiar task of the geographer. Then, secondly, he must discuss in a fitting manner the several parts of the inhabited world, both land and sea, noting in passing wherein the subject has ben treated inadequately by those of our predecessors whom we have believed to be the best authorities on these matters.

Now let us take as hypothesis that the earth together with the sea is sphere-shaped and that the surface of the earth is one and the same with that of the high seas; for the elevations on the earth's surface would disappear from consideration, because they are small in comparison with the great size of the earth and admit of being overlooked; and so we use "sphere-shaped" for figures of this kind, not as though they were turned on a lathe, nor yet as the geometrician uses the sphere for demonstration, but as an aid to our conception of the earth—and that, too, a rather rough conception. Now let us conceive of a sphere with five zones, and let the equator be drawn as a circle upon that sphere, and let a second circle be drawn parallel thereto, bounding the frigid zone in the northern hem-isphere, and let a third circle be drawn through the poles, cutting the other two circles at right angles. Then, since the northern hemisphere contains two-fourths of the earth, which are formed by the equator with the circle that passes through the poles, a quadrilateral area is cut off in each of the two fourths. The northern side of the quadrilateral is half of the parallel next to the pole; the southern side is half of the equator; and the two remaining sides are segments of the circle that runs through the poles, these segments lying opposite to each other and being equal in length. Now in one of these two quadrilaterals (it would seem to make no difference in which one) we say that our inhabited world lies, washed on all sides by the sea and like an island; for, as I have already said above, the evidence of our senses and of reason prove this. But if anyone disbelieves the evidence of reason, it would make no difference, from the point of view of the geographer, whether we make the inhabited world an island, or merely admit what experience has taught us, namely, that it is possible to sail round the inhabited world on both sides, from the east as well as from the west, with the exception of a few intermediate stretches. And, as to these stretches, it makes no difference whether they are bounded by sea or by uninhabited land; for the geographer undertakes to describe the known parts of the inhabited world, but he leaves out of consideration the unknown parts of it—just as he does what is outside of it. And it will suffice to fill out and complete the outline of what we term "the island" by joining with a straight line the extreme points reached on the coasting-voyages made on both sides of the inhabited world.

On India

We made it clear in our former discussion that the most trustworthy account of the country regarded as India when Alexander invaded it is the summary set forth by Eratosthenes in the third book of his geographical work. The Indus was then the boundary between this land in Ariana . . . The following is the purport of Eratosthenes's account: India is bounded on the north by the farthest parts of the Taurus stretching from Ariana to the eastern sea; the natives give different names to different parts of the mountains—Paropamisus, Emodus, Imaus, and so on, but the Macedonians named them Caucasus; on the west the boundary is the river Indus, while the southern and eastward sides, which are much bigger than the others, jut out into the Atlantic Sea (*here, the whole Ocean*), and the shape of the country becomes rhomboidal, each of the greater sides exceeding the opposite side by as much as 3,000 stades, this being the distance which the headland, common to both the eastern and the southern coasts, juts out beyond the rest of the shore equally on the east and south. The length of the western side from the Caucasian mountains (*Hindu Kush and Himalayas*) to the southern sea is said to be about 13,000 stades, measured along the river Indus to its mouths, so that the eastern side opposite, with the addition of the 3,000 stades of the headland, will measure like 6,000 stades. The total length where it is shortest will be 16,000 stades of the country, the length being the measurement from west to east. About this as far as Palibothra we may speak with more certainty, for it has been measured by measuring lines, and there is a royal road of 10,000 stades. The regions beyond are measured by conjecture from sailings from the sea up the Ganges as far as Palibothra; the figure would be something like 6,000 stades. The total length where it is shortest will be 16,000 stades according to Eratosthenes, who says he took it from the register of Resting-stations which is looked on as trustworthy. Megasthenes's statement in this case agrees with him, but Patrocles makes it 1,000 stades less. But by adding to this distance that of the headland which juts out far towards the east, these further 3,000 stades will represent the greatest length, which is taken as from the mouths of the river Indus along the coast immediately following, as far as the aforesaid headland and its eastern limits.

On the Alps

Polybius, in speaking of the size and height of the Alps, compares them with the biggest mountains among the Greeks—Taygetus, Lycaeus, Parnassus, Olympus, Pelion, Ossa; and in Thrace Haemus, Rhodope, and Dunax. And he says that each of these latter can be climbed in about a day if you travel light, and can be walked round in a day, but nobody could climb the Alps even in five days. The length of that part which stretches along the plains is 2,200 stades. He names four passes only, one through Liguria, nearest to the Tyrrhenian Sea, another through the Taurini, through which Hannibal passed, another (*Great or Little St. Bernard?*) through the Salassi, and a fourth through the Rhaeti (*in Tyrol*), all very steep. He says that there are in the mountains several lakes, three of them large; of these Benacus (*Garda*) has a length of 500 stades, and a breadth of 30; from it flows the river Mincius (*Mincio*). Next in order comes Larius (*Como*), 400 long, but in breadth narrower than the former; it discharges the river Adus (*Adda*). The third is Verbanus (*Maggiore*), in length nearly 300 stades, and breadth 30; it sends out a big river the Ticinus (*Ticino*). All these rivers flow into the Padus (*Po*).

On the Iberian Peninsula

Iberia is shaped much like a stripped ox-hide, of which the parts forming as it were a neck project over into Celtice; these are the regions which lie towards the east, and within these the range called Pyrene runs in and divides off the eastern side... Iberia is washed all round by the sea, the southern part by our sea as far as the Pillars, and the rest by the Atlantic as far as the northern extremity of Pyrene. The greatest length of this country is round about 6,000 stades, and the breadth 5,000.

Most of this land is but poorly inhabited. The people dwell in mountains and oak forests and plains which have fine friable earth, most of this being not evenly irrigated. The part towards the north is exceedingly cold, be-

sides being rugged, and borders on the Ocean, and has the further dis-
advantage of lack of intercourse and communication with other regions
. . . But nearly all the southern part is fertile . . . There are parts where the
breadth is much less than 3,000 stades, especially towards Pyrene, which
forms the eastern side; this continuous range, stretching from south to
north, forms the boundary between Iberia and Celtice. Both Celtice and
Iberia vary in breadth, the narrowest breadth of either from our sea to
the Ocean being near Pyrene, especially on either side of it; this produces
gulfs, some on the Ocean, some on our sea. The biggest are the Celtic,
which they also call Galatic (*Gulfs of Lyons and Gascony*), making the
Celtic isthmus narrower than the Iberic. The eastern side then of Iberia
is formed by Pyrene, the southern by our sea from Pyrene as far as the
Pillars and the adjoining outer sea up to the promontory called Sacred
(*Cape St. Vincent*). Third comes the western side, roughly parallel to
Pyrene, from the Sacred Promontory to the headland of the Artabri, which
is called Nerium (*Cape Finisterre*). The fourth side stretches thence as
far as the northern extremity of Pyrene . . . The Sacred Promontory is the
westernmost point not only of Europe, but also of the whole inhabited
earth . . .

On the Dead Sea

The Lake Sirbonis is of great extent. Some say that it is 1000 stadia in circumference. It stretches along the coast, to the distance of a little more
than 200 stadia. It is deep, and the water is exceedingly heavy, so that no
person can dive into it; if any one wades into it up to the waist, and attempts
to move forward, he is immediately lifted out of the water. It abounds with
asphaltus, which rises, not however at any regular season, in bubbles,
like boiling water, from the middle of the deepest part. The surface is
convex, and presents the appearance of a hillock. Together with the
asphaltus, there ascends a great quantity of sooty vapour, not perceptible
to the eye, which tarnishes copper, silver, and everything bright—even gold.
The neighbouring people know by the tarnishing of their vessels that the
asphaltus is beginning to rise, and they prepare to collect it by means of
rafts composed of reeds. The asphaltus is a clod of earth, liquified by heat;
the air forces it to the surface, where is spreads itself. It is again changed
into so firm and solid a mass by cold water, such as the water of the lake,
that it requires cutting or chopping (for use). It floats upon the water,
which, as I have described, does not admit of diving or immersion, but
lifts up the person who goes into it. Those who go on rafts for the asphaltus
cut it in pieces, and take away as much as they are able to carry.

Such are the phenomena. But Posidonius says, that the people being
addicted to magic, and practising incantations, (by these means) consolidate
the asphaltus, pouring upon it urine and other fetid fluids, and then cut

it into pieces. (Incantations cannot be the cause), but perhaps urine may have some peculiar power (in effecting the consolidation) in the same manner that chrysocolla is formed in the bladders of persons who labour under the disease of the stone, and in the urine of children.

It is natural for these phenomena to take place in the middle of the lake, because the source of the fire is in the centre, and the greater part of the asphaltus comes from thence. The bubbling up, however, of the asphaltus is irregular, because the motion of fire, like that of many other vapours, has no order perceptible to observers. There are also phenomena of this kind at Apollonia in Epirus.

Many other proofs are produced to show that this country is full of fire. Near Moasada are to be seen rugged rocks, bearing the marks of fire; fissures in many places; a soil like ashes; pitch falling in drops from the rocks; rivers boiling up, and emitting a fetid odour to a great distance; dwellings in every direction overthrown; when we are inclined to believe the common tradition of the natives, that thirteen cities once existed there, the capital of which was Sodom, but that a circuit of about 60 stadia around it escaped uninjured; shocks of earthquakes, however, eruptions of flames and hot springs, containing asphaltus and sulphur, caused the lake to burst its bounds, and the rocks took fire; some of the cities were swallowed up, others were abandoned by such of the inhabitants as were able to make their escape.

But Eratosthenes asserts, on the contrary, that the country was once a lake, and that the greater part of it was uncovered by the water discharging itself through a breach, as was the case in Thessaly.

On Egypt

We must, however, enter into a further detail of particulars. And first, we must speak of the parts about Egypt, proceeding from those that are better known to those which follow next in order.

The Nile produces some common effects in this and the contiguous tract of country, namely, that of the Ethiopians above it, in watering them at the time of its rise, and leaving those parts only habitable which have been covered by the inundation; it intersects the higher lands, and all the tract elevated above its current on both sides, which however are uninhabited and a desert, from an absolute want of water. But the Nile does not traverse the whole of Ethiopia, nor alone, nor in a straight line, nor a country which is well inhabited. But Egypt it traverses both alone and entirely, and in a straight line, from the lesser cataract above Syene and Elephantina, (which are the boundaries of Egypt and Ethiopia,) to the mouths by which it discharges itself into the sea. The Ethiopians at present lead for the most part a wandering life, and are destitute of the means of subsistence, on account of the barrenness of the soil, the disadvantages of climate, and their great distance from us.

Now the contrary is the case with the Egyptians in all these respects. For they have lived from the first under a regular form of government, they were a people of civilized manners, and were settled in a well-known country; their institutions have been recorded and mentioned in terms of praise, for they seemed to have availed themselves of the fertility of their country in the best possible manner by the partition of it (and by the classification of persons) which they adopted, and by their general care.

When they had appointed a king, they divided the people into three classes, into soldiers, husbandmen, and priests. The latter had the care of everything relating to sacred things (of the gods), the others of what related to man; some had the management of warlike affairs, others attended to the concerns of peace, the cultivation of the ground, and the practice of the arts, from which the king derived his revenue.

The priests devoted themselves to the study of philosophy and astronomy, and were companions of the kings.

The country was at first divided into nomes. The Thebaïs contained ten, the Delta ten, and the intermediate tract sixteen. But according to some writers, all the nomes together amounted to the number of chambers in the Labyrinth. Now these were less than thirty [six]. The nomes were again divided into other sections. The greater number of the nomes were distributed into toparchies, and these again into other sections; the smallest portions were the arouræ.

An exact and minute division of the country was required by the frequent confusion of boundaries occasioned at the time of the rise of the Nile, which takes away, adds, and alters the various shapes of the bounds, and obliterates other marks by which the property of one person is distinguished from that of another. It was consequently necessary to measure the land repeatedly. Hence it is said geometry originated here, as the art of keeping accounts and arithmetic originated with the Phoenicians, in consequence of their commerce.

As the whole population of the country, so the separate population in each nome, was divided into three classes; the territory also was divided into three equal portions.

The attention and care bestowed upon the Nile is so great as to cause industry to triumph over nature. The ground by nature, and still more by being supplied with water, produces a great abundance of fruits. By nature also a greater rise of the river irrigates a larger tract of land; but industry has completely succeeded in rectifying the deficiency of nature, so that in seasons when the rise of the river has been less than usual, as large a portion of the country is irrigated by means of canals and embankments, as in seasons when the rise of the river has been greater.

Before the times of Petronius there was the greatest plenty, and the rise of the river was the greatest when it rose to the height of fourteen cubits; but when it rose to eight only, a famine ensued. During the government of Petronius, however, when the Nile rose twelve cubits only, there was a most abundant crop; and once when it mounted to eight only, no famine followed. Such then is the nature of this provision for the physical state of the country. We shall now proceed to the next particulars.

The Nile, when it leaves the boundaries of Ethiopia, flows in a straight line towards the north, to the tract called the Delta, then "cloven at the head," (according to the expression of Plato), makes this point the vertex, as it were, of a triangle, the sides of which are formed by the streams, which separate on each side, and extend to the sea, one on the right hand to Pelusium, the other on the left to Canobus and the neighbouring Heracleium, as it is called; the base is the coast lying between Pelusium and the Heracleium.

An island was therefore formed by the sea and by both streams of the river, which is called Delta from the resemblance of its shape to the letter (Δ) of that name. The spot at the vertex of the triangle has the same appellation, because it is the beginning of the above-mentioned triangular figure. The village, also, situated upon it is called Delta.

These then are two mouths of the Nile, one of which is called the Pelusiac, the other the Canobic and Heracleiotic mouth. Between these are five other outlets, some of which are considerable, but the greater part are of inferior importance. For many others branch off from the principal streams, and are distributed over the whole of the island of the Delta, and form many streams and islands; so that the whole Delta is accessible to boats, one canal succeeding another, and navigated with so much ease, that some persons make use of rafts floated on earthen pots, to transport them from place to place.

The whole island is about 3000 stadia in circumference, and is called, as also the lower country, with the land on the opposite sides of the streams, the Delta.

But at the time of the rising of the Nile, the whole country is covered, and resembles a sea, except the inhabited spots, which are situated upon natural hills or mounds; and considerable cities and villages appear like islands in the distant prospect.

The water, after having continued on the ground more than forty days in summer, then subsides by degrees, in the same manner as it rose. In sixty days the plain is entirely exposed to view, and dries up. The sooner the land is dry, so much the sooner and ploughing and sowing are accomplished, and it dries earlier in those parts where the heat is greater.

The country above the Delta is irrigated in the same manner, except that the river flows in a straight line to the distance of about 4000 stadia in one channel, unless where some island intervenes, the most considerable of which comprises the Heracleiotic Nome; or, where it is diverted by a canal into a large lake, or a tract of country which it is capable of irrigating, as the lake Mœris and the Arsinoïte Nome, or where the canals discharge themselves into the Mareotis.

In short, Egypt, from the mountains of Ethiopia to the vertex of the Delta, is merely a river tract on each side of the Nile, and rarely if anywhere comprehends in one continued line a habitable territory of 300 stadia in breadth. It resembles, except the frequent diversions of its course, a bandage rolled out.

The mountains on each side (of the Nile), which descend from the parts about Syene to the Egyptian Sea, give this shape to the river tract of which I am speaking, and to the country. For in proportion as these mountains extend along that tract, or recede from each other, in the same degree is the river contracted or expanded, and they impart to the habitable country its variety of shape. But the country beyond the mountains is in a great measure uninhabited.

The ancients understood more by conjecture than otherwise, but persons in later times learnt by experience as eye-witnesses, that the Nile owes its rise to summer rains, which fall in great abundance in Upper Ethiopia, particularly in the most distant mountains. On the rains ceasing, the fulness of the river gradually subsides. This was particularly observed by those who navigated the Arabian Gulf on their way to the Cinnamon country, and by those who were sent out to hunt elephants, or for such other purposes as induced the Ptolemies, kings of Egypt, to despatch persons in that direction. These sovereigns had directed their attention to objects of this kind, particularly Ptolemy surnamed Philadelphus, who was a lover of science, and on account of bodily infirmities always in search of some new diversion and amusement. But the ancient kings paid little attention to such inquiries, although both they and the priests, with whom they passed the greater part of their lives, professed to be devoted to the study of philosophy. Their ignorance therefore is more surprising, both on this account and because Sesostris had traversed the whole of Ethiopia as far as the Cinnamon country, of which expedition monuments exist even to the present day, such as pillars and inscriptions. Cambyses also, when he was in possession of Egypt, had advanced with the Egyptians as far even as Meroë; and it is said that he gave this name both to the island and to the city, because his sister, or according to some writers his wife, Meroë died there. For this reason therefore he conferred the appellation on the island, and in honour of a woman. It is surprising how, with such opportunities of obtaining information, the history of these rains should not have been clearly known to persons living in those times, especially as the priests registered with the greatest diligence in the sacred books all extraordinary facts, and preserved records of everything which seemed to contribute to an increase of knowledge. And, if this had been the case, would it be necessary to inquire what is even still a question, what can possibly be the reason why rain falls in summer, and not in winter, in the most southerly parts of the country, but not in the Thebaïs, nor in the country about Syene? nor should we have to examine whether the rise of the water of the Nile is occasioned by rains, nor require such evidence for these facts as Poseidonius adduces. For he says, that Callisthenes asserts that the cause of the rise of the river is the rain of summer. This he borrows from Aristotle, who borrowed it from Thrasyalces the Thasian (one of the ancient writers on physics), Thrasyalces from some other person, and he from Homer, who calls the Nile "heaven-descended:"

"back to Egypt's heaven-descended stream."

On Arabia and the Red Sea

In speaking of the northern and desert region which lies between Arabia the Blest and Hollow Syria and Judaea, as far as the recess of the Arabian Gulf, he says that it measures 5,600 stades from Heroopolis, which forms a recess of the Arabian Gulf near the Nile, by way of Petra (*in the Wadi Musa*) of the Nabataei to Babylon; the whole journey lies towards the summer sunrise (*north-east*), and passes through the adjacent nations the Nabataei, the Chaulotaei, and the Agraei. Above (*south of*) these is Arabia the Blest, stretching out for a distance of 12,000 stades southwards to the Atlantic Sea (*the Ocean in general*). The first people who occupy it after the Syrians and the Judaeans are husbandmen; after these the land is sand-bound and poor, producing a few date-palms, mimosa, and tamarisk, and providing water by excavation only, like Gedrosia. It is occupied by Arabians who are tent-dwellers and camel-breeders. The farthest parts towards the south, which rise under the same parallel as Ethiopia, are watered by summer rains, and are sown twice like India and have rivers which exhaust in supplying plains and lakes and waters brought down . . .

The four greatest nations inhabit the farthest part of the country mentioned. The Minaeans dwell in the part towards the Red Sea; their biggest city is Carna or Carnana; next to these come the Sabaeans, whose chief city is Mariaba (*Marib*); third come the Cattabaneis (*of Kataban*), who reach down to the straits and the crossing of the Arabian Gulf; their royal seat is called Tamna: Farthest towards the east are the Chatramotitae (*of Hadramut*), who have a city Sabata (*Sawa*). Each of these cities is ruled by a monarch and is flourishing . . . and the four districts occupy a region larger than the Delta in Egypt . . .

Frankincense is produced by Cattabania, myrrh by Chatramotitis; these and other aromatics are exchanged by the merchants. Men come to them —to Minaea from Aelana in seventy days . . . while Gerrhaeans reach Chatramotitis in forty days.

That part of the Arabian Gulf which runs along the side of Arabia, if we begin from the Aelanitic recess (*Gulf of Akabah*), measures, according to those who were with Alexander and Anaxicrates, 14,000 stades, but this figure gives too much. The part opposite Tryglodytice (*western coast-land of Red Sea*), which lies on the right as we sail from Heroopolis, measures as far as Ptolemais and its elephant-hunting ground 9,000 stades towards the south, and a little towards the east; thence as far as the straits (*of Bab el-Mandeb*) it measures about 4,500 and turns more towards the east. The straits on the Ethiopian side are formed by a headland (*Ras Bir*) called Deire, and a small city of the same name. It is inhabited by Fish-Eaters . . . The straits at Deire are contracted to 60 stades; however, it is not this passage which is nowadays called the straits; instead, when we have sailed farther on, to the part where the sea-crossing between the continents is one of 200 stades, and six islands, coming one closely after another, fill up the crossing and leave very narrow passages for ships— this is the part which they speak of as the straits; through them the men convey their loads of wares from one side to the other on rafts. From these islands onwards navigation consists of coasting round the bays of the Myrrh-bearing country (*Guban*), both southwards and eastwards as far as the Cinnamon-bearing country (*roughly from Berbera to Guardafui*) over a distance of about 5,000 stades. Beyond this land, they say no one has gone up to now, and that while the cities on the coast are not many, inland there are many which are well inhabited.

30. Ptolemy on the field of geography and on divisions of the earth

Of Claudius Ptolemy little is known except the approximate dates of his birth and death (c. 100–c. 170 A.D.) and that he was a native of Hellenistic Egypt, where he spent his working life. He was—and is —equally famous as an astronomer, a geographer, and a cartographer; his are among the few works of late classical writers to have been preserved in their entirety. Ptolemy's astronomical observations have been identified as having been made between 127 and 141 A.D., confirming the fact that he worked during the second Christian century; his influence, however, endured for more than a thousand years. When "rediscovered" by the Christian West, he was known as a great astronomer, and it was not until the appearance of his *Geography* in Latin translation, in the early 1400s, that his ideas on mapmaking and on the shape and size of the earth began to exert direct influ-

The first of these selections is from Ptolemy, *Tetrabiblos*, trans. F. E. Robbins, Loeb Classical Library (London: William Heinemann, 1940), pp. 35–36. The others are from *The Geography of Ptolemy*, trans. Edward Luther Stevenson (New York: New York Public Library, 1932), pp. 25–28, 18–19, 29–32, 41. Reprinted by permission of the New York Public Library.

ence. The first of the excerpts from Ptolemy's writings describes the effects of location on physiology and character and is typical of his general views on geography. This is followed by specific discussions on the scope of geography, on measurements, and on divisions of the earth by parallels, an essential part of Ptolemy's views and prescriptions for mapmaking.

Of the characteristics of the inhabitants of the general climes

The demarcation of national characteristics is established in part by entire parallels and angles, through their position relative to the ecliptic and the sun. For while the region which we inhabit is in one of the northern quarters, the people who live under the more southern parallels, that is, those from the equator to the summer tropic, since they have the sun over their heads and are burned by it, have black skins and thick, woolly hair, are contracted in form and shrunken in stature, are sanguine of nature, and in habits are for the most part savage because their homes are continually oppressed by heat; we call them by the general name Ethiopians. Not only do we see them in this condition, but we likewise observe that their climate and the animals and plants of their region plainly give evidence of this baking by the sun.

Those who live under the more northern parallels, those, I mean, who have the Bears over their heads, since they are far removed from the zodiac and the heat of the sun, are therefore cooled; but because they have a richer share of moisture, which is most nourishing and is not there exhausted by heat, they are white in complexion, straight-haired, tall and well-nourished, and somewhat cold by nature; these too are savage in their habits because their dwelling-places are continually cold. The wintry character of their climate, the size of their plants, and the wildness of their animals are in accord with these qualities. We call these men, too, by a general name, Scythians.

The inhabitants of the region between the summer tropic and the Bears, however, since the sun is neither directly over their heads nor far distant at its noon-day transits, share in the equable temperature of the air, which varies, to be sure, but has no violent changes from heat to cold. They are therefore medium in colouring, of moderate stature, in nature equable, live close together, and are civilized in their habits. The southernmost of them are in general more shrewd and inventive, and better versed in the knowledge of things divine because their zenith is close to the zodiac and to the planets revolving about it. Through this affinity the men themselves are characterized by an activity of the soul which is sagacious, investigative, and fitted for pursuing the sciences specifically called mathematical. Of them, again, the eastern group are more masculine, vigorous of soul, and frank in all things, because one would reasonably assume that the orient partakes of the nature of the sun. This region therefore is diurnal, masculine, and right-handed, even as we observe that among the animals too their right-hand parts are better fitted for strength and vigour. Those to the west are more feminine, softer of soul, and secretive, because this region, again, is lunar, for it is always in the west that the moon emerges and makes its appearance after conjunction. For this reason it appears to be a nocturnal clime, feminine, and, in contrast with the orient, left-handed.

And now in each of these general regions certain special conditions of character and customs naturally ensue. For as likewise, in the case of the climate, even within the regions that in general are reckoned as hot, cold, or temperate, certain localities and countries have special peculiarities of excess or deficiency by reason of their situation, height, lowness, or adjacency; and again, as some peoples are more inclined to horsemanship because theirs is a plain country, or to seamanship because they live close to the sea, or to civilization because of the richness of their soil, so also would one discover special traits in each arising from the natural familiarity of their particular climes with the stars in the signs of the zodiac. These traits, too, would be found generally present, but not in every individual. We must, then, deal with the subject summarily, in so far as it might be of use for the purpose of particular investigations.

In what geography differs from chorography

Geography is a representation in picture of the whole known world together with the phenomena which are contained therein.

It differs from Chorography in that Chorography, selecting certain places from the whole, treats more fully the particulars of each by themselves—even dealing with the smallest conceivable localities, such as harbors, farms, villages, river courses, and such like.

It is the prerogative of Geography to show the known habitable earth as a unit in itself, how it is situated and what is its nature; and it deals with those features likely to be mentioned in a general description of the earth, such as the larger towns and the great cities, the mountain ranges and the principal rivers. Besides these it treats only of features worthy of special note on account of their beauty.

The end of Chorography is to deal separately with a part of the whole, as if one were to paint only the eye or the ear by itself. The task of Geography is to survey the whole in its just proportions, as one would the entire head. For as in an entire painting we must first put in the larger features, and afterward those detailed features which portraits and pictures may require, giving them proportion in relation to one another so that their correct measure apart can be seen by examining them, to note whether they form the whole or a part of the picture. Accordingly therefore it is not unworthy of Chorography, or out of its province, to describe the smallest details of places, while Geography deals only with regions and their general features.

The habitable parts of the earth should be noted rather than the parts which are merely of equal size, especially the provinces or regions and their divisions, the differences between these being rather the more important. Chorography is most concerned with what kind of places those are which it describes, not how large they are in extent. Its concern is to paint a true likeness, and not merely to give exact position and size. Geography looks at the position rather than the quality, noting the relation of distances every-where, and emulating the art of painting only in some of its major descriptions. Chorography needs an artist, and no one presents it rightly unless he is an artist. Geography does not call for the same requirements, as any one, by means of lines and plain notations can fix positions and draw general outlines. Moreover Chorography does not have need of mathematics, which is an important part of Geography. In Geography one must contemplate the extent of the entire earth, as well as its shape, and its position under the heavens, in order that one may rightly state what are the peculiarities and proportions of the part with which one is dealing, and under what parallel of the celestial sphere it is located, for so one will be able to discuss the length of its days and nights, the stars which are fixed overhead, the stars which move above the horizon, and the stars which never rise above the horizon at all; in short all things having regard to our earthly habitation.

It is the great and the exquisite accomplishment of mathematics to show all these things to the human intelligence so that the sky, too, having a representation of its own character, which, although it can not be seen as moving around us, yet we can look upon it by means of an image as we look upon the earth itself, for the earth being real and very large, and neither wholly nor in part moving around us, yet it can be mapped by the same means as is the sky.

What presuppositions are to be made use of in geography

What Geography aims at, and wherein it differs from Chorography, we have definitely shown in our preceding chapter. But now as we propose to describe our habitable earth, and in order that the description may correspond as far as possible with the earth itself, we consider it fitting at the outset to put forth that which is the first essential, namely, a reference to the history of travel, and to the great store of knowledge obtained from the reports of those who have diligently explored certain regions; whatever concerns either the measurement of the earth geometrically or the observation of the phenomena of fixed localities; whatever relates to the measurement of the earth that can be tested by pure distance calculations to determine how far apart places are situated; and whatever relations to fixed positions can be tested by meteorological instruments for recording shadows. This last is a certain method, and is in no respect doubtful. The other method is less perfect and needs other support, since first of all it is necessary to know in determining the distance between two places, in what direction each place lies from the other; to know how far this place is distant from that, we must also know under what part of the sky each is located, that is, whether each extends toward the north, or, so to speak, toward the rising of the sun (the east), or in some other particular direction. And these facts it is impossible to ascertain without the use of the instruments to which we refer. By the use of these instruments, anywhere and at any time, the position of the meridian line can easily be found, and from this we can ascertain the distances that have been traveled. But when this has been done, the measurement of the number of stadia does not give us sure information, because journeys very rarely are made in a straight line. There being many deviations from a straight course both in land and in sea journeys, it is necessary to conjecture, in the case of a land journey, the nature and the extent of the deviation, and how far it departs from a straight course, and to subtract something from the number of stadia to make the journey a straight one.

Even in sailing the sea the same thing happens, as the wind is never constant throughout the whole voyage. Thus although the distance of the places noted is carefully counted, it does not give us a basis for the determination of the circumference of the whole earth; nor do we ascertain an exact position for the equatorial circle or for the location of the poles.

Distance which is ascertained from an observation of the stars shows accurately all these things, and in addition shows how much of the circumference is intercepted in turn by the parallel circles, and by the meridian circles which are drawn through the places themselves; that is to say, what part of the circumference of parallel circles and of the equatorial circle is intercepted by the meridians, or what part of the meridian circles are intercepted by the parallels and equatorial circle.

After this it will readily be seen how much space lies between the two places themselves on the circumference of the large circle which is drawn through them around the earth. This measurement of stadia obtained from careful calculations does not require a consideration of the parts of the earth traversed in a described journey; for it is enough to suppose that the circuit of the earth itself is divided into as many parts as one desires, and that some of these parts are contained within distances noted on the great circles that gird the earth itself. Dividing the whole circuit of the earth or any part of it noted by our measurements which are known as stadia measurements, is a method not equally convincing. Therefore because of this fact alone it has been found necessary to take a certain part of the circumference of a very large celestial circle, and by determining the ratio of this part to the whole of the circle, and by counting the number of stadia contained in the given distance on the earth, one can measure the stadia circumference of the globe.

When we grant that it has been demonstrated by mathematics that the surface of the land and water is in its entirety a sphere, and has the same center as the celestial globe, and that any plane which passes through the center makes at its surface, that is, at the surface of the earth and of the sky, great circles, and that the angles of the planes, which angles are at the center, cut the circumferences of the circles which they intercept proportionately, it follows that in any of the distances which we measure on the earth the number of the stadia, if our measurements are correct, can be determined, but the proportion of this distance to the whole circumference of the earth can not be found, because no proportion to the whole earth can thus be derived, but from the similar circumference of the celestial globe that proportion can be derived, and the ratio of any similar part on the earth's surface to the great circle of the earth is the same.

How, from measuring the stadia of any given distance, although not on the same meridian, we may determine the number of stadia in the circumference of the earth, and vice versa

Those geographers who lived before us sought to fix correct distance on the earth, not only that they might determine the length of the greatest circle, but also that they might determine the extent which a region occupied in one plane on one and the same meridian. After observing therefore, by means of the instrument of which I have spoken, the points which were directly over each terminus of the given distance, they calculated from the intercepted part of the circumference of the meridian, distances on the earth.

As we have said, they assume the location of the points to be in one plane, and the lines passing through the terminals of the distance, to the points which are directly overhead, must necessarily meet, and the points where they meet would be the common center of the circles. Therefore if the circle drawn through the poles were intercepted by lines drawn from the two points that were marked overhead, it would be understood that it formed the total extent of the intercepted circumference compared with the whole circuit of the earth.

If a distance of this kind is not on the circle drawn through the poles, but on another of the great circles, the same thing can be shown by observing in like manner the elevation of the pole from the extremities of the distance, and noting simultaneously the position which the same distance has on the other meridian. This we have clearly shown by an instrument which we ourselves have constructed for measuring shadows, by which instrument we can easily ascertain a great many other useful things. For on any day or night we have the elevation of the north pole, and at any hour we have the meridian position of the given distance by performing a single measurement, that is, by measuring the angle that the greatest circle drawn through the line of the distance makes with the meridian circle at the vertical point: in this way we can show the required circumference by means of this instrument, and the circumference of the equatorial circle which is intercepted between the two meridians, these meridians being parallel and circles like the equator. According to this demonstration, if we measure only one straight distance on the surface of the earth, then the number of stadia of the whole circuit of the earth can be ascertained. And as a result of this we can obtain the measurements of all distances, even when they are not exactly on the same meridian or parallel, by observing carefully the elevation of the pole, and the inclination of the distance to the meridian, and vice versa. From the ratio of the given part of the circumference to the great circle, the number of the stadia can be calculated from the known number of stadia in the circuit of the whole earth.

The opinions of Marinus relating to the earth's latitude are corrected by observed phenomena

First of all, Marinus places Thule as the terminus of latitude on the parallel that cuts the most northern part of the known world. And this parallel, he shows as clearly as is possible, at a distance of sixty-three degrees from the equator, of which degrees a meridian circle contains three hundred and sixty. Now the latitude he notes as measuring 31,500 stadia, since every degree, it is accepted, has 500 stadia. Next, he places the country of the *Ethiopians,* Agisymba by name, and the promontory of Prasum on the same parallel which terminates the most southerly land known to us, and this parallel he places below the winter solstice.

Between Thule and the southern terminus he inserts altogether about eighty-seven degrees which is 43,500 stadia, and he tries to prove the correctness of this southern termination of his by certain observations (which he thinks to be accurate) of the fixed stars and by certain journeys made both on land and on sea. Concerning this we will make a few observations.

In his observation concerning the fixed stars, in the third volume of his work, he uses these words: "The Zodiac is considered to lie entirely above the torrid zone and therefore in that zone the shadows change, and all the fixed stars rise and set. Ursa Minor begins to be entirely above the horizon from the north shore of Ocele which is 5,500 stadia distant. The parallel through Ocele is elevated eleven and two-fifths degrees.

"We learn from Hipparchus that the star in Ursa Minor which is the most southerly or which marks the end of the tail, is distant from the pole twelve and two-fifths degrees, and that in the course of the sun from the equinoctial to the summer solstice, the north pole continually rises above the horizon while the south pole is correspondingly depressed, and that on the contrary in the course of the sun from the equator to the winter solstice the south pole rises above the horizon while the north pole is depressed."

In these statements Marinus narrates only what is observed (on) the equator, or between the tropics. But what, after being learned from the records or from accurate observations of the fixed stars, are the happenings in places south of the equator, he in no wise informs us, as if one should place southern stars rather than equatorial directly overhead, or assert that mid-day shadows over the equator incline south, or show all the stars of Ursa Minor risen or set, or some of them visible at the time when the south pole is raised above the horizon.

In what he adds later he tells us of certain observations, of which, nevertheless, he is not entirely certain in his own mind.

He says that those who sail from India to Limyrica, as did Diodorus the Samian, which is related in his third book, tell us that Taurus is in a higher position in the mid-heavens than in reality it is and that the Pleiades are seen in the middle of the masts, and he continues, "those who sail from Arabis to Azania sail straight to the south, and toward the star Canopus, which there is called Hippos, that is the Horse, and which is far south. Stars are seen there which are not known to us by name, and the Dog Star rises before Procyon and Orion, and before the time when the sun turns back toward the summer solstice."

For these observations concerning the stars Marinus clearly states that some places are located more northerly than the equator, as when he says that Taurus and the Pleiades are directly over the heads of the sailors. As a matter of fact these stars are near the equator. He indeed shows some stars to be no further south than north, for Canopus can be seen by those who dwell a long distance north of the summer solstice; and several of the fixed stars, never seen by us, can be seen above the horizon in places south of us, and in places more toward the equatorial region than those in the north, as around Meroë. They can be seen as is Canopus itself, which, when appearing above the horizon is never visible to those who dwell north of us. Those who dwell toward the south call this star Hippos, that is the Horse, nor is any other star of those known to us called by that name.

Marinus infers that he himself determined by mathematical proofs that Orion is entirely visible, before the summer solstice, to those who dwell below the equatorial circle; also that with them the Dog Star rises before Procyon, which he says is observed as far south even as Syene. In these conclusions of Marinus there is nothing appropriate or of value to us because he extends the position of his inhabited countries too far south of the equator.

They are also corrected by measuring journeys on land

In computing the days one by one, occupied in journeying from Leptis Magna to Agisymba, Marinus shows that the latter locality is 24,680 stadia south of the equator. By adding together the days occupied in sailing from Ptolemais Trogloditica to Prasum he concluded that Prasum is 27,800 stadia south of the equator, and from these data he infers that the promontory of Prasum and the land of Agisymba, which, as he himself expresses it, belongs to Ethiopia (and is not the end of Ethiopia), lies on the south coast in the frigid zone opposite to ours. In a southerly direction 27,800 stadia make up fifty-five and three-fourths degrees, and this number of degrees in an opposite direction (i. e. north) marks a like temperate climate, and the region of the swamp Meotis, which the *Scythians* and *Sarmatians* inhabit.

Marinus then reduces the stated number of his stadia by half or less than half, that is to 12,000 which is about the distance of the winter solstice from the equatorial circle. The only reason for this reduction that he gives us is the deviation from a straight line of the journeys and their daily variations in length. After he has stated these reasons, it seemed to us necessary not only to show that he was mistaken, but also to reduce his figures by the required one-half.

At the outset, when writing of the journey from Garama to Ethiopia he says that Septimius Flaccus, having set out from Libya with his army, came to the land of the *Ethiopians* from the land of the *Garamantes* in the space of three months by journeying continuously southward. He says furthermore that Julius Meternus, setting out from Leptis Magna and Garama with the king of the *Garamantes,* who was beginning an expedition against the *Ethiopians,* by bearing continuously southward came within four months to Agisymba, the country of the *Ethiopians* where the rhinoceros is to be found.

Each of these statements, on the face of it, is incredible, first, because the *Ethiopians* are not so far distant from the *Garamantes* as to require a three months' journey, seeing that the *Garamantes* are themselves for the most part Ethiopians, and have the same king; secondly, because it is ridiculous to think that a king should march through regions subject to him only in a southerly direction when the inhabitants of those regions are scattered widely east and west, and ridiculous also that he should never have made a single halt that would alter the reckoning. Wherefore we conclude that it is not unreasonable to suppose that those men either spoke in hyperbole, or else, as rustics say, "To the south," or "Toward Africa" to those who prefer to be deceived by them, rather than take the pains to ascertain the truth.

They are also corrected by measuring journeys by water

Concerning the voyage from Aromata to Rhapta, Marinus tells us that a certain Diogenes, one of those who were accustomed to sail to India, having been driven out of his course, and being off the coast of Aromata, was caught by the north wind and, after having sailed with Trogloditica on his right, came in twenty-five days to the lake from which the Nile flows, to the south of which lies the promontory of Rhaptum. He tells us also that a certain Theophilus, one of those who were accustomed to sail to Azania, driven from Rhapta by the south wind came to Aromata on the twentieth day. In neither of these cases does he tell us how many days were occupied in actual sailing, but merely states that Theophilus took twenty days, and Diogenes, who sailed along the coast of Trogloditica, took twenty-five days.

He only tells us how many days they were on the voyage, and not the exact sailing time, nor the changes of the wind in strength and direction, which must have taken place during a voyage of such long duration. Moreover he does not say that the sailing was continuously south or north, but merely says that Diogenes was carried along by the north wind while Theophilus sailed with the south wind. That the wind kept the same strength and direction during the whole voyage is related in neither case, and it is incredible that for the space of so many days in succession, it should have done so. Therefore although Diogenes sailed from Aromata to the swamps, to the south of which lies the promontory of Rhaptum, in twenty-five days, and Theophilus from Rhapta to Aromata, a greater distance, in twenty days, and although Theophilus tells us that a single day's sailing under favorable circumstances is calculated at 1,000 stadia (and this computation Marinus himself approves) Disocorus nevertheless says that the voyage from Rhapta to the promontory of Prasum, which takes many days, as computed by Diogenes is only 5,000 stadia. The wind, he says, varies very suddenly at the equator, and squalls around the equator on either side of the line are more dangerous.

From these considerations we thought we ought not to assent to the numbering of the days, because it is plain to all that on the reckoning made by Marinus, the *Ethiopians* and the haunts of the rhinoceros should be moved to the cold zone of the earth, that is, opposite to ours. Reason herself asserts that all animals, and all plants likewise, have a similarity under the same kind of climate or under similar weather conditions, that is, when under the same parallels, or when situated at the same distance from either pole.

Marinus has shortened the measures of latitude around the winter solstice, but has given no sufficient reason for his contraction. Even should we admit the number of days occupied in the series of voyages that he relates, he has shortened the number of daily stadia and has reasoned contrary to his customary measure in order to reach the desired and correct parallel. He should have done exactly the opposite, for it is easy to believe the same daily distance traveled as possible, but in the even course of the journeys, or voyages or that they were wholly made in a straight line, he ought not to have believed. From them it was not possible to ascertain the distance but it was correct to assert that in latitude the places in question extended beyond the equator. Even this could be known with greater certainty from astronomical observations. Any one could have ascertained exactly the required distances if he had, with more skill in mathematics, considered what takes place in those localities. Since this observation was not made, it remains that we follow what reason dictates, that is, we must ascertain how far the distance extends beyond the equator. We can also ascertain what we require to know through information concerning the kinds, the forms, and the colors of the animals living there, from which we draw the conclusion that the parallel of the region of Agisymba is the same as that of the *Ethiopians* and extends from the winter solstice to the equator; although with us in places opposite to that region, that is, in the summer solstice, they do not have the color of the *Ethiopians,* nor is the rhinoceros and elephant to be found, yet in places not far south of us the inhabitants are moderately black such as from the same cause are the *Garamantes,* whom Marinus himself describes, and whom he places neither under the summer solstice nor north of it, but much too far to the south. In the regions around Meroë the inhabitants are very black, and closely resemble the *Ethiopians,* and there we find that elephants and other kinds of monstrous animals are bred.

Ethiopia should not be placed more to the south than the parallel which is opposite the parallel passing through Meroe

In agreement therefore with this information, viz., that the inhabitants are *Ethiopians,* as those who have sailed there have told us, Marinus describes the region of Agisymba and the promontory of Prasum, and the other places lying on the same parallel, as situated all on one parallel, which is opposite the parallel passing through Meroë. That would place them on a parallel distant from the equator in a southerly direction 16°25' or about 8,200 stadia, and by the same reckoning the whole width of the habitable world amounts to 79°25' or altogether 40,000 stadia.

Now the distance between Leptis Magna and Garama, according to Flaccus and Maternus, is placed at 5,400 stadia. The time of their second journey was twenty days, a more nearly correct time than the first because it was directly north, while the first journey of thirty days had many deviations. The travelers who several times made the voyage kept the reckoning of each day's distance, and this was not only properly done, but done of necessity on account of the changes of the water and the weather. Just as we should have doubts with regard to distances that are great in extent, and rarely traveled, and not fully explored, so in regard to those that are not great and not rarely but frequently gone over, it seems right to give credit to the reports of the voyagers.

Explanation of the meridians and parallels used in our delineation

The meridians, according to what we have already shown, will embrace the space of twelve hours. The parallel that bounds the most southern limit of the habitable world will be distant from the equator in a southerly direction only as far as the parallel passing through Meroë is distant in a northerly direction.

It has seemed proper to us to put in the meridians at a distance from each other the third part of an equinoctial hour, that is, through five of the divisions marked on the equator. The parallels that are not of the equator we have inserted so that:

The first parallel is distant from the equator the fourth part of an hour, and is distant from it geometrically about 4°15′.

The second parallel we make distant half an hour from the equator, and geometrically distant 8°25′.

The third parallel we make distant from the equator three-fourths of an hour, and geometrically 12°30′.

The fourth parallel is distant one hour and is 16°25′. This is the parallel through Meroë.

The fifth parallel is distant one and one-fourth hours, and 20°15′.

The sixth parallel, which is under the summer solstice, is distant one and one-half hours, and 23°50′, and is drawn through Syene.

The seventh parallel is distant one and three-fourths hours, and 27°10′.

The eighth parallel is distant two hours, and 30°20′.

The ninth parallel is distant two and one-fourth hours, and 33°20′.

The tenth parallel is distant two and one-half hours, and 36°, and is drawn through Rhodes.

The eleventh parallel is distant two and three-fourths hours, and 38°35′.

The twelfth parallel is distant three hours, and 40°55′.

The thirteenth parallel is distant three and one-fourth hours, and 43°05′.

The fourteenth parallel is distant three and one-half hours, and 45°.

The fifteenth parallel is distant four hours, and 48°30′.

The sixteenth parallel is distant four and one-half hours, and 51°30′.

The seventeenth parallel is distant five hours, and 54°.

The eighteenth parallel is distant five and one-half hours, and 56°10′.

The nineteenth parallel is distant six hours, and 58°.

The twentieth parallel is distant seven hours, and 61°.

The twenty-first parallel is distant eight hours, and 63°, and is the parallel drawn through Thule.

Besides these, one other parallel must be drawn south of the equator with the time difference of half an hour. It should pass through Rhaptum promontory and Cattigara, and should be about the same length as the parallel in the opposite part of the earth which is distant 8°25′ north of the equator.

Latin Encyclopedists

The writings of the Roman encyclopedists were based on the work of their Greek predecessors. During the late Empire and through much of the Middle Ages, the Latin-speaking West, unable to read the original sources, relied entirely upon these Roman authors of handbooks. Varro, Pliny, Pomponius Mela, Solinus, and Macrobius were read, copied, and later still, printed—while Aristotle, Hipparchus, and Strabo were only names in a long list of authorities quoted in geographical writings.

31. Pliny: from the Natural History

Pliny, or Gaius Plinius Secundus (23–79 A.D.), author of the *Natural History,* was the most popular Roman writer on all manner of natural phenomena and the one most quoted by later writers on these subjects. High official of the Roman Empire, voracious reader, and writer of many books, Pliny was also an avid observer of nature: he died (while commander of the Roman naval base northwest of Naples) observing the great eruption of Vesuvius in 79 A.D. that buried Pompeii and Herculaneum.

Pliny offered his readers views on the general characteristics of the earth; on earthquakes, tides, and eclipses; on flora and fauna; and on lands and peoples both within the Roman Empire and far removed from its Mediterranean center.

From Pliny, *Natural History,* trans. H. Rackham, Loeb Classical Library (Cambridge, Mass.: Harvard University Press, 1938 and 1949), I, 171–175; II, 5; I, 309–315, 325–327, 343–345; II, 41–43, 125, 339, 187–188, 521–525.

Pliny not only was quoted by subsequent writers on natural history but was also the authoritative source for maps of his own time (which are now lost) and for later, mediaeval maps as well. The strange beings described in the passage quoted here on India appear, for example, as the real inhabitants of distant lands on the Ebstorf map, made in the second half of the thirteenth century.

The world—its nature

The world and this—whatever other name men have chosen to designate the sky whose vaulted roof encircles the universe, is fitly believed to be a deity, eternal, immeasurable, a being that never began to exist and never will perish. What is outside it does not concern men to explore and is not within the grasp of the human mind to guess. It is sacred, eternal, immeasurable, wholly within the whole, nay rather itself the whole, finite and resembling the infinite, certain of all things and resembling the uncertain, holding in its embrace all things that are without and within, at once the work of nature and nature herself . . .

Its shape has the rounded appearance of a perfect sphere. This is shown first of all by the name of 'orb' which is bestowed upon it by the general consent of mankind. It is also shown by the evidence of the facts: not only does such a figure in all its parts converge upon itself; not only must it sustain itself, enclosing and holding itself together without the need of any fastenings, and without experiencing an end or a beginning at any part of itself; not only is that shape the one best fitted for the motion with which, as will shortly appear, it must repeatedly revolve, but our eyesight also confirms this belief, because the firmament presents the aspect of a concave hemisphere equidistant in every direction, which would be impossible in the case of any other figure . . .

The world thus shaped then is not at rest but eternally revolves with indescribable velocity, each revolution occupying the space of 24 hours: the rising and setting of the sun have left this not doubtful. Whether the sound of this vast mass whirling in unceasing rotation is of enormous volume and consequently beyond the capacity of our ears to perceive, for my own part I cannot easily say—any more in fact than whether this is true of the tinkling of the stars that travel round with it, revolving in their own orbits; or whether it emits a sweet harmonious music that is beyond belief charming. To us who live within it the world glides silently alike by day and night.

The circuit of the earth

The whole circuit of the earth is divided into three parts, Europe, Asia and Africa. The starting point is in the west, at the Straits of Gibraltar, where the Atlantic Ocean bursts in and spreads out into the inland seas. On the right as you enter from the ocean is Africa and on the left Europe, with Asia between them; the boundaries are the river Don and the river Nile. The ocean straits mentioned are fifteen miles long and five miles broad, from the village of Mellaria in Spain to the White Cape in Africa, as given by Turranius Gracilis, a native of the neighbourhood, while Livy and Cornelius Nepos state the breadth at the narrowest point as seven miles and at the widest as ten miles: so narrow is the mouth through which pours so boundless an expanse of water. Nor is it of any great depth, so as to lessen the marvel, for recurring streaks of whitening shoal-water terrify passing keels, and consequently many have called this place the threshold of the Mediterranean. At the narrowest part of the Straits stand mountains on either side, enclosing the channel, Ximiera in Africa and Gibraltar in Europe; these were the limits of the labours of Hercules, and consequently the inhabitants call them the Pillars of that deity, and believe that he cut the channel through them and thereby let in the sea which had hitherto been shut out, so altering the face of nature.

The shape of the earth and the motion of the sun

That the earth is at the centre of the universe is proved by irrefragable arguments, but the clearest is the equal hours of day and night at the equinox. For if the earth were not at the centre, it can be realized that it could not have the days and nights equal; and binoculars confirm this very powerfully, since at the season of the equinox sunrise and sunset are seen on the same line, whereas sunrise at midsummer and sunset at midwinter fall on a line of their own. These things could not occur without the earth's being situated at the centre.

But the three circles intertwined between the zones aforesaid are the cause of the differences of the seasons: the Tropic of Cancer on the side of the highest part of the zodiac to the northward of us, and opposite to it the Tropic of Capricorn towards the other pole, and also the equator that runs in the middle circuit of the zodiac.

The cause of the remaining facts that surprise us is found in the shape of the earth itself, which together with the waters also the same arguments prove to resemble a globe. For this is undoubtedly the cause why for us the stars of the northern region never set and their opposites of the southern region never rise, while on the contrary these northern stars are not visible to the antipodes, as the curve of the earth's globe bars our view of the tracts between. Cave-dweller Country and Egypt which is adjacent to it do not see the Great and Little Bear, and Italy does not see Canopus and the constellation called Berenice's Hair, also the one that in the reign of his late Majesty Augustus received the name of Caesar's Throne, constellations that are conspicuous there. And so clearly does the rising vault curve over that to observers at Alexandria Canopus appears to be elevated nearly a quarter of one sign above the earth, whereas from Rhodes it seems practically to graze the earth itself, and on the Black Sea, where the North Stars are at their highest, it is not visible at all. Also Canopus is hidden from Rhodes, and still more from Alexandria; in Arabia in November it is hidden during the first quarter of the night and shows itself in the second; at Meroe it appears a little in the evening at midsummer and a few days before the rising of Arcturus is seen at daybreak. These phenomena are most clearly disclosed by the voyages of those at sea, the sea sloping upward in the direction of some and downward in the direction of others, and the stars that were hidden behind the curve of the ball suddenly becoming visible as it were rising out of the sea. For it is not the fact, as some have said, that the world rises up at this higher pole—or else these stars would be visible everywhere; but these stars are believed to be higher the nearer people are to them, while they seem low to those far away, and just as at present this pole seems lofty to those situated on the declivity, so when people pass across to yonder downward slope of the earth those stars rise while the ones that here were high sink, which could not happen except with the conformation of a ball.

Consequently inhabitants of the East do not perceive evening eclipses of the sun and moon, nor do those dwelling in the West see morning eclipses, while the latter see eclipses at midday later than we do. The victory of Alexander the Great is said to have caused an eclipse of the moon at Arbela at 8 p.m. while the same eclipse in Sicily was when the moon was just rising. An eclipse of the sun that occurred on April 30 in the consulship of Vipstanus and Fonteius a few years ago was visible in Campania between 1 and 2 p.m. but was reported by Corbulo commanding in Armenia as observed between 4 and 5: this was because the curve of the globe discloses and hides different phenomena for different localities. If the earth were flat, all would be visible to all alike at the same time; also the nights would not vary in length, because corresponding periods of 12 hours would be visible equally to others than those at the equator, periods that as it is do not exactly correspond in every region alike.

Consequently also although night and day are the same thing all over the world, it is not night and day at the same time all over the world, the intervention of the globe bringing night or its revolution day. This has been discovered by many experiments—that of Hannibal's towers in Africa and Spain, and in Asia when piratical alarms prompted the precaution of watchtowers of the same sort, warning fires lit on which at noon were often ascertained to have been seen by the people farthest to the rear at 9 p.m. Alexander above mentioned had a runner named Philonides who did the 1200 stades from Sicyon to Elis in 9 hours from sunrise and took till 9 p.m. for the return journey, although the way is downhill; this occurred repeatedly. The reason was that going his way lay with the sun but returning he was passing the sun as it met him travelling in the opposite direction. For this reason ships sailing westward beat even in the shortest day the distances they sail in the nights, because they are going with the actual sun.

On earthquakes

Earthquakes occur in a variety of ways, and cause remarkable consequences, in some places overthrowing walls, in others drawing them down into a gaping cleft, in others thrusting up masses of rock, in others sending out rivers and sometimes even fires or hot springs, in others diverting the course of rivers. They are however preceded or accompanied by a terrible sound, that sometimes resembles a rumble, sometimes the lowing of cattle or the shouts of human beings or the clash of weapons struck together, according to the nature of the material that receives the shock and the shape of the caverns or burrows through which it passes, proceeding with smaller volume in a narrow channel but with a harsh noise in channels that bend, echoing in hard channels, bubbling in damp ones, forming waves in stagnant ones, raging against solid ones. Accordingly even without any movement occurring a sound is sometimes emitted. And sometimes the earth is not shaken in a simple manner but trembles and vibrates. Also the gap sometimes remains open, showing the objects that it has sucked in, while sometimes it hides them by closing its mouth and drawing soil over it again in such a way as to leave no traces; it being usually cities that are engulfed, and a tract of farmland swallowed, although seaboard districts are most subject to earthquakes, and also mountainous regions are not free from disaster of the kind: I have ascertained that tremors have somewhat frequently occurred in the Alps and Apennines.

Earthquakes are more frequent in autumn and spring, as is lightning. Consequently the Gallic provinces and Egypt suffer very little from them, as in the latter the summer is the cause that prevents them and in the former the winter. Similarly they are more frequent by night than in the daytime. The severest earthquakes occur in the morning and the evening, but they are frequent near dawn and in the daytime about noon. They also occur at an eclipse of the sun or moon, since then storms are lulled, but particularly when heat follows rain or rain heat.

On tides

About the nature of bodies of water a great deal has been said. But the rise and fall of the tides of the sea is extremely mysterious, at all events in its irregularity; however the cause lies in the sun and moon. Between two risings of the moon there are two high and two low tides every 24 hours, the tide first swelling as the world moves upward with the moon, then falling as it slopes from the midday summit of the sky towards sunset, and again coming in as after sunset the world goes below the earth to the lowest parts of the heaven and approaches the regions opposite to the meridian, and from that point sucking back until it rises again; and never flowing back at the same time as the day before, just as if gasping for breath as the greedy star draws the seas with it at a draught and constantly rises from another point than the day before; yet returning at equal intervals and in every six hours, not of each day or night or place but equinoctial hours, so that the tidal periods are not equal by the space of ordinary hours whenever the tides occupy larger measures of either diurnal or nocturnal hours and only equal everywhere at the equinox. It is a vast and illuminating proof, and one of even divine utterance, that those are dull of wit who deny that the same stars pass below the earth and rise up again, and that they present a similar appearance to the lands and indeed to the whole of nature in the same processes of rising and setting, the course or other operation of a star being manifest beneath the earth in just the same way as when it is travelling past our eyes.

Moreover, the lunar difference is manifold, and to begin with, its period is seven days: inasmuch as the tides, which are moderate from new moon to half-moon, therefrom rise higher and at full moon are at their maximum; after that they relax, at the seventh day being equal to what they were at first; and they increase again when the moon divides on the other side, at the union of the moon with the sun being equal to what they were at full moon. When the moon is northward and retiring further from the earth the tides are gentler than when she has swerved towards the south and exerts her force at a nearer angle. At every eighth year the tides are brought back at the hundredth circuit of the moon to the beginnings of their motion and to corresponding stages of increase. They make all these increases owing to the yearly influences of the sun, swelling most at the two equinoxes and more at the autumn than the spring one, but empty at midwinter and more so at midsummer. Nevertheless this does not occur at the exact points of time I have specified, but a few days after, just as it is not at full or new moon but afterwards, and not immediately when the world shows or hides the moon or slopes it in the middle quarter, but about two equinoctial hours later, the effect of all the occurrences in the sky reaching the earth more slowly than the sight of them, as is the case with lightning, thunder and thunder-bolts.

The Tiber River

The Tiber, the former name of which was Thybris, and before that Albula, rises in about the middle of the Apennine in the territory of Arezzo. At first it is a narrow stream, only navigable when its water is dammed by sluices and then discharged, in the same way as its tributaries, the Tinia and the Chiana, the waters of which must be so collected for nine days, unless augmented by showers of rain. But the Tiber, owing to its rugged and uneven channel, is even so not navigable for a long distance, except for rafts, or rather logs of wood; in a course of 150 miles it divides Etruria from the Umbrians and Sabines, passing not far from Tifernum, Perugia and Ocriculum, and then, less than 16 miles from Rome, separates the territory of Veii from that of Crustumium, and afterwards that of Fidenae and Latium from Vaticanum. But below the confluence of the Chiana from Arezzo it is augmented by forty-two tributaries, the chief being the Nera and the Severone (which latter is itself navigable, and encloses Latium in the rear), while it is equally increased by the aqueducts and the numerous springs carried through to the city; and consequently it is navigable for vessels of whatever size from the Mediterranean, and is a most tranquil trafficker in the produce of all the earth, with perhaps more villas on its banks and overlooking it than all the other rivers in the whole world. And no river is more circumscribed and shut in on either side; yet of itself it offers no resistance, though it is subject to frequent sudden floods, the inundations being nowhere greater than in the city itself. But in truth it is looked upon rather as a prophet of warning, its rise being always construed rather as a call to religion than as a threat of disaster.

The isthmus of Corinth

The Peloponnese, which was previously called Apia and Pelasgia, is a peninsula inferior in celebrity to no region of the earth. It lies between two seas, the Aegean and the Ionian, and resembles in shape the leaf of a plane-tree; on account of the angular indentations the circuit of its coast-line, according to Isidore, amounts to 563 miles, and nearly as much again in addition, measuring the shores of the bays. The narrow neck of land from which it projects is called the Isthmus. At this place the two seas that have been mentioned encroach on opposite sides from the north and east and swallow up all the breadth of the peninsula at this point, until in consequence of the inroad of such large bodies of water in opposite directions the coasts on either side have been eaten away so as to leave a space between them of only five miles, with the result that the Morea is only attached to Greece by a narrow neck of land. The inlets on either side are called the Gulf of Lepanto and the Gulf of Egina, the former ending in Lecheae and the latter in Cenchreae. The circuit of the Morea is a long and dangerous voyage for vessels prohibited by their size from being carried across the isthmus on trolleys, and consequently successive attempts were made by King Demetrius, Caesar the dictator and the emperors Caligula and Nero, to dig a ship-canal through the narrow part—an undertaking which the end that befell them all proves to have been an act of sacrilege. In the middle of this neck of land which we have called the Isthmus is the colony of Corinth, the former name of which was Ephyra; its habitations cling to the side of a hill, 7½ miles from the coast on either side, and the top of its citadel, called the Corinthian Heights, on which is the spring of Pirene, commands views of the two seas in opposite directions. The distance across the Isthmus from Leucas to Patras on the Gulf of Corinth is 88 miles. The colony of Patras is situated on the longest projection of the Peloponnese opposite to Aetolia and the river Evenus, separated from them at the actual mouth of the gulf by a gap of less than a mile, as has been said; but in length the Gulf of Corinth extends 85 miles from Patras to the Isthmus.

The Black Sea

The Euxine or Black Sea, formerly because of its inhospitable roughness called the Axine, owing to a peculiar jealousy on the part of Nature, which here indulges the sea's greed without any limit, actually spreads into Europe and Asia. The Ocean was not content to have encircled the earth, and with still further cruelty to have reft away a portion of her surface, nor to have forced an entrance through a breach in the mountains and rent Gibraltar away from Africa, so devouring a larger area than it left remaining, nor to have swallowed up a further space of land and flooded the Sea of Marmara through the Dardanelles; even beyond the Straits of Constantinople also it widens out into another desolate expanse, with an appetite unsatisfied until the Sea of Azov links on its own trespass to its encroachments. That this event occurred against the will of the earth is proved by the number of narrows, and by the smallness of the gaps left by Nature's resistance, measuring at the Dardanelles 875 paces, at the Straits of Constantinople and Kertsch the passage being actually fordable by oxen— which fact gives both of them their name;—and also by a certain harmonious affinity contained in their disseverance, as the singing of birds and barking of dogs on one side can be heard on the other, and even the interchange of human speech, conversation going on between the two worlds, save when the actual sound is carried away by the wind.

The Hyperboreans

Then come the Ripaean Mountains . . . it is a part of the world that lies under the condemnation of nature and is plunged in dense darkness, and occupied only by the work of frost and the chilly lurking-places of the north wind. Behind these mountains and beyond the north wind there dwells (if we can believe it) a happy race of people called the Hyperboreans, who live to extreme old age and are famous for legendary marvels. Here are believed to be the hinges on which the firmament turns and the extreme limits of the revolutions of the stars, with six months' daylight and a single day of the sun in retirement, not as the ignorant have said, from the spring equinox till autumn: for these people the sun rises once in the year, at midsummer, and sets once, at midwinter. It is a genial region, with a delightful climate and exempt from every harmful blast. The homes of the natives are the woods and groves; they worship the gods severally and in congregations; all discord and all sorrow is unknown. Death comes to them only when, owing to satiety of life, after holding a banquet and anointing their old age with luxury, they leap from a certain rock into the sea: this mode of burial is the most blissful.

On India

The Ganges is said by some people to rise from unknown sources like the Nile and to irrigate the neighbouring country in the same manner, but others say that its source is in the mountains of Scythia, and that it has nineteen tributaries, among which the navigable ones besides those already mentioned are the Crenacca, Rhamnumbova, Casuagus and Sonus. Others state that it bursts forth with a loud roar at its very source, and after falling over crags and cliffs, as soon as it reaches fairly level country finds hospitality in a certain lake, and flows out of it in a gentle stream with a breadth of 8 miles where narrowest, and 12½ miles as its average width, and nowhere less than 100 feet deep, the last race situated on its banks being that of the Gangarid Calingae: the city where their king lives is called Pertalis. This monarch has 60,000 infantry, 1000 cavalry and 700 elephants always equipped ready for active service. For the peoples of the more civilised Indian races are divided into many classes in their mode of life: they cultivate the land, others engage in military service, others export native merchandise and import goods from abroad, while the best and wealthiest administer the government and serve as judges and as counsellors of the kings. There is a fifth class of persons devoted to wisdom, which is held in high honour with these people and almost elevated into a religion; those of this class always end their life by a voluntary death upon a pyre to which they have previously themselves set light. There is one class besides these, half-wild people devoted to the laborious task—from which the classes above mentioned are kept away—of hunting and taming elephants; these they use for ploughing and for transport, these are their commonest kind of cattle, and these they employ when fighting in battle and defending their country: elephants to use in war are chosen for their strength and age and size. . .

It is known that many of the inhabitants are more than seven feet six inches high, never spit, do not suffer from headache or toothache or pain in the eyes, and very rarely have a pain in any other part of the body—so hardy are they made by the temperate heat of the sun; and that the sages of their race, whom they call Gymnosophists, stay standing from sunrise to sunset, gazing at the sun with eyes unmoving, and continue all day long standing first on one foot and then on the other in the glowing sand. Megasthenes states that on the mountain named Nulus there are people with their feet turned backwards and with eight toes on each foot, while on many of the mountains there is a tribe of human beings with dogs' heads, who wear a covering of wild beasts' skins, whose speech is a bark and who live on the produce of hunting and fowling, for which they use their nails as weapons; he says that they numbered more than 120,000 when he published his work. Ctesias writes that also among a certain race of India the women bear children only once in their life-time, and the children begin to turn grey directly after birth; he also describes a tribe of men called the Monocoli who have only one leg, and who move in jumps with surprising speed; the same are called the Umbrella-foot tribe, because in the hotter weather they lie on their backs on the ground and protect themselves with the shadow of their feet; and that they are not far away from the Cave-dwellers; and again westward from these there are some people without necks, having their eyes in their shoulders. There are also satyrs in the mountains in the east of India (it is called the district of the Catarcludi); this is an extremely swift animal, sometimes going on all fours and sometimes standing upright as they run, like human beings; because of their speed only the old ones or the sick are caught. Tauron gives the name of Choromandae to a forest tribe that has no speech but a horrible scream, hairy bodies, keen grey eyes and the teeth of a dog. Eudoxus says that in the south of India men have feet eighteen inches long and the women such small feet that they are called Sparrow-feet. Megasthenes tells of a race among the Nomads of India that has only holes in the place of nostrils, like snakes, and bandy-legged; they are called the Sciritae. At the extreme boundary of India to the East, near the source of the Ganges, he puts the Astomi tribe, that has no mouth and a body hairy all over; they dress in cottonwool and live only on the air they breathe and the scent they inhale through their nostrils; they have no food or drink except the different odours of the roots and flowers and wild apples, which they carry with them on their longer journeys so as not to lack a supply of scent; he says they can easily be killed by a rather stronger odour than usual. Beyond these in the most outlying mountain region we are told of the Three-span men and Pygmies, who do not exceed three spans, *i.e.* twenty-seven inches, in height; the climate is healthy and always spring-like, as it is protected on the north by a range of mountains; this tribe Homer has also recorded as being beset by cranes. It is reported that in springtime their entire band, mounted on the backs of rams and she-goats and armed with arrows, goes in a body down to the sea and eats the cranes' eggs and chickens, and that this outing occupies three months; and that otherwise they could not protect themselves against the flocks of cranes that would grow up; and that their houses are made of mud and feathers and egg-shells.

32. Varro on soils

Marcus Terentius Varro (116–37 B.C.) has been described as "the prototype for a host of pedestrian Latin scholars who, through reading in many fields and digesting or excerpting the works of earlier authorities in the production of their own compendia, expected to gain a reputation for learning."[1] Varro's *De re rustica* offers sound advice on many aspects of farming. The following selection is concerned with the qualities of various kinds of soil.

It is expedient then, as I was saying, to study each kind of soil to determine for what it is, and for what it is not, suitable. The word *terra* is used in three senses: general, particular and mixed. It is a general designation when we speak of the orb of the earth, the land of Italy or any other country. In this designation is included rock and sand and other such things. In the second place, *terra* is referred to particularly when it is spoken of without qualification or epithet. In the third place, which is the mixed sense, when one speaks of *terra* as soil—that in which seeds are sown and developed; as for example, clay soil or rocky soil or others. In this sense there are as many kinds of earth as there are when one speaks of it in the general sense, on account of the mixtures of substances in it in varying quantities which make it of different heart and strength, such as rock, marble, sand, loam, clay, red ochre, dust, chalk, gravel, carbuncle (which is a condition of soil formed by the burning of roots in the intense heat of the sun); from which each kind of soil is called by a particular name, in accordance with the substances of which it is composed, as a chalky soil, a gravelly soil, or whatever else may be its distinguishing quality. And as there are different varieties of soil so each variety may be subdivided according to its quality, as, for example, a rocky soil is either very rocky, moderately rocky or hardly rocky at all. So three grades may be made of other mixed soils. In turn each of these three grades has three qualities: some are very wet, some very dry, some moderate. These distinctions are of the greatest importance in respect of the crops, for the skilled husbandman plants spelt rather than wheat in wet land, and on dry land barley rather than spelt, in medium land both. Furthermore there are still more subtle distinctions to be made in respect of all these kinds of soil, as for example it must be considered in respect of loam, whether it is white loam or red loam, because white loam is unfit for nursery beds, while red loam is what they require. But the three great dis-

---

From A Virginia Farmer, *Roman Farm Management: the Treatises of Cato and Varro,* (New York: Macmillan, 1913), pp. 88–91.

1. William H. Stahl, *Roman Science: Origins, Development and Influence to the Later Middle Ages* (Madison: University of Wisconsin Press, 1962), p. 74.

tinctions of quality of soil are whether it is lean or fat, or medium. Fat soils are apparent from the heavy growth of their vegetation, and the lean lie bare; as witness the territory of Pupinia (in Latium), where all the foliage is meagre and the vines look starved, where the scant straw never stools, nor the fig tree blooms, while for the most part the trees are as covered with moss as are the arid pastures. On the other hand, a rich soil like that of Etruria reveals itself heavy with grain and forage crops and its umbrageous trees are clean of moss. Soil of medium strength, like that near Tibur, which one might say is rather hungry than starved, repays cultivation in proportion as it takes on the quality of rich land.

"Diophanes of Bithynia," said Stolo, "was very much to the point when he wrote that the best indication of the suitability of soil for cultivation can be had either from the soil itself or from what grows in it: so one should ascertain whether it is white or black, if it is light and friable when it is dug, whether its consistency is ashy, or too heavy: or it can be tested by evidence that the wild growth upon it is heavy and fruitful after its kind."

33. Pomponius Mela on the earth, on Europe, and on Africa

De situ orbis, a short geographical treatise by Pomponius Mela, (fl. middle of 1st century A.D.), remained a popular text for nearly fifteen hundred years. It is the oldest extant work on geography in Latin.

The land Mela calls Sarmatia is the Scythia of Herodotus.

The division of the earth

All of this great whole, which we call Earth and Heaven, is but a single unit that encompasses all things within it. Yet it is divided into parts: that region where the sun rises is called Orient, where it sets is known as Occident, where it reaches its highest point in the heavens is South, the opposite is North. The earth, rising in the midst of all this sublime universe, is surrounded by the sea, which also divides it, from east to west, into two hemispheres; within these we can distinguish five zones. The central zone is cursed by heat, the two extreme zones by cold; the remaining zones are inhabitable and have the same seasons, though at opposite times of the year. We inhabit this hemisphere, the antipodes the opposite one, but their dwelling place, due to the heat of the zone separating us, is unknown, therefore we shall only deal with our own . . .

From Pomponius Mela, De situ orbis, in Macrobe–Varron–Pomponius Mela (Paris: Firmin-Didot, 1883). Editor's translation.

Our hemisphere extends from east to west, its length is somewhat greater then its greatest width. It is completely surrounded by the ocean, whence it receives four seas: one from the north [the Caspian], two from the south [The Persian Gulf and the Arabian Sea], and one from the west [the Mediterranean] . . . This sea [the Mediterranean] and two great rivers, the Don and the Nile, divide our hemisphere into three parts. The Don, running from north to south, ends in the Sea of Azov. . . The Nile, running in the opposite direction, ends in our sea. All of the lands located on one side of our sea, between the strait [of Gibraltar] and the rivers, are called Africa, those on the opposite side, Europe; Africa extending as far as the Nile, Europe as far as the Don. All lands lying beyond these rivers are part of Asia.

Europe

Europe is bordered in the east by the Don, the Sea of Azov, and the Black Sea; in the south, by the Mediterranean; in the west by the Atlantic; in the north, by the British Sea. . .

The coasts of Europe form three great gulfs between the Dardanelles and the Strait of Gibraltar. . . The first of these is the Aegean, the second, the Ionian, extended towards the interior of the continent by the Adriatic; the third is the one we call the Etruscan Sea; the Greeks call it the Tyrrhenian Sea.

The first of the European countries is Scythia, extending from the Don to about the middle of the Black Sea coast. Hence, and extending into the Aegean, lie Macedonia and Thrace. Further on, Greece projects itself between the Aegean and the Ionian seas. Illyria lies on one of the coasts of the Adriatic, while Italy occupies a position between the Adriatic and the Etruscan seas. Adjoining Italy, and lying on the shore of the Etruscan sea is Gaul and beyond it, Spain. The coasts of Spain also face westward and northward. Beyond them one encounters Gaul once more, it extends a great distance inland from the Mediterranean. Beyond Gaul is Germany and beyond it Sarmatia which, in turn, borders on Asia.

Cyrenaica

In this province, located between the Catabathmon and the limits of Africa, we find the oracle of Ammon, famous for its accurate prophecies; a fountain called the fountain of the sun; and a rock sacred to the south wind. If a man touches this rock with his hand, the wind, irritated, rises in full fury, stirring up waves of sand raging in the same manner as a storm at sea. The waters of the fountain of the sun are boiling hot in the middle of the night, then cool off slowly and, already fresh at daybreak, continue to turn colder until, at midday, are frigid. Past noon, the water begins to warm up gradually: being lukewarm at sunset, it warms up until, at midnight, it is boiling once more. . .

Egypt

Egypt, the first part of Asia, . . . extends a considerable distance inland, being contiguous with Ethiopia. Though a dry land, it is wondrously fertile, a prodigious producer of men and other animals. It is the Nile, largest of all rivers draining to the Mediterranean, which is responsible for this. This river comes out of the African desert, being neither navigable, nor bearing the name of Nile. Having traversed as a single river a large area, it divides in Ethiopia into two branches, surrounding the large island of Meroe. One of these branches is called Astaboras (White Nile), the other Astape (Blue Nile). The name Nile is applied to the river starting at the point where the two branches merge into one stream . . .

The customs of the Egyptians are quite different from those of other peoples. At funerals they daub themselves with mud and wail. They regard the burning or burying of the dead as impieties, instead they embalm bodies with care and bury them in underground caves. They write from right to left. They handle dung with their hands and trash grain with their feet. Women discuss public affairs on the forum, while the men stay home and look after household chores. Women must look after their parents, men do not have to. They eat outside their houses, but to obey the call of nature they return home. They worship the images of all sorts of animals, and even more the animals themselves, but these practices differ from place to place, and it is a capital crime to kill these animals, even accidentally. If an animal dies of illness or by accident, they hold a solemn funeral and indulge in mourning.

Sarmatia

Sarmatia widens out south of the sea (the Baltic); it is separated from its westerly neighbors by the Vistula, and extends southward to the Danube. Its inhabitants resemble, in their dress and in the arms they bear, the Parthians, but living in a more rigorous climate they have a rougher character. They have neither cities, nor fixed abode. Whether they are in search of good pasture or in pursuit of an enemy, they carry all their possessions with them and always dwell in fortified camps. They are warlike, free and indomitable, and so fierce and cruel that even the women go to war together with the men. To enable the women to fight, they burn the girl-babies' right breast at birth: having thus a man-like chest, they can deliver blows easily. To shoot the bow, to ride horses, to hunt, these are the duties of girls; when they grow to adulthood they are to kill enemies, and until they have accomplished this their punishment is to remain virgins . . .

Ethiopia

Ever since, in the days of our ancestors, a certain Eudoxus . . . sailed out of the Arabian Gulf into the Ocean and finally reached Gades [Cadiz], we do have some information on the coasts of Africa.

Beyond the desert shores of which we spoke live people who are mute, who cannot make themselves understood except by signs. Some do have a tongue but cannot articulate sounds; some do not have a tongue; some, whose lips are closed, possess under their nostrils an opening through which they can drink, using straws, and when they feel the need to eat, they smell the grains of growing plants they find here and there. Until the landing of Eudoxus, fire was unknown to some of these people, to the extent that, when they saw it, they were so astonished as to hold burning coals in their arms and against their bodies; thus the fire that pleased them so much gave them much pain.

Beyond these lands the coast becomes a vast gulf, in it there is a large island, inhabited only by women. Their bodies are covered by hair, and they bear children without having intercourse with men. Besides, they are so wild that the strongest bonds can hardly restrain them. Hanno tells this story; its truth is attested by the skins of some of these women whom he had killed and brought back.

Beyond this gulf there is a high mountain the Greeks call "the chariot of the Gods," which always omits fire. Beyond it, the coast offers an endless vista of green hills and meadows that appear as though they were peopled by Pans and satyrs. What supports this view is that the land is without trace of cultivation, nor does one see even a vestige of man and his habitations. During the day this is a land of great solitude and even greater silence, but at night, with fires burning everywhere, it resembles a vast camp, where one hears the crash of drums and cymbals, and the sound of flutes played louder than any human being is capable of.

34. Solinus describes Italy, Thrace, the Hyperboreans; the crocodile, China, and India

Virtually nothing is known of the life of Caius Julius Solinus, nicknamed "Polyhistor," master of everything. He was the author of one of the most successful Latin compilations describing the world and its wonders and it is assumed that he lived during the third century

From *The Excellent and Pleasant Worke of Iulius Solinus Polyhistor—Containing the noble actions of human creatures, the secrets & providence of nature, the description of Countries, the manners of the people . . .* , trans. Arthur Golding, (London, 1587).

of the Christian era. His book, *Collectanea rerum memorabilium,* was translated into English in 1587 for use in English schools. The following excerpts, taken from Golding's 1587 text, were transcribed according to modern English spelling.

Italy

Now to the intent that we return to our determined purpose, our style is to be directed to the Recital of places: chiefly and principally to Italy, the beauty where of we have already touched upon lightly, in the City of Rome. But Italy has been written of so thoroughly by all men, especially by Marcus Cato, that there cannot be found that thing which the diligence of former authors have not presented; for the country is so excellent that it receives abundant praise. The notable writers consider the healthfulness of the places, the temperateness of the air; the fruitfulness of the soil; the open prospects of the hills; the cool meadows of the woods; the unhurtful low grounds; the plentiful increase of wines and olives; the sheeps' courses; the pasture grounds; so many rivers; such great lakes; places that bear flowers twice a year; together with the mountain Vesuvius, casting up a flaming breath of air as if it had a soul; the baths with their springs of warm water; the continual beautifying of the land with new cities, so goodly a site of ancient towns which first the aborigines, Arunks, Pelasgians, Arcadians, Sicilians, and lastly the inhabitants of all parts of Greece, and above all others, the victorious Romans, have built. Besides this, it has shores full of havens and coasts with large bays and harbors, that are the crossroads for traffic from all places of the world.

Thrace

Now it is time to take our journey into Thrace, and to set sail towards the most powerful nations of Europe: which whoever will consider looking upon, shall easily find that there is a contempt of life in the barbarous Thracians, through a certain discipline of mother wit. They all agree to die willingly: some of them believing that the souls of those who die return again, and others thinking that they die not, but are in a more happy and blissful state. Among most of them, birth days are sorrowful, and contrariwise, the burials are joyful, in as much that the fathers and mothers fall to weeping when their children are newborn and rejoice when they are dead. The men glory in the number of their wives, and count it an honor to have many bed-fellows. Such women as are careful of their chastity, leap into the fires where their dead husbands are burned, and (which they think is the greatest possible token of chastity) run headlong into the flame. When women come to the time of marriage, they do not take husbands at the appointment of their parents: but those that excel others in beauty set themselves forth for sale, marrying who will give most; they marry not to him that is of the best disposition, but to him that is the best merchant. Those who are foul or deformed, bring dowries with them to their husbands. When they feast, both sexes of them go about the hearths and cast the seeds of certain herbs growing among them into the fire. The fume of these herbs so strikes their heads, that it wounds their senses and makes them like drunken folk, whereat they have a good sport. This much concerning their customs.

The Hyperboreans

Sundry things that have been reported of the Hyperboreans have been but a fable and a flying tale if the things, that have come from thence unto us, had been rashly believed. But seeing the best authors, and such as are of sufficient credibility, do agree in one constant report, no man needs to fear any falsehood. Of the Hyperboreans, they speak in this way. They inhabit most of the Pteropheron, which we hear lies beyond the North Pole, a most blessed nation. They ascribe it rather unto Asia than unto Europe, and some do place it midway between the sun's rising and the sunset: that is to say, between the west of the Antipodes and our East; a thing which reason reproves when considering what a vast sea runs between the two worlds. They are therefore in Europe, and among them are thought to be the poles of the world; and the uttermost circuit of the stars; and the half year night, lacking the sun but one day. However, there are those who think the sun rises not day by day to them as it does to us, but that it rises in the springtime, and goes not down again before the fall of the leaf, so that they have continued day within the space of five months and continual night within the space of the other five months. The air is very mild, the blasts wholesome, and the wind not hurtful. Their houses are the wild fields or the woods and the trees yield them food from day to day. They know no debate, they are not troubled with diseases, all men have one desire which is to live innocently. They have death, and by the wilfull doing away with themselves, prevent the long tarriance of their decease. For when they have lived as long as they desire, then feasting and anointing themselves, they throw themselves headlong from some known rock into the deep sea, believing this to be the best kind of burial.

The crocodile

The crocodile, a four footed mischief, has force both upon land and water alike. He has no tongue, and he moves the upper jaw. Where he bites, he takes horrible fast hold, his teeth shutting checkerwise one within another. For the most part he grows to the size of twenty fathoms, and they lay eggs like geese eggs. He chooses a place to build a nest in, where the water of the Nile cannot come when it is at its fullest. In raising their young, the male and female take their turns. Besides his wide chops, he is also armed with outrageously long tails. At night he keeps in the waters, and at day he rests upon the land. He is clad in marvelously strong hide, in so much as a piece of artillery shot at him out of any engine rebounds again from his skin. There is a little pretty bird called Trochylos, which in seeking to feed upon the flesh that sticks in the crocodile's teeth, does little by little scrape his mouth, and so delighting him easily with his soft tickling, makes him gape, that he may stand between his chops.

China

As ye turn from the Scythian Ocean and Caspian Sea towards the East Ocean: from the beginning of this coast; first deep snows, then long deserts, beyond that the Cannibals, a most cruel kind of people, and lastly, places full of most outrageously wild beasts, make almost one half of the way impassable. These distresses have their end at a Mount, which the barbarous people call Tabis, that butts upon the sea, beyond which the wilderness does nevertheless continue a great way on still. So in that coast which faces the Northeast, beyond those vast and uninhabitable countries, the first men that we have heard of are the Seres, who, sprinkling water upon the leaves of their trees, do, with the help of that liquor, comb certain fleeces, and with moisture so card that fine cotton, that they make what they will thereof. This is that silk admitted to be worn commonly, to the hindrance of gravity, and wherewith the lust of excess has persuaded first women, and now men to apparel themselves, rather to set out the bodies to sale, then to clothe it.

India

No cost is bestowed in burial of the dead. Furthermore (as is expressed in the books of King Juba and King Archelaus) as the people disagree in manners and conditions, so also is their great difference in their attire. Some wear linen garments, some woolen, some go all naked, some cover but their private members, and many go clad in barks of trees. Some people are so tall, that they can easily bolt over elephants as if they were horses. Many think it good neither to kill any living thing, nor to eat any flesh. Some eat only fish and live by the sea.

But among those who have some more care to live according to reason, many women are married to one man, and when the husband is deceased, each of them pleads before most grave judges concerning her deserts and she, who by the sentence of the judges is deemed to have been more dutiful and fervent than the rest, receives this reward of her victory: that at her pleasure she may leap into the fire where her husband is burning and offer herself as a sacrifice upon his corpse. The rest live with infamy.

35. Macrobius: a late Roman geographer

Macrobius, who flourished during the first part of the fifth century A.D., remained a popular author on matters regarding geography throughout the Middle Ages. His *Commentary on the Dream of Scipio* was among the important incunabula, and maps illustrating his concepts exist in both manuscript and early printed versions. The following excerpts are typical of his use of earlier sources and his methods of presentation.

The earth and the sun, their dimensions

We must recognize this fact, too, that the earth's shadow, which the sun after setting and passing into the lower hemisphere sends upwards, creating on earth the darkness called night, is sixty times the diameter of the earth; its full extent reaches the circle of the sun's path and, by cutting off the light, causes darkness to fall back upon the earth. Now we must reveal the length of the earth's diameter in order to determine the distance that would be sixty times as long, and thus fulfill our promise.

By most obvious and accurate methods of measurement we know that the earth's circumference, including the habitable and uninhabitable areas, is 252,000 stades. This being the circumference, the diameter will of course be 80,000 stades, or slightly more, according to the method explained above of tripling the diameter with the addition of a seventh part to get the circumference. Now, since it is the diameter of the earth and not the circumference which is multiplied sixty times to obtain the measurement of the earth's shadow—inasmuch as this is the dimension that casts the shadow—sixty times 80,000 stades (the earth's diameter) gives the distance of 4,800,000 stades from the earth to the sun's path, the point we said that the earth's shadow reaches. The earth, moreover, is at the midpoint of the orbit in which the sun revolves, and so the length of the earth's shadow is equal to half the diameter of the sun's orbit; and if an equal distance is measured in the opposite direction from the earth to the sun's orbit, we have the diameter of the circle in which the sun moves. Double 4,800,000 stades and you get 9,600,000 stades, or the diameter of the sun's orbit, after which it is an easy matter to find the circumference. You must multiply the length of the diameter by 3 1/7, as we have often stated, and thus you will find that the circumference of the circle in which the sun moves is roughly 30,170,000 stades.

At this point we have obtained the measurements of the earth's diameter and circumference, and of the sun's orbit and the diameter of that orbit;

From Macrobius, *Commentary on the Dream of Scipio*, trans. William Stahl (New York: Columbia University Press, 1952), pp. 171–174, 182–184, 201–205, 214–216. By permission of the publisher.

now let us reveal the diameter of the sun and how the skillful Egyptians ascertained it. As it was possible to learn the measurement of the orbit in which the sun moves from the earth's shadow, so the dimensions of the sun have been found by means of that orbit in the following ingenious manner. On the day of the equinox before sunrise place in an exactly level position a stone vessel that has been hollowed until its cavity forms a perfect hemisphere [in Greek: *scaphē*]. On the bottom there must be lines representing the twelve hours of the day which the shadow cast by a stylus will mark off as the sun passes along in the sky. Moreover, we know that with a vessel of this kind, in the exact amount of time that it takes for the shadow to go from one lip of the vessel to the other, the sun will traverse half the sky, from rising to setting, the distance of one hemisphere. A complete circuit of the sky requires a day and a night, and so we may be sure that the shadow in the vessel advances in proportion to the space traversed by the sun. Shortly before sunrise let the observer closely fix his eyes upon this bowl—be sure that it is in a perfectly level position—and then when the first ray of the sun, as its top just steals over the horizon, has cast a shadow of the top of the stylus upon the edge of the bowl, let him carefully mark the place where the shadow first fell. Then he must watch until the moment when the full orb comes into view so that the bottom seems to be just resting on the horizon. Again he must mark the point which the shadow has reached in the bowl. The measurement between the two shadow marks, corresponding to the full orb or diameter of the sun, will be found to be a ninth part of the space in the bowl between the line marking sunrise and the line marking the first hour.

Thus it becomes clear that nine times the sun's diameter is equal to the distance that the sun progresses in one hour at the time of the equinox. Twelve hours complete the sun's course through one hemisphere, and 12 times 9 is 108; so the sun's diameter must be equal to 1/108 of the equinoctial hemisphere, or 1/216 of the whole equinoctial orbit. We have proved that the sun's orbit is 30,170,000 stades long, and therefore the sun's diameter will be 1/216 of this. This will be found to be a little less than 140,000 stades; and this, you see, is almost twice the diameter of the earth.

Now geometry teaches us that when the diameter of one sphere is twice as great as the diameter of another sphere, the sphere with the larger diameter is really eight times as great as the other; so we must agree that the sun is eight times greater than the earth. This discussion of the sun's size is only a brief summary of the many things written on the subject.

On gravity

Of all the matter that went into the creation of the universe, that which was purest and clearest took the highest position and was called ether; the part that was less pure and had some slight weight became air and held second place; next came that part which was indeed still clear but which the sense of touch demonstrates to be corporeal, that which formed the bodies of water; lastly, as a result of the downward rush of matter, there was that vast, impenetrable solid, the dregs and off-scourings of the purified elements, which had settled to the bottom, plunged in continual and oppressing chill, relegated to the last position in the universe, far from the sun. Because this became so hardened it received the name *terra.* Compact air, closer to the chill of the earth than to the heat of the sun, supports the earth on all sides and keeps it in its place by its very density. The power of this belt of atmosphere, supporting on every side with equal force, or perhaps the spherical nature of the sky itself, keeps the earth from moving in one direction or another. If it should deviate from the midpoint it would move nearer to the surface and leave the bottom since only the midpoint is equidistant from any point on the surface of a sphere. To this point which is the bottom and middle, so to speak, and is stationary because it is the center, all weights must be drawn since the earth itself, like a weight, has fallen to this place.

We might cite countless arguments as proof, but particularly convincing is that of the rains that fall upon the earth from every region of the atmosphere. Not only do they fall upon this portion of the earth that we inhabit, but on the slopes that give the earth its sphericity; and, what is more, there is the same sort of rainfall in the region that we consider the underside. If air is condensed by the chill of the earth and forms a cloud which sends down a shower, and if, moreover, air surrounds the whole of the earth, then, assuredly, rain falls from all regions of the air except in those places parched by constant heat. From all quarters it drops upon the earth, which is the only resting place for objects possessing weight. Anyone who rejects this theory will be forced to admit that any precipitation of snow, rain, or hail falling outside the region inhabited by us would continue on down from the air into the celestial sphere. The sky is of course equidistant from the earth in every direction and is as far beyond the sloping regions and that portion which we regard the underside as it is beyond us. If all weights were not drawn to the earth, therefore, the rain that falls beyond the sides of the earth would not fall upon the earth but upon the celestial sphere, an assumption too ridiculous to consider. Let a circle represent the earth; upon it inscribe the letters ABCD. About this draw another circle with the letters EFGLM inscribed, representing the belt of the atmosphere. Divide both circles by drawing a line from E to L. The upper section will be the one we inhabit, the lower the one beneath us. If it were not true that all weights are drawn towards the earth, then the earth would receive a very small portion of the rainfall, that which falls from A to C. The atmosphere between F and E and G and L would send its moisture into the air and sky. Furthermore, the rain from the lower half of the celestial sphere would have to continue on into the outer regions, unknown to our world, as we see from the diagram.

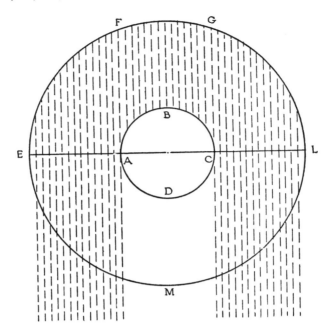

To refute this notion would not befit a sober treatise; it is so absurd that it collapses without discussion. In conclusion, we may therefore say that we have clearly proved that all weights are drawn to the earth by their own inclination . . .

On zones

As regards the five belts, I beg you not to think that the two founders of Roman eloquence, Virgil and Cicero, disagree in their views because the latter says that the belts *encircle the earth* and the former that the belts, which he calls by their Greek name zones, "holds the sky"; a later discussion will prove that they are both correct and still not contradictory. But in order to clarify all the difficulties we have undertaken to explain at this time, we must first discuss these belts, for when we have the location of each set before our eyes we shall more easily understand the rest.

We must first explain how they girdle the earth and then how they hold the sky. The earth is the ninth and lowest sphere. The horizon—that is, the circular boundary to which we have previously referred—divides this equally: one half, a part of which we inhabit, lies beneath that portion of the sky which is above the horizon, and the other half lies beneath that portion of the sky which in its revolutions has descended to the regions that seem to be below us. The earth, fixed in the middle of the universe, looks out upon the sky from every direction. Insignificant as it is in comparison with the sky—it is only a point in comparison, though a vast sphere to us—

it is divided into regions of excessive cold or heat, with two temperate zones between the hot and cold regions. The northern and southern extremities are frozen with perpetual cold, two belts, so to speak, that go around the earth but are small since they encircle the extremities. Neither zone affords habitation, for their icy torpor withholds life from animals and vegetation; animal life thrives upon the same climate that sustains plant life. The belt in the middle and consequently the greatest, scorched by an incessant blast of heat, occupies an area more extensive in breadth and circumference, and is uninhabited because of the raging heat. Between the extremities and the middle zone lie two belts which are greater than those at the poles and smaller than the one in the middle, tempered by the extremes of the adjoining belts; in these alone has nature permitted the human race to exist . . .

On the Antipodes

Of the four cardinal points in our region, the east, west, and north are referred to by their proper names because we know about them from their beginnings; the northern extremity, though uninhabited, is still not far removed from us. But the fourth cardinal point received another name, not being known as *australis,* "southern," but rather as *meridies,* "midday," for two reasons: first, because only that region is properly called southern which originates in the other extremity, lying opposite the north pole; second, the region where the south wind is first felt by the inhabitants of our quarter is the *mid*-part of the earth on which the light of *day* falls; hence the name *medidies,* and then by the substitution of a letter *meridies.* You must realize that Auster, the wind that reaches us from that direction, is at its starting point cold, as is our north wind, which pleases us with its refreshing chill, but that as it passes through the burning torrid zone, it gathers heat and reaches us as a warm wind though it was originally cold. Indeed, it would be contrary to reason and nature for winds originating at the two poles that are subjected to the same degree of cold to differ much in temperature. It is also obvious for the same reason that our north wind reaches the inhabitants of the south temperate zone as a hot wind, and that the south wind agreeably affects their bodies with its innate chill. It is then impossible to doubt that over the portion of the earth's surface which we think of as the underside the continuation and complete circuit of the zones that above are temperate are there also to be considered temperate, and accordingly, that the same two zones lie as far apart there as here, and are likewise inhabited. If anyone refuses to believe in this let him state his objections.

If life is possible for us in this quarter of the earth that we inhabit because we tread on the ground and look up at the sky overhead, and thrive upon the abundant air that we inhale, and because the sun rises and sets for us, why should we not assume that there are men living there as well where similar conditions always obtain? We must agree that the men who are supposed to be dwelling there breathe the same air as we because both zones have the same moderate temperature over their entire circuit; the same sun will of course be setting for them when it is rising for us and will be rising for them when setting for us; they will tread the ground as well as we and above their heads will always see the sky. Have no fear that they will fall off the earth into the sky, for nothing can ever fall upwards. If for us *down* is the earth and *up* is the sky—to affirm it is to be jesting—then for those people as well *up* will be what they see above them, and there is no danger of their falling upwards. I can assure you that the uninformed among them think the same thing about us and believe that it is impossible for us to be where we are; they, too, feel that anyone who tried to stand in the region beneath them would fall. But just as there has never been anyone among us who was afraid he might fall into the sky, so no one in their quarter is going to fall upwards since we saw from a previous discussion that all weights are borne by their own inclination towards the earth . . .

On the ocean and the oekumene

Let us now confirm the statement we made about Ocean, that the whole earth is girt about not by a single but by a twofold body of water whose true and original course man in his ignorance has not yet determined.

That Ocean which is generally supposed to be the only one is really a secondary body, a great circle which was obliged to branch off from the original body. The main course actually flows around the earth's torrid zone, girdling our hemisphere and the underside, and follows the circumference of the equator. In the east it divides, one stream flowing off to the northern extremity, the other to the southern; likewise, in the west, streams flow to the north and south, where they meet the streams from the east at the poles. As they rush together with great violence and impetus and buffet each other, the impact produces the remarkable ebb and flow of Ocean; and wherever our sea extends, whether in narrow straits or open coast, it shares in the tidal movement of Ocean's streams. These we now speak of as Ocean proper because of the fact that our sea is filled from Ocean's streams. But the truer bed of Ocean, if I may call it that, keeps to the torrid zone; it follows the circuit of the equator as the streams originating in it follow the circuit of the horizon in their course, thus dividing the whole earth into four parts and making each inhabited quarter, as we previously stated, an island.

Separating us from the people of the southern hemisphere, Ocean flows along the whole extent of the equator; again, as its streams branch out at the extremities of both regions, it forms two islands on the upper face of the earth and two on the underside. Cicero, wishing to imply this, did not say, "The whole earth is a small island," but rather, *The whole of the por-*

tion that you inhabit is a small island, since each of the four inhabited quarters becomes an island, with Ocean flowing about them all in two great circles, as we have explained.

The accompanying diagram will lay everything before our eyes; the origin of our sea, one part of the whole, and the sources of the Red Sea and the Indian Ocean will be evident; here, too, you will see where the Caspian Sea rises, although I am aware that there are some who deny that it has any connection with Ocean. It is certain, too, that in the temperate zone of the southern hemisphere there is a sea comparable to ours flowing in from Ocean, but we do not have the evidence for marking this off since its location continues to be unknown.

From our diagram we shall also understand Cicero's statement that our quarter is *narrow at the top and broad at the sides.* As the tropical circle is greater than the arctic circle, so our zone is narrower at the top than at the sides, for the top is pressed together by the smallness of the northern circle, whereas the sides extend in either direction over the broad expanse of the tropics. Indeed, the ancients remarked that the whole of our inhabited quarter was like an outspread chlamys.

Furthermore, since the whole earth, including Ocean, is but a point in comparison with any circle in the celestial sphere, Cicero was obliged to add, when speaking about Ocean, *But you can see how small it is despite its name!* Although we call it the Atlantic Sea and the Great Sea, it cannot seem great to those who behold it from the sky, since in comparison with the sky the earth is a mark that cannot be divided. His reason for emphasizing the earth's minuteness was that worthy men might realize that the quest for fame should be considered unimportant since it could not be great in so small a sphere.

Landscape in Latin Prose and Poetry

Geography, as its name indicates, includes descriptions of the earth's surface and of the several landscapes that compose it. Yet some of the greatest descriptions of landscape, and of earth features in general, have come not from the practitioners of the art and science of geography but from poets and prose writers. Examples of these descriptions are legion; those selected for this section are taken from the writings of the great poets and historians of the Golden Age of Latin literature: Virgil, Horace, Caesar, and Tacitus. These men could discuss with equal fluency the familiar features of their own Italy and the peculiarities of lands unknown to their contemporaries such as Britain, Germany and Judaea.

36. A victorious general reports: Caesar on Gaul, Britain, and Germany

Having completed the conquest of Gaul, Caius Julius Caesar (101–44 B.C.) wrote his own report of this triumph, including in it observations not only about that large province and its inhabitants but also about two neighboring countries unknown to most Romans: Britain and Germany.

Gaul is a whole divided into three parts, one of which is inhabited by the Belgae, another by the Aquitani, and a third by a people called in their

---

From Caesar, *Gallic Wars*, trans. H. J. Edwards, Loeb Classical Library (London: William Heinemann, 1939), pp. 3–5, 181–183, 191, 249–253, 345–347, 351–355.

own tongue Celtae, in the Latin Galli. All these are different one from another in language, institutions, and laws. The Galli (Gauls) are separated from the Aquitani by the river Garonne, from the Belgae by the Marne and the Seine. Of all these peoples the Belgae are the most courageous, because they are farthest removed from the culture and the civilization of the Province and least often visited by merchants introducing the commodities that make for effeminacy; and also because they are nearest to the Germans dwelling beyond the Rhine, with whom they are continually at war. For this cause the Helvetii also excel the rest of the Gauls in valour, because they are struggling in almost daily fights with the Germans, either endeavouring to keep them out of Gallic territory or waging an aggressive warfare in German territory. The separate part of the country which, as has been said, is occupied by the Gauls, starts from the river Rhone, and is bounded by the river Garonne, the Ocean, and the territory of the Belgae; moreover, on the side of the Sequani and the Helvetii, it touches the river Rhine; and its general trend is northward. The Belgae, beginning from the edge of the Gallic territory, reach to the lower part of the river Rhine, bearing towards the north and east. Aquitania, starting from the Garonne, reaches to the Pyrenees and to that part of the Ocean which is by Spain: its bearing is between west and north.

The Suebi are by far the largest and the most warlike nation among the Germans. It is said that they have a hundred cantons, from each of which they draw one thousand armed men yearly for the purpose of war outside their borders. The remainder, who have stayed at home, support themselves and the absent warriors; and again, in turn, are under arms the following year, while the others remain at home. By this means neither husbandry nor the theory and practice of war is interrupted. They have no private or separate holding of land, nor are they allowed to abide longer than a year in one place for their habitation. They make not much use of corn for food, but chiefly of milk and of cattle, and are much engaged in hunting; and this, owing to the nature of the food, the regular exercise, and the freedom of life—for from boyhood up they are not schooled in a sense of duty or discipline, and do nothing whatever against their wish— nurses their strength and makes men of immense bodily stature. Moreover, they have regularly trained themselves to wear nothing, even in the coldest localities, except skins, the scantiness of which leaves a great part of the body bare, and they bathe in the rivers.

They give access to traders rather to secure purchasers for what they have captured in war than to satisfy any craving for imports. And, in fact, the Germans do not import for their use draught-horses, in which the Gauls take the keenest delight, procuring them at great expense; but they take their home-bred animals, inferior and ill-favoured, and by regular exercising they render them capable of the utmost exertion. In cavalry combats they often leap from their horses and fight on foot, having trained their horses to remain in the same spot, and retiring rapidly upon them at need; and their tradition regards nothing as more disgraceful or more indolent than the use of saddles. And so, however few in number, they dare approach any party, however large, of saddle-horsemen. They suffer no importation of wine whatever, believing that men are thereby rendered soft and womanish for the endurance of hardship . . .

The Meuse flows from the range of the Vosges, in the territory of the Lingones, and, receiving from the Rhine a certain tributary called the Waal, forms the island of the Batavi; then, no more than eighty miles from the Ocean, it flows into the Rhine. The Rhine rises in the land of the Lepontii, who inhabit the Alps; in a long, swift course it runs through the territories of the Nantuates, Helvetii, Sequani, Mediomatrices, Triboci, and Treveri, and on its approach to the Ocean divides into several streams, forming many large islands (a great number of which are inhabited by fierce barbaric tribes, believed in some instances to live on fish and birds' eggs); then by many mouths it flows into the Ocean.

The inland part of Britain is inhabited by tribes declared in their own tradition to be indigenous to the island, the maritime part by tribes that migrated at an earlier time from Belgium to seek booty by invasion. Nearly all of these latter are called after the names of the states from which they sprang when they went to Britain; and after the invasion they abode there and began to till the fields. The population is innumerable; the farm-buildings are found very close together, being very like those of the Gauls; and there is great store of cattle. They use either bronze, or gold coins, or instead of coined money tallies of iron, of a certain standard of weight. In the midland districts of Britain tin is produced, in the maritime iron, but of that there is only a small supply; the bronze they use is imported. There is timber of every kind, as in Gaul, save beech and pine. They account it wrong to eat of hare, fowl, and goose; but these they keep for pastime or pleasure. The climate is more temperate than in Gaul, the cold seasons more moderate.

The natural shape of the island is triangular, and one side lies opposite to Gaul. Of this side one angle, which is in Kent (where almost all the ships from Gaul come in to land), faces the east, the lower angle faces south. This side stretches about five hundred miles. The second side bears towards Spain and the west, in which direction lies Ireland, smaller by one half, as it is thought, than Britain; the sea-passage is of equal length to that from Gaul to Britain. Here in mid-channel is an island called Man;

in addition, several smaller islands are supposed to lie close to land, as touching which some have written that in midwinter night there lasts for thirty whole days. We could discover nothing about this by inquiries; but, by exact water measurements, we observed that the nights were shorter than on the Continent. The length of this side, according to the belief of the natives, is seven hundred miles. The third side bears northwards, and has no land confronting it; the angle, however, of that side faces on the whole towards Germany. The side is supposed to be eight hundred miles long. Thus the whole island is two thousand miles in circumference.

Of all the Britons the inhabitants of Kent, an entirely maritime district, are by far the most civilised, differing but little from the Gallic manner of life. Of the inlanders most do not sow corn, but live on milk and flesh and clothe themselves in skins. All the Britons, indeed, dye themselves with woad, which produces a blue colour, and makes their appearance in battle more terrible. They wear long hair, and shave every part of the body save the head and the upper lip. Groups of ten or twelve men have wives together in common, and particularly brothers along with brothers, and fathers with sons; but the children born of the unions are reckoned to belong to the particular house to which the maiden was first conducted.

The Germans have no Druids to regulate divine worship, no zeal for sacrifices. They reckon among the gods those only whom they see and by whose offices they are openly assisted—to wit, the Sun, the Fire-god, and the Moon; of the rest they have learnt not even by report. Their whole life is composed of hunting expeditions and military pursuits; from early boyhood they are zealous for toil and hardship. Those who remain longest in chastity win greatest praise among their kindred; some think that stature, some that strength and sinew are fortified thereby. Further, they deem it a most disgraceful thing to have had knowledge of a woman before the twentieth year; and there is no secrecy in the matter, for both sexes bathe in the rivers and wear skins or small cloaks of reindeer hide, leaving great part of the body bare.

For agriculture they have no zeal, and the greater part of their food consists of milk, cheese, and flesh. No man has a definite quantity of land or estate of his own: the magistrates and chiefs every year assign to tribes and clans that have assembled together as much land and in such place as seems good to them, and compel the tenants after a year to pass on elsewhere. They adduce many reasons for that practice—the fear that they may be tempted by continuous association to substitute agriculture for their warrior zeal; that they may become zealous for the acquisition of broad territories, and so the more powerful may drive the lower sort from their holdings; that they may build with greater care to avoid the extremes of cold and heat; that some passion for money may arise to be the parent of parties and of quarrels. It is their aim to keep common people in contentment, when each man sees that his own wealth is equal to that of the most powerful . . .

The breadth of this Hercynian forest . . . is as much as a nine days' journey for an unencumbered person; for in no other fashion can it be determined, nor have they means to measure journeys. It begins in the borders of the Helvetii, the Nemetes, and the Rauraci, and, following the direct line of the river Danube, it extends to the borders of the Daci and the Anartes; thence it turns leftwards, through districts apart from the river, and by reason of its size touches the borders of many nations. There is no man in the Germany we know who can say that he has reached the edge of that forest, though he may have gone forward a sixty days' journey, or who has learnt in what place it begins. It is known that many kinds of wild beasts not seen in any other places breed therein, of which the following are those that differ most from the rest of the animal world and appear worthy of record.

There is an ox shaped like a stag, from the middle of whose forehead between the ears stands forth a single horn, taller and straighter than the horns we know. From its top branches spread out just like open hands. The main features of female and of male are the same, the same the shape and the size of the horns.

There are also elks so-called. Their shape and dappled skin are like unto goats, but they are somewhat larger in size and have blunted horns. They have legs without nodes or joints, and they do not lie down to sleep, nor, if any shock has caused them to fall, can they raise or uplift themselves. Trees serve them as couches; they bear against them, and thus, leaning but a little, take their rest. When hunters have marked by their tracks the spot to which they are wont to betake themselves, they either undermine all the trees in that spot at the roots or cut them so far through as to leave them just standing to outward appearance. When the elks lean against them after their fashion, their weight bears down the weakened trees and they themselves fall along with them.

A third species consists of the ure-oxen so-called. In size these are somewhat smaller than elephants; in appearance, colour, and shape they are as bulls. Great is their strength and great their speed, and they spare neither man nor beast once sighted. These the Germans slay zealously, by taking them in pits; by such work the young men harden themselves and by this kind of hunting train themselves, and those who have slain most of them bring the horns with them to a public place for a testimony thereof, and win great renown. But even if they are caught very young, the animals cannot be tamed or accustomed to human beings. In bulk, shape, and appearance their horns are very different from the horns of our own oxen. The natives collect them zealously and encase the edges with silver, and then at their grandest banquets use them as drinking-cups.

37. Vergil on the Creation, on zones of the earth, and on winds

These selections from the works of Vergil (c. 70–19 B.C.) are taken from "Silenus," one of his pastorals, and from the *Georgics*, his poem dealing with agriculture. They offer us a brief description of Creation, of the nature of the earth's zones, and of winds and their effect.

He sung the secret seeds of Nature's frame;
How seas, and earth, and air, and active flame,
Fell through the mighty void, and, in their fall,
Were blindly gather'd in this goodly ball.
The tender soil, then stiff'ning by degrees,
Shut from the bounded earth the bounding seas.
Then earth and ocean various forms disclose;
And a new sun to the new world arose;
And mists, condens'd to clouds, obscure the sky
And clouds, dissolv'd, the thirsty ground supply.
The rising trees the lofty mountains grace:
The lofty mountains feed the savage race,
Yet few, and strangers, in th' unpeopled place.
From thence the birth of man the song pursu'd,
And how the world was lost, and how renew'd.

Five girdles bind the skies: the torrid zone
Glows with the passing and repassing sun:
Far on the right and left th' extremes of heav'n
To frosts and snows and bitter blasts are given:
Betwixt the midst and these, the gods assign'd
Two habitable seats for human kind,
And 'cross their limits, cut a sloping way,
Which the twelve signs in beauteous order sway.
Two poles turn round the globe; one seen to rise
O'er Scythian hills, and one in Libyan skies;
The first sublime in heav'n, the last is whirl'd
Below the regions of the nether world.
Around our pole the spiry Dragon glides,
And, like a winding stream, the Bears divides—
The less and greater, who by Fate's decree
Abhor to dive beneath the northern sea.

From *The Works of Virgil*, trans. J. Dryden (New York, 1883).

There, as they say, perpetual night is found,
In silence brooding on th' unhappy ground:
Or, when Aurora leaves our northern sphere,
She lights the downward heav'n, and rises there;
And, when on us she breathes the living light,
Red Vesper kindles there the tapers of the night.
From hence uncertain season we may know:
And when to reap the grain, and when to sow;
Or when to fell the furzes: when 'tis meet
To spread the flying canvas for the fleet.
Observe what stars arise or disappear;
And the four quarters of the rolling year

. . . . .

And that by certain signs we may presage
Of heats and rains, and wind's impetuous rage,
the sov'reign of the heav'ns has set on high
The moon, to mark the changes of the sky;
When southern blasts should ease, and when the swain
Should near their fold his feeding flocks restrain.
For, ere the rising winds begin to roar,
The working seas advance to wash the shore:
Soft whispers run along the leafy woods;
And mountains whistle to the murm'ring floods.
E'en then the doubtful billows scarce abstain
From the toss'd vessel on the troubled main;
When crying cormorants forsake the sea,
And, stretching to the covert, wing their way;
When sportful coots run skimming o'er the strand,
When watchful herons leave their wat'ry stand,
And, mounting upward with erected flight,
Gain on the skies, and soar above the sight.
And oft, before tempestuous winds arise,
The seeming stars fall headlong from the skies,
And, shooting through the darkness, gild the night
With sweeping glories, and long trails of light;
And chaff with eddy-winds if whirl'd around,
And dancing leaves are lifted from the ground;
And floating feathers on the waters play.
But, when the winged thunder takes his way
From the cold north, and east and west engage,
And at their frontiers meet with equal rage,
The clouds are crash'd: a glut of gather'd rain
The hollow ditches fills, and floats the plain,
And sailors furl their dropping sheets amain.

38. Horace describes the Italian landscape

Horace (Quintus Horatius Flaccus, 65–9 B.C.) was happiest in his rustic retreat, a villa in the hills near Rome. It is that setting which he describes in these excerpts from his *Epistles* and *Epodes*. In the third excerpt, from his "Carmina," Horace uses a set of glimpses of the Italian landscape to underline the passing value of all material things in the face of death.

> My dear friend Quinctius, you might ask about my farm—
> if it enriches its master with ploughland or with olives,
> with fruit or pastures or with elm-trees clothed in vines:
> so let me lovingly sketch its shape and situation.
> Unbroken hills, just once divided by a dark
> valley: the sun approaching looks on the right side,
> departing, warms the left with the rays of his vanishing chariot.
> You'd praise the temperate air. And think of this!—the kindly
> bushes bear wild plums and cherries, the oak and ilex
> delight the cattle with their acorns, me with their shade.
> Tarentum might have been brought north with all its foliage.
> There is a spring besides, worthy to name a river:
> Hebrus, wandering through Thrace, is no purer or colder.
> Its flow will cure distempered heads and sickly stomachs.
> That is my hidden home: dear, yes, and beautiful.

'Happy the man who far from schemes of business, like the early generations of mankind, ploughs and ploughs again his ancestral land with oxen of his own breeding, with no yoke of usury on his neck! He is not wakened like a soldier by the fierce clarion; he dreads no angry sea. He avoids the Forum and the insolent portals of the great. And so he is either wedding the tall poplar to the full-grown vine-plant, or looking forth on his herds of lowing cattle as they stray in the shady valley, or cutting off with the pruning-hook useless boughs and grafting in those of happier fruit. He is either storing in clean pitchers the squeezed honey, or shearing the unresisting sheep. Or when Autumn has lifted over the land his head wreathed with mellow fruitage, what joy it is to gather the pear from the tree he grafted, and the grape that vies with the purple dye, to present to thee,

The first selection is from Horace, *Epistles,* as translated by Gilbert Highet in *Poets in a Landscape* (Harmondsworth: Penguin Books, 1959), p. 137. Reprinted by permission of Curtis Brown, Ltd. Copyright © 1957 by Gilbert Highet. The other selections, from the *Epodes,* are taken from E. C. Wickham, *Horace for English Readers* (London: Oxford University Press, 1930), pp. 138–140, 62–63, by permission of the Oxford University Press.

Priapus, and thee, Sire Silvanus, guardian of his bounds! Now it is his fancy to lie under some aged holm-oak, now on the soft deep grass, whilst the streams slide along in brimming courses, birds make moan in the woods, and springs babble with gushing water, sounds to invite light slumbers. And then when the wintry months of Jove the Thunderer gather storms and snow, he either drives this way and that with his pack of dogs the wild boars into the toils set for them, or spreads on smooth pole the wide-meshed nets to catch the greedy fieldfares, or sets snares for the timid hare and the crane from over seas, sweet prizes. In such a life who does not forget any evil cares which belong to love? But if a chaste wife do her part and grace his house with its sweet children (such as is a Sabine spouse or the sunburnt partner of the sturdy Apulian); pile the sacred hearth with old logs against the return of her wearied lord; and, as she shuts the glad cattle in the wattled fold, drain dry their full udders, and broaching the sweet cask's wine of the year make ready the unbought banquet, no Lucrine shellfish could give me more delight, no turbot or scar, if the storm that burst in thunder on Eastern waves should direct any to our waters. The bird of Africa would not cross my palate, nor the woodcock of Ionia, with more pleasant flavour than the olive gathered from the tree's richest boughs, or the sorrel plant that loves the meadow land and the mallows that give health to the laden body, or, it may be, the lamb slain on the festivals of Terminus, or the kid snatched from the wolf's jaws. While we sit at such a banquet, what delight to look out and see the well-fed sheep hastening home, to see the wearied bullocks dragging with tired neck and reversed plough, and the home-born slaves, the swarm that makes a wealthy home, all gathered round the glowing images of the home gods!'

So spake the usurer Alfius, on the point of turning farmer. He got in on the Ides all his money that was on interest: next Kalends he is seeking to put it out again.

Remember when life's path is steep to keep your mind even, and as much when things are prosperous to keep it chastened from extravagant joy. O Dellius, who have still to die, whether you have spent all your days in mourning, or flung at your length on some quiet grass-plat from holiday to holiday have made yourself happy with the oldest bin of Falernian! To what end, think you, huge pine and white poplar love to join in partnership their hospitable shade? Why frets the runaway water in its haste to escape down its tortuous channel? This is the spot; bid bring wine and unguents and the two shortlived flowers of the lovely rose, while fortune and our years and the black threads of the three sisters allow us. You will leave the forest pastures that you have bought up, and town mansion, and country house by yellow Tiber's stream, you will leave them, and the riches piled so high will belong to an heir. It matters nothing whether rich and born of old Inachus' line, or poor and of the dregs of the people, you linger out your span under heaven, the victim due to Orcus who has no pity. We all are driven one road. For all alike the lot is being shaken in the urn and, be it sooner or later, will leap out and place us in Charon's bark for everlasting exile.

39. Tacitus on Germany, Britain, and Judaea

Cornelius Tacitus (c. 55–c. 120 A.D.) was the leading historian of the Golden Age of Rome. His writings contain many valuable descriptions of the world of the Roman Empire, particularly his work on *The Geography, the Manners and Customs, and the Tribes of Germany*—the full title of *Germania*. The passages included here concern Germany, Britain, and Judaea and are taken, respectively, from *Germania, Agricola,* and *Historiae*. An interesting comparison may be made between the concise and factual narrative style of Tacitus and the tendency of Pliny, his contemporary, to indulge the reader with tales both wondrous and unbelievable.

On Germany

Germany is separated from the Galli, the Rhæti, and Pannonii, by the rivers Rhine and Danube; mountain ranges, or the fear which each feels for the other, divide it from the Sarmatæ and Daci. Elsewhere ocean girds it, embracing broad peninsulas and islands of unexplored extent, where certain tribes and kingdoms are newly known to us, revealed by war. The Rhine springs from a precipitous and inaccessible height of the Rhætian Alps, bends slightly westward, and mingles with the Northern Ocean. The Danube pours down from the gradual and gently rising slope of Mount Abnoba, and visits many nations, to force its way at last through six channels into the Pontus; a seventh mouth is lost in marshes...

For my own part, I agree with those who think that the tribes of Germany are free from all taint of intermarriages with foreign nations, and that they appear as a distinct, unmixed race, like none but themselves. Hence, too, the same physical peculiarities throughout so vast a population. All have fierce blue eyes, red hair, huge frames, fit only for a sudden exertion. They are less able to bear laborious work. Heat and thirst they cannot in the least endure; to cold and hunger their climate and their soil inure them.

From Tacitus, *The Complete Works,* trans. A. J. Church and W. J. Brodribb, ed. Moses Hadas (New York: Random House, 1942), pp. 709–711, 682–684, 660–661, by permission of the publisher. Copyright © 1942, by Random House, Inc.

Their country, though somewhat various in appearance yet generally either bristles with forests or reeks with swamps; it is more rainy on the side of Gaul, bleaker on that of Noricum and Pannonia. It is productive of grain, out unfavourable to fruit-bearing trees; it is rich in flocks and herds, but these are for the most part undersized, and even the cattle have not their usual beauty or noble head. It is number that is chiefly valued; they are in fact the most highly prized, indeed the only riches of the people. Silver and gold the gods have refused to them, whether in kindness or in anger I cannot say. I would not, however, affirm that no vein of German soil produces gold or silver, for who has ever made a search? They care but little to possess or use them. You may see among them vessels of silver, which have been presented to their envoys and chieftains, held as cheap as those of clay. The border population, however, value gold and silver for their commercial utility, and are familiar with, and show preference for, some of our coins. The tribes of the interior use the simpler and more ancient practice of the barter of commodities. They like the old and well-known money, coins milled, or showing a two-horse chariot. They likewise prefer silver to gold, not from any special liking, but because a large number of silver pieces is more convenient for use among dealers in cheap and common articles.

On Britain

Britain, the largest of the islands which Roman geography includes, is so situated that it faces Germany on the east, Spain on the west; on the south it is even within sight of Gaul; its northern extremities, which have no shores opposite to them, are beaten by the waves of a vast open sea. The form of the entire country has been compared by Livy and Fabius Rusticus, the most graphic among ancient and modern historians, to an oblong shield or battle-axe. And this no doubt is its shape without Caledonia, so that it has become the popular description of the whole island. There is, however, a large and irregular tract of land which juts out from its furthest shores, tapering off in a wedge-like form. Round these coasts of remotest ocean the Roman fleet then for the first time sailed, ascertained that Britain is an island, and simultaneously discovered and conquered what are called the Orcades, islands hitherto unknown. Thule too was descried in the distance, which as yet had been hidden by the snows of winter. Those waters, they say, are sluggish, and yield with difficulty to the oar, and are not even raised by the wind as other seas. The reason, I suppose, is that lands and mountains, which are the cause and origin of storms, are here comparatively rare, and also that the vast depths of that unbroken expanse are more slowly set in motion. But to investigate the nature of the ocean and the tides is no part of the present work, and many writers have discussed the subject. I would simply add, that nowhere has the sea a wider dominion, that it has many currents running in every direction, that it does not merely flow and ebb within the limits of the shore, but penetrates and winds far inland, and finds a home among hills and mountains as though in its own domain . . .

Their sky is obscured by continual rain and cloud. Severity of cold is unknown. The days exceed in length those of our part of the world; the nights are bright, and in the extreme north so short that between sunlight and dawn you can perceive but a slight distinction. It is said that, if there are no clouds in the way, the splendour of the sun can be seen throughout the night, and that he does not rise and set, but only crosses the heavens. The truth is, that the low shadow thrown from the flat extremities of the earth's surface does not raise the darkness to any height, and the night thus fails to reach the sky and stars.

With the exception of the olive and vine, and plants which usually grow in warmer climates, the soil will yield, and even abundantly, all ordinary produce. It ripens indeed slowly, but is of rapid growth, the cause in each case being the same, namely, the excessive moisture of the soil and of the atmosphere. Britain contains gold and silver and other metals, as the prize of conquest. The ocean, too, produces pearls, but of a dusky and bluish hue. Some think that those who collect them have not the requisite skill, as in the Red Sea the living and breathing pearl is torn from the rocks, while in Britain they are gathered just as they are thrown up. I could myself more readily believe that the natural properties of the pearls are in fault than our keenness for gain.

On Judaea

Eastward the country is bounded by Arabia; to the south lies Egypt; on the west are Phœnicia and the Mediterranean. Northward it commands an extensive prospect over Syria. The inhabitants are healthy and able to bear fatigue. Rain is uncommon, but the soil is fertile. Its products resemble our own. They have, besides, the balsam-tree and the palm. The palm-groves are tall and graceful. The balsam is a shrub; each branch, as it fills with sap, may be pierced with a fragment of stone or pottery. If steel is employed, the veins shrink up. The sap is used by physicians. Libanus is the principal mountain, and has, strange to say amidst these burning heats, a summit shaded with trees and never deserted by its snows. The same range supplies and sends forth the stream of the Jordan. This river does not discharge itself into the sea, but flows entire through two lakes, and is lost in the third. This is a lake of vast circumference; it resembles the sea, but is more nauseous in taste; it breeds pestilence among those who live near by its noisome odour; it cannot be moved by the wind, and it affords no home either to fish or water-birds. These strange waters support what is thrown upon them, as on a solid surface, and all persons, whether they can swim or no, are equally buoyed up by the waves. At a certain season of the year the lake throws up bitumen, and the method of collecting it has been taught by that experience which teaches all other arts. It is naturally a fluid of dark colour; when vinegar is sprinkled upon it, it coagulates and floats upon the surface. Those whose business it is take it with the hand, and draw it on to the deck of the boat; it then continues of itself to flow in and lade the vessel till the stream is cut off. Nor can this be done by any instrument of brass or iron. It shrinks from blood or any cloth stained by the menstrua of women. Such is the account of old authors; but those who know the country say that the bitumen moves in heaving masses on the water, that it is drawn by hand to the shore, and that there, when dried by the evaporation of the earth and the power of the sun, it is cut into pieces with axes and wedges just as timber or stone would be.

Christian Geography

During the reign of Constantine, early in the fourth century, Christianity became the established religion of the Roman Empire. The Emperor's religious decision affected all aspects of daily life and left its mark on the arts and sciences. Geography was no exception, and many of the geographical writings that survive from the first seven centuries after Constantine's conversion may well be called examples of Christian geography.

The earliest genre of Christian geography is that of guidebooks for pilgrims bound for the Holy Land, exemplified here by the *Bordeaux Itinerary* and by Bishop Eucherius' *Epitome . . . about certain holy places.* A second type of Christian geography, based entirely on scripture, is illustrated by excerpts from the *Christian Topography* by Cosmas. The third, and perhaps most popular, genre was the encyclopedia, a form already well established in Roman times; excerpts from Isidore of Seville's *Etymologiae* are used here as examples. Finally, passages from the works of Bede and Dicuil—an English and an Irish monk who lived and worked in Britain in the eighth and ninth centuries—illustrate the spread of Christianity across Europe and the kind of geographical description taught in centers of Christian learning during the early Middle Ages.

40. The Bordeaux Itinerary: a pilgrim's guide to the Holy Land

The itinerary from Bordeaux to Jerusalem, or *Itinerarium Burdigalense*, is the first of the Christian guidebooks and marks the beginning of

From *The Bordeaux Pilgrim* (London, Palestine Pilgrims Text Society, 1889), I, pp. 15–17, 24–25, 26–27.

geographical writings with a distinct ecclesiastic flavor. It is not only the earliest extant description of the route taken by pilgrims to the Holy Land, but also the most detailed surviving example of an important genre of Latin geographical work, the *itinerarium scriptum,* or written description of the great roads of the Roman Empire. That there also existed *itineraria picta,* or road maps, is known only from a very late (thirteenth century) copy of such a work, the *Peutinger Table.* The date of the Bordeaux itinerary is also significant: it was written in the year 333, shortly after the journey of Constantine's mother, Helena, to the Holy Land, and only eight years after the first of the great councils, that of Nicaea. Distances in the itinerary are given in miles, changes, and stations or halts, and while the figures are obviously not always correct, the spatial sequence of the stops made on the way is exact. The brief references to places of significance highlight the purpose of the work: to be a guidebook to holy places.

Here is a city in the sea, two miles from the shore.

Change at Spiclis	xii
Change at Basiliscum	xii
Halt at Arcæ (Arca, Cæsarea, *'Arka*)	viii
Change at Bruttus	iv
City of Tripolis (*Tarabulus*)	xii
Change at Triclis	xii
Change at Bruttus alius	xii
Change at Alcobilis	xii
City of Berytus (*Beirút*)	xii
Change at Heldua (*Khan Khulda*)	xii
Change at Porphyrion (Porphyreon)	viii
City of Sidon (*Saida*)	viii
Thence to Sarepta	ix

Here Helias went up to the widow and begged food for himself.

Change at 'ad Nonum' (at the ninth milestone?)	iv
City of Tyre (*Sûr*)	xii

Total from *Antioch* to *Tyre* 174 miles, 20 changes, 11 halts.

Change at Alexandroschene	xii
Change at Ecdeppa (Achzib, *ez Zîb*)	xii
City of Ptolemais (Accho, *'Akka, St. Jean d'Acre*)	ix
Change at Calamon	xii
Halt at Sycaminos (Sycaminon)	iii

Here is the Mount Carmel, where Helias offered sacrifice.

Change at Certa	viii

Frontier of Syria, Phœnïcia, and Palestine.

City of Cæsarea Palæstina (*Kaisarieh*), that is, Judæa   viii

Total from *Tyre* to Cæsarea Palæstina (*Kaisarieh*) 73 miles, 2 changes, 3 halts.

Here is the bath (balneus) of Cornelius the centurion who gave many alms.

At the third milestone from thence is the mountain Syna, where there is a fountain, in which, if a woman bathes, she becomes pregnant.

City of Maximianopolis   xviii

City of Stradela   x

Here reigned King Achab (Ahab), and here Helias prophesied. Here is the field in which David slew Goliath . . .

Also as one goes from Jesusalem to the gate which is to the eastward, in order to ascend the Mount of Olives, is the valley called that of Josaphat. Towards the left, where are vineyards, is a stone at the place where Judas Iscariot betrayed Christ; on the right is a palm-tree, branches of which the children carried off and strewed in the way when Christ came. Not far from thence, about a stone's-throw, are two notable (monubiles) tombs of wondrous beauty; in the one, which is a true monolith, lies Isaiah the prophet, and in the other Hezekiah, King of the Jews.

From thence you ascend to the Mount of Olives, where before the Passion, the Lord taught His disciples. There by the orders of Constantine a basilica of wonderous beauty has been built. Not far from thence is the little hill which the Lord ascended to pray, when he took Peter and John with Him, and Moses and Elias were beheld. A mile and a half to the eastward is the village (villa) called Bethany. There is a vault (crypta) in which Lazarus, whom the Lord raised, was laid . . .

From Jericho to the Dead sea   ix

The water of it is very bitter, and in it there is no kind of fish whatever, nor any vessel; and if a man casts himself into it in order to swim, the water turns him over. From thence to the Jordan, where the Lord was baptized by John   v

There is a place by the river, a little hill upon the further (left) bank, from which Elijah was caught up into heaven.

From Jerusalem going to Bethlehem   vi

On the road, on the right hand, is a tomb, in which lies Rachel, the wife of Jacob. Two miles from thence, on the left hand, is Bethlehem, where our Lord Jesus Christ was born. A basilica has been built there by the orders of Constantine. Not far from thence is the tomb of Ezekiel, Asaph, Job, Jesse, David, and Solomon, whose names are inscribed in Hebrew letters upon the wall as you go down into the vault itself.

41. Bishop Eucherius on the holy places

The guide to the holy places of Palestine by Eucherius, Bishop of Lyon, is a brief account of the Holy Land, probably written in the middle of the fifth century. The date of the work has been established partly by our knowledge of the author, a churchman of distinction, and partly by evidence provided through the mention, or omission, in the text of certain landmarks whose dates of construction are known. The *Epitome* has a geographical flavor mainly because it indicates the physical characteristics of certain sites and supplies the reader with accurate directions and distances.

The letter of Bishop Eucherius to Faustinus the Priest

Bishop Eucherius wishes health to the Priest Faustinus the Islander.

I have briefly put together an account of the position of the city of Jerusalem and of Judæa, as I have learned it either by conversation or by reading, and have described it in a brief preface, because it is by no means fitting that a work of no length should have a long preface. Fare thee well in Christ, thou in whom I glory and put my trust.

I. Jerusalem is called Ælia from Ælius Hadrianus; for, after its destruction by Titus, it received the name together with the works of its founder, Ælius. The place, they say, is naturally lofty, so that one has to ascend to it from every side; it rises by a long yet gentle slope.

The site of the city itself is almost circular, enclosed within a circuit of walls of no small extent, whereby it now receives within itself Mount Sion, which was once outside, and which, lying on the south side, overhangs the city like a citadel. The greater part of the city stands on the level ground of the lower hill below the mount.

II. Mount Sion on one side, that which faces north, is set apart for the dwellings of priests and monks; the level ground on its summit is covered by the cells of monks surrounding a church, which, it is said, was built there by the apostles out of reverence for the place of our Lord's Resurrection; because, as promised before by the Lord, they were filled with the Holy Ghost.

III. The most frequented gates (of the city) are three in number; one on the west, another on the east, and the third on the north side of the city.

IV. First, about the Holy Places.

In consequence of the direction of the streets, one must visit the basilica, which is called the "Martyrium" (Church of the Testimony), built by Constantine with great splendour. Adjacent to this, upon the west side, are to be seen Golgotha and the "Anastasis" (Church of the Resurrection).

From *The Epitome of S. Eucherius about certain holy places,* trans. Aubrey Stewart (London, Palestine Pilgrims Text Society, 1890), pp. 7–12.

Now, the "Anastasis" is at the place of the Resurrection, and Golgotha, which is between the "Anastasis" and the "Martyrium," is the place of the Lord's Passion, wherein may be seen the very rock which once supported the cross itself on which the Lord's body hung. These things are to be seen beyond Mount Sion, where there is a slight swell in the ground which slopes away towards the north.

V. The Temple, which was situated in the lower part of the city near the city wall on the east side and was splendidly built, was once a world's wonder, but out of its ruins there stands only the pinnacle of one wall, the rest being destroyed to their very foundations. A few cisterns (*cisternæ*) to collect water are to be seen on the north side of the city in the neighbourhood of the Temple.

VI. Bethesda is to be seen, remarkable for its twin pools, one of which is generally filled by the winter rain, while the water of the other is coloured red.

VII. On that side of the Mount Sion which looks from a precipitous rock towards the east, below the city walls and at the foot of the hill, the fountain of Siloam gushes forth, not running continuously, but on certain days and hours welling through caverns in the rock, and flows with an intermittent stream towards the south. Opposite the city wall of Jerusalem and on the east of the Temple is Geenon, the valley of Josaphat, stretching from the north towards the south, through which a torrent flows whenever it receives rain-water.

VIII. The neighbourhood of the city of Jerusalem looks rough and mountainous, and the Mount of Olives is to be seen at a distance of a mile to the eastward of it. Thereon are two very famous churches, one of which is built on the very place where we are told that the Lord talked with His disciples, the other on the place from which He is said to have ascended to heaven.

IX. Six miles from Jerusalem, on the south side, stands Bethlehem, which is a very narrow space contained within a low wall without towers, wherein the Lord's manger, covered with decorations of silver and gold, is contained in a splendid building.

X. Jericho is eighteen miles distant from Jerusalem in a north-easterly direction. The Jordan also bounds Judæa in the same quarter, distant fourteen or fifteen miles from Ælia; which Jordan first passes through the Lake of Tiberias, and then emerging from it, flows with an ample and calm stream into the Dead Sea; which sea is about twenty miles distant from Jerusalem. It is called Jordan, because two fountains, one of which is named Jor, and the other Dan, join their waters together.

XI. Hebron, once a city of giants, is distant about twenty-two miles from Jerusalem in a southerly direction. The "Great" (Mediterranean) Sea nowhere approaches nearer (to Jerusalem) than at the town of Joppa, which is distant forty miles to the north-west.

XII. The longest part of the (Holy) land reaches from Dan to Bersabee, stretching from the north to the south. Dan is a little village on that frontier of Judæa which looks towards the north, at the fourth milestone from Paneas as you go towards Tyre. (The Hebrews call a river Jor.) Now, Bersabee, as we have stated above, is a large village towards the south, twenty miles from Hebron.

42. The Christian Topography of Cosmas Indicopleustes

Cosmas the Traveler, or *Indicopleustes* ("who had been to India"), was a businessman of Alexandria who, in his later years, turned to the religious life. His *Christian Topography,* written probably in the middle of the sixth century, provides the clearest example of a geographical world view based squarely on Scripture. Manuscripts of the *Topography* contain several of the earliest known maps of distinctly Christian origin, which reflect this same world view. The first passage quoted here sums up Cosmas's concept of the structure of the universe; the second, his refutation of the existence of the Antipodes. The third speaks of his experience when sailing between Egypt and India.

It is written: *In the beginning God made the heaven and the earth.* We therefore first depict along with the earth, the heaven which is vaulted and which has its extremities bound together with the extremities of the earth. To the best of our ability we have endeavoured to delineate it on its western side and its eastern; for these two sides are walls, extending from below to the vault above. There is also the firmament which, in the middle, is bound together with the first heaven, and which, on its upper side, has the waters according to divine scripture itself. The position and figure are such as here sketched. To the extremities on the four sides of the earth the heaven is fastened at its own four extremities, making the figure of a cube, that is to say, a quadrangular figure, while up above it curves round in the form of an oblong vault and becomes as it were a vast canopy. And in the middle the firmament is made fast to it, and thus two places are formed.

From the earth to the firmament is the first place, this world, namely, in which are the angels and men and all the present state of existence. From the firmament again to the vault above is the second place—the

---

From J. W. McCrindle, *The Christian Topography of Cosmas, An Egyptian Monk* (London: Hakluyt Society, no. 98, 1897), pp. 129–130, 17, 39–40. © The Hakluyt Society, London.

Kingdom of Heaven, into which Christ, first of all, entered, after his ascension, having prepared for us a new and living way.

On the western side and the eastern the outline presented is short, as in the case of an oblong vault, but on its north and south sides it shows its length.

On the Antipodes

But should one wish to examine more elaborately the question of the Antipodes, he would easily find them to be old wives' fables. For if two men on opposite sides placed the soles of their feet each against each, whether they chose to stand on earth, or water, or air, or fire, or any other kind of body, how could both be found standing upright? The one would assuredly be found in the natural upright position, and the other, contrary to nature, head downward. Such notions are opposed to reason, and alien to our nature and condition. And how, again, when it rains upon both of them, is it possible to say that the rain falls down upon the two, and not that it falls down to the one and falls up to the other, or falls against them, or towards them, or away from them. For to think that there are Antipodes compels us to think also that rain falls on them from an opposite direction to ours; and any one will, with good reason, deride these ludicrous theories, which set forth principles incongruous, ill-adjusted, and contrary to nature.

Between Egypt and India

Once on a time, when we sailed in these gulfs, bound for Further India and had almost crossed over to Barbaria, beyond which there is situated Zingium, as they term the mouth of the ocean, I saw there to the right of our course a great flight of the birds which they call Souspha, which are like kites, but somewhat more than twice their size. The weather was there so very unsettled that we were all in alarm; for all the men of experience on board, whether passengers or sailors, all began to say that we were near the ocean and called out to the pilot: "Steer the ship to port and make for the gulf, or we shall be swept along by the currents and be carried into the ocean and be lost." For the ocean rushing into the gulf was swelling into billows of portentous size, while the currents from the gulf were driving the ship into the ocean, and the outlook was altogether so dismal that we were kept in a state of great alarm. A great flock, all the time, of the birds called Souspha followed us, flying generally high over our heads, and the presence of these was a sign that we were near the ocean.

43. The Etymologiae of Isidore of Seville: an early Christian encyclopedia

Isidore, Bishop of Seville and primate of Spain (c. 560–636), was one of the most popular encyclopedists of Christian Europe. His fame is reflected in inscriptions such as those found on the great medieval *mappaemundi*: the Ebstorf Map (c. 1270) admonishes the reader: "If you wish to know more, read Isidore!"

On the earth

The earth is placed in the middle region of the universe, being situated like a center at an equal interval from all parts of heaven; in the singular number it means the whole circle; in the plural the separate parts; and reason gives different names for it; for it is called *terra* from the upper part where it suffers attrition (*teritur*); *humus* from the lower and *humid* part, as for example, under the sea; again, *tellus,* because we take (*tollimus*) its fruits; it is also called *ops* because it brings opulence. It is likewise called *arva,* from ploughing (*arando*) and cultivating.

Earth in distinction from water is called dry; since the Scripture says that "God called the dry land, earth." For dryness is the natural property of earth. Its dampness it gets by its relation to water. As to its motion (earthquakes) some say it is wind in its hollow parts, the force of which causes it to move.

On the circle of lands

The circle of lands (*orbis*) is so called from its roundness, which is like that of a wheel, whence a small wheel is called *orbiculus.* For the Ocean flowing about on all sides encircles its boundaries. It is divided into three parts; of which the first is called Asia; the second, Europe; the third, Africa.

These three parts the ancients did not divide equally; for Asia stretches from the South through the East to the North, and Europe from the North to the West, and thence Africa from the West to the South. Whence plainly the two, Europe and Africa, occupy one-half, and Asia alone the other. But the former were made into two parts because the Great Sea enters from the Ocean between them and cuts them apart. Wherefore if you divide the circle of lands into two parts, East and West, Asia will be in one, and in the other, Europe and Africa.

From Isidore of Seville's *Etymologiae,* as translated in Ernest Brehaut, *An Encylopedist of the Dark Ages* (New York: Longmans Green & Co., 1912), pp. 237, 243–245, 238–239, 239–242, 245–246.

On Europe

Europe, which was parted off to form a third part of the circle, begins at the river Tanais, passing to the west along the Northern ocean as far as the limits of Spain. Its Eastern and Southern parts begin at the Pontus, extend along the whole Mare Magnum, and end at the island of Gades.

On the air and the clouds

Air is emptiness, having more rarity mixed with it than the other elements. Of it Virgil says: *"Longum per inane secutus."* Air (*aer*) is so called from αἴρειν (*to raise*), because it supports the earth or, it may be, is supported by it. This belongs partly to the substance of heaven, partly to that of the earth. For yonder thin air where windy and gusty blasts cannot come into existence, belongs to the heavenly part; but this more disordered air which takes a corporeal character because of dank exhalations, is assigned to earth, and it has many subdivisions; for being set in motion it makes winds; and being vigorously agitated, lightnings and thunderings; being contracted, clouds; being thickened, rain; when the clouds freeze, snow; when thick clouds freeze in a more disordered way, hail; being spread abroad, it causes fine weather; for it is known that thick air is a cloud and that a cloud that thins and melts away, is air.

On the rainbow and the causes of clouds

The rainbow is so called from its resemblance to a bent bow. Its proper name is Iris and it is called Iris, as it were *aeris* (of the air), because it comes down through the air to earth. It comes from the radiance of the sun when hollow clouds receive the sun's ray full in front, and they create the appearance of a bow, and rarified water, bright air, and a misty cloud under the beams of the sun create those varied hues.

Rains (*pluviae*) are so called because they flow, as if *fluviae*. They arise by exhalation from earth and sea, and being carried aloft they fall in drops on the lands, being acted upon by the heat of the sun or condensed by strong winds.

On the winds

There are four chief winds. The first of these is from the east, *Subsolanus,* and *Auster* from the south, *Favonius* from the west, and from *Septentrio* (north) a wind of the same name blows. These winds have kindred winds one on each side.

Subsolanus has on its right *Vulturnus,* on its left *Eurus; Auster* has on its right *Euroauster,* on its left *Austroafricus; Favonius* on its right *Africus,* on its left *Corus*. Further, *Septentrio* has on its right *Circius*, on its left *Aquilo*. These twelve winds surround the globe of the universe with their blasts.

On the waters

The two most powerful elements of human life are fire and water, whence they who are forbidden fire and water are seriously punished.

The element of water is master of all the rest. For the waters temper the heavens, fertilize the earth, incorporate air in their exhalations, climb aloft and claim the heavens; for what is more marvelous than the waters keeping their place in the heavens!

On the sea

The depth of the sea varies; still the level of its surface is invariable.

Moreover that the sea does not increase, though it receives all streams and all springs, is accounted for in this way; partly that its very greatness does not feel the waters flowing in; secondly, because the bitter water consumes the fresh that is added, or that the clouds draw up much water to themselves, or that the winds carry it off, and the sun partly dries it up; lastly, because the water leaks through certain secret holes in the earth, and turns and runs back to the sources of rivers and to the springs.

On the Mediterranean Sea

The *Mare Magnum* is that which flows from the west out of the Ocean and extends toward the South, and then stretches to the North. And it is called *Magnum* because the rest of the seas are smaller in comparison with it. It is also called Mediterranean because it flows through the midst of the land (*per mediam terram*) as far as the Orient, separating Europe and Africa and Asia.

On rivers

Certain of the rivers have received their names from causes peculiar to them, and of these some which are told of as famous in history should be mentioned.

Geon is a river issuing from Paradise and surrounding the whole of Ethiopia, being called by this name because it waters the land of Egypt by its flood, for γῆ in the Greek means *terra* in the Latin. This river is called Nile by the Egyptians, on account of the mud which it brings, which gives fertility.

The river Ganges, which the holy Scriptures call Phison, issuing from

Paradise, takes its course toward the regions of India. . . It is said to rise in the manner of the Nile and overflow the lands of the East.

The Tigris, a river of Mesopotamia, rises in Paradise, and flows opposite the Assyrians (*contra Assyrios*), and after many windings flows into the Dead Sea. And it is called by this name because of its velocity, like a wild beast that runs with great speed.

The Euphrates, a river of Mesopotamia, greatly abounding in gems, rises in Paradise and flows through the midst of Babylonia . . . It irrigates Mesopotamia in certain places just as the Nile does Alexandria. Sallust, however, a most reliable author, asserts that the Tigris and the Euphrates arise from one source in Armenia, and going by different ways are far separated, an intervening space of many miles being left, and the land which is enclosed by them is called Mesopotamia. Therefore as Hieronymous noted, there must be a different explanation of the rivers of Paradise.

Tanus was the first king of the Scythians, from whom the river *Tanais* is said to have been named. It rises in the Riphaean forest, and separates Europe from Asia, flowing in the midst between two divisions of the world, and emptying into the Pontus.

On Libya (Africa)

It begins at the boundaries of Egypt, extending along the South through Ethiopia as far as Mt. Atlas. On the north it is bounded by the Mediterranean Sea, and it ends at the strait of Gades, having the provinces Libya Cyrenensis, Pentapolis, Tripolis, Byzacium, Carthago, Numidia, Mauritania Stifensis, Mauritania Tingitana, and in the neighborhood of the sun's heat, Ethiopia.

Ethiopia is so called from the color of its people, who are scorched by the nearness of the sun. The color of the people betrays the sun's intensity, for there is never-ending heat here. Whatever there is of Ethiopia is under the south pole. Towards the west it is mountainous, sandy in the middle, and toward the eastern region, a desert. Its situation extends from the Atlas Mts. on the west to the bounds of Egypt on the east. It is bounded on the south by the ocean, on the north by the river Nile. It has many peoples, of diverse appearance and fear-inspiring because of their monstrous aspect.

Besides the three parts of the circle there is a fourth part across the Ocean on the South, which is unknown to us on account of the heat of the sun, in whose boundaries, according to story, the Antipodes are said to dwell.

On islands

Britannia, an island of the Ocean, completely separated from the circle of lands by the sea that flows between, is called by the name of its people. It lies in the rear of the Gauls and looks toward Spain. Its circuit is 4,875 miles; there are many large rivers in it and hot springs, and an abundant and varied supply of metals. Jet is very common there, and pearls.

Thanatos, an island of the Ocean in the Gallic sea, separated from Britain by a narrow strait, with fields rich in grain and a fertile soil. It is called Thanatos from the death of snakes, for it is destitute of them itself, and earth taken thence to any part of the world kills snakes at once.

Thyle is the furthest island in the ocean, between the region of North and that of West, beyond Britain, having its name from the sun, because there the sun makes its summer halt, and there is no day beyond it; whence the sea there is sluggish and frozen.

Scotia, the same as Hibernia, an island very near Britain, narrower in the extent of its lands but more fertile; this reaches from Africa towards Boreas, and Iberia and the Cantabrian ocean are opposite to the first part of it. Whence, too, it is called Hibernia. It is called Scotia because it is inhabited by the tribes of Scots. There are no snakes there, few birds, no bees; and so if any one scatters among beehives stones or pebbles brought thence, the swarms desert them . . .

Taprobana is an island lying close to India on the Southeast, where the Indian Ocean begins, extending in length eight hundred and seventy-five miles, in width, six hundred and twenty-five. It is separated [from India] by a river that flows between. It is all full of pearls and gems. Part of it is full of wild beasts and elephants, but men occupy part. In this island they say that there are two summers and two winters in one year, and that the place blooms twice with flowers.

44. Britain in the eighth century: the Venerable Bede on the situation of Britain and Ireland

The Venerable Bede, (672/673–735), an English religious, lived most of his active life in the monastery of Jarrow, in northern England. Besides his *Ecclesiastical History of the English Nation,* from which the following excerpt is taken, Bede wrote extensively on many other subjects, both religious and secular. His most important contributions are in the field of *computus,* the computation of significant time cycles. He was the first scholar known to have recorded the basis of the Metonic, or perpetual, Easter cycle, a vital part of the calendar of the Church; the first to have articulated the "establishment of port," a device used to calculate tidal periods; and the first to use the term *anno Domini.* His determination of the beginning of the year, his

From Bede, *Ecclesiastical History of the English Nation,* Everyman's Library (New York: E. P. Dutton & Co.), pp. 4–7.

calculation of the movable feast of Easter, and his calculation of years in terms of B.C. and A.D. are all still in use.

Britain, an island in the ocean, formerly called Albion, is situated between the north and west, facing, though at a considerable distance, the coasts of Germany, France, and Spain, which form the greatest part of Europe. It extends 800 miles in length towards the north and is 200 miles in breadth, except where several promontories extend further in breadth, by which its compass is made to be 3675 miles. To the south, as you pass along the nearest shore of the Belgic Gaul, the first place in Britain which opens to the eye is the city of Rutubi Portus, by the English corrupted into Reptacestir. The distance from hence across the sea to Gessoriacum, the nearest shore of the Morini, is fifty miles, or as some writers say, 450 furlongs. On the back of the island, where it opens upon the boundless ocean, it has the islands called Orcades. Britain excels for grain and trees, and is well adapted for feeding cattle and beasts of burden. It also produces vines in some places, and has plenty of land and water-fowls of several sorts; it is remarkable also for rivers abounding in fish, and plentiful springs. It has the greatest plenty of salmon and eels; seals are also frequently taken, and dolphins, as also whales; besides many sorts of shell-fish, such as mussles, in which are often found excellent pearls of all colours, red, purple, violet, and green, but mostly white. There is also a great abundance of cockles, of which the scarlet dye is made; a most beautiful colour, which never fades with the heat of the sun or the washing of the rain; but the older it is, the more beautiful it becomes. It has both salt and hot springs, and from them flow rivers which furnish hot baths, proper for all ages and sexes, and arranged according. For water, as St. Basil says, receives the heating quality, when it runs along certain metals, and becomes not only hot but scalding. Britain has also many veins of metals, as copper, iron, lead, and silver; it has much and excellent jet, which is black and sparkling, glittering at the fire, and when heated, drives away serpents; being warmed with rubbing, it holds fast whatever is applied to it, like amber. The island was formerly embellished with twenty-eight noble cities, besides innumerable castles, which were all strongly secured with walls, towers, gates, and locks. And, from its lying almost under the North Pole, the nights are light in summer, so that at midnight the beholders are often in doubt whether the evening twilight still continues, or that of the morning is coming on; for the sun, in the night, returns under the earth, through the northern regions at no great distance from them. For this reason the days are of a great length in summer, as, on the contrary, the nights are in winter, for the sun then withdraws into the southern parts, so that the nights are eighteen hours long. Thus the nights are extraordinarily short in summer, and the days in winter, that is, of only six equinoctial hours. Whereas, in Armenia,

Macedonia, Italy, and other countries of the same latitude, the longest day or night extends but to fifteen hours, and the shortest to nine.

This island at present, following the number of the books in which the Divine law was written, contains five nations, the English, Britons, Scots, Picts, and Latins, each in its own peculiar dialect cultivating the sublime study of Divine truth. The Latin tongue is, by the study of the Scriptures, become common to all the rest. At first this island had no other inhabitants but the Britons, from whom it derived its name, and who, coming over into Britain, as is reported, from Armorica, possessed themselves of the southern parts thereof. When they, beginning at the south, had made themselves masters of the greatest part of the island, it happened, that the nation of the Picts, from Scythia, as is reported, putting to sea, in a few long ships, were driven by the winds beyond the shores of Britain, and arrived on the northern coast of Ireland, where, finding the nation of the Scots, they begged to be allowed to settle among them, but could not succeed in obtaining their request. Ireland is the greatest island next to Britain, and lies to the west of it; but as it is shorter than Britain to the north, so, on the other hand, it runs out far beyond it to the south, opposite to the northern parts of Spain, though a spacious sea lies between them. The Picts, as has been said, arriving in this island by sea, desired to have a place granted them in which they might settle. The Scots answered that the island could not contain them both; but "We can give you good advice," said they, "what to do; we know there is another island, not far from ours, to the eastward, which we often see at a distance, when the days are clear. If you will go thither, you will obtain settlements; or, if they should oppose you, you shall have our assistance." The Picts, accordingly, sailing over into Britain, began to inhabit the northern parts thereof, for the Britons were possessed of the southern. Now the Picts had no wives, and asked them of the Scots; who would not consent to grant them upon any other terms, than that when any difficulty should arise, they should choose a king from the female royal race rather than from the male: which custom, as is well known, has been observed among the Picts to this day. In process of time, Britain, besides the Britons and the Picts, received a third nation, the Scots, who, migrating from Ireland under their leader Reuda, either by fair means, or by force of arms, secured to themselves those settlements among the Picts which they still possess. From the name of their commander, they are to this day called Dalreudins; for, in their language, Dal signifies a part.

Ireland, in breadth, and for wholesomeness and serenity of climate, far surpasses Britain; for the snow scarcely ever lies there above three days: no man makes hay in the summer for winter's provision, or builds stables for his beasts of burden. No reptiles are found there, and no snake can live there; for, though often carried thither out of Britain, as soon as the ship comes near the shore, and the scent of the air reaches them, they die. On the contrary, almost all things in the island are good against poison. In short, we have known that when some persons have been bitten by serpents, the scrapings of leaves of books that were brought out of Ireland, being put into water, and given them to drink, have immediately expelled the spreading poison, and assuaged the swelling. The island abounds in milk and honey, nor is there any want of vines, fish, or fowl; and it is remarkable for deer and goats. It is properly the country of the Scots, who, migrating from thence, as has been said, added a third nation in Britain to the Britons and the Picts. There is a very large gulf of the sea, which formerly divided the nation of the Picts from the Britons; which gulf runs from the west very far into the land, where, to this day, stands the strong city of the Britons, called Alcluith. The Scots, arriving on the north side of this bay, settled themselves there.

45. From Dicuil's De mensura orbis terrae

Dicuil was an Irish monk who lived and worked during the latter part of the eighth and the early part of the ninth century. In *De Mensura Orbis Terrae* (Of the Dimensions of the World), he speaks of "our island Ireland" (VII)[1], signs the work with his name, and indicates the date of its composition, 825 (IX). Using information contained in the work, Dicuil has been identified with one of several monks, abbots of the monasteries of Innis Muredaich, Kilmor, and Pahlacht—all of whom were active during the 9th century.

As indicated by Gustav Parthey in the preface to his edition (Berlin, 1870), the work itself depends heavily on earlier authorities, and especially on Pliny and on Solinus. It does have significance of its own, however, since the author refers to firsthand sources of information: his contemporaries.

Dicuil relates a story told by Fidelis, a fellow religious, concerning a pilgrimage of Irish clerics and laymen to Jerusalem. As was the custom of many pilgrims, they visited Egypt as well as the Holy Land and "sailing on the Nile, admired the seven barns that Joseph built, according to the number of years of plenty: there were four of these in one place, three in another," (VI). (The "barns" are the pyramids.)

Fidelis adds that when several of the pilgrims landed on the bank of the river to observe the pyramids more closely, eight of them, both men and women, were killed by a lion. The lion was finally killed by the other pilgrims with lances and swords. All of this happened, says Fidelis, because the pyramids are built in lonely places.

From Dicuil, *De Mensura Orbis Terrae,* in A. Letronne's edition (Paris, 1814). Editor's translation.

1. Numbers in parentheses refer to chapters of Dicuil's work.

Reimbarking on their vessels, the pilgrims sailed on a canal from the Nile down to the Red Sea, to the port near where Moses led his people across the Sea. Dicuil adds a touch of verisimilitude concerning the curiosity characteristic of pilgrims and tourists in any age. One of the pilgrims had actually measured the sides of a pyramid; arriving at the port of embarcation on the Red Sea, that same "measurer," as the narrator calls him, wanted to go into the water to look for the tracks made by Pharaoh's chariots in pursuit of Moses and his people. Reason prevailed, however, for the sailors of the pilgrims' ship would not tolerate such goings-on and proceeded to sail across the Gulf of Suez, to the shores of Sinai. There is evidence that the waterway connecting the Nile and the Red Sea was still in use at the time of the pilgrimage of Brother Fidelis.

Dicuil refers obliquely to another contemporary event—the arrival of an embassy from the Caliph Harun-al-Rashid to Charlemagne in Germany in the year 802. Dicuil disputes the assertion of Solinus that elephants cannot lie down by saying: "He can certainly lie down, just like an ox, as the people of the kingdom of the Franks observed, seeing an elephant in the days of the Emperor Karolus" (VII).

The most interesting information Dicuil provides concerns observations made by his contemporaries on the world of the North Atlantic —the Orkneys, the Shetlands, and Iceland—which was then being explored by Dicuil's fellow Irishmen. Iceland is called "Thule" although neither Dicuil nor his contemporaries, nor anyone down to our time, has ever provided a satisfactory reason for identifying Iceland with the "Thule" of the ancients.

After quoting Solinus about the long summer days and winter nights at "Ultima Thule," Dicuil furnishes the following information:

It was thirty years ago (795) that priests who stayed on that island from the first of February to the first of August spoke of it in this manner. Not only at the time of the summer solstice, but for several days before and after, the sun during the evening seemed merely to be hidden, as it were, behind a tiny mound, in such a manner that there was no darkness, not even for the shortest time; a man could do whatever he wished, such as picking lice from his shirt, in full daylight. Had they been on top of a mountain, perhaps the sun would not have been hidden from their sight at all . . . I believe that at the time of the winter solstice, and for a few days before and after, the opposite would occur, i.e. there would be only the shortest period of daylight . . . These priests then sailed hence and, in a day's sail, did reach the frozen sea to the north" (VII).

One of Dicuil's most attractive traits is his refusal to accept the word of the "authorities" (Pliny and Solinus) without question. Pliny, for example, attributes to one Fabianus the statement that the greatest depth of the ocean is fifteen stadia. Dicuil duly reports that figure, but cannot help adding: "But who believes that Fabianus could know of the depths of the ocean?" (VIII)

Solinus, too, comes in for his share of criticism. Dicuil finds him guilty of making two completely contradictory statements about Mount Atlas, in the first saying that its summit extends above the clouds, and in the second, that it is always covered with snow. "But if the peak is always covered with snow," writes Dicuil, "it could not possibly rise above the clouds, for truly snow and hail and rain as well as thunder and lightning do not rise up from the clouds but descend from them" (IX).

A typical scholar of the early Middle Ages, Dicuil respects authority and is particularly impressed by facts and figures. He reports the height of the highest Alpine peaks to be fifty miles, a figure he had read somewhere, but adds in a refreshingly honest line "I cannot remember the book where I had read this" (IX).

In the eighth chapter of *De Mensura* Dicuil offers the reader a thumbnail resumé of the "Seven Chief Things" in this world, to wit, the number of important features found on the earth's surface. He first provides a breakdown of these statistics for each of the four quarters, East, South, West, and North, then sums it up in these words: "There are in the whole world 28 seas, 72 islands, 40 mountains, 65 provinces, 281 cities, 55 rivers, and 116 nations." C. R. Beazley felt that "this curious table . . . of things in the visible world stands for all the invisible things that are not expressly mentioned . . . quite in the same spirit of studious credulity which had nearly as strong a hold over Dicuil as it had over his models, Solinus, Isidore, and the rest."[2]

46. Ohthere's report on northernmost Europe

Paulus Orosius (fl. 390–417), a cleric, was a contemporary and friend of Saint Augustine. He wrote one of the most popular historical-geographical treatises of the early Middle Ages, the *Historia adversum paganos,* which enjoyed renown for at least seven centuries. Because of its reputation, Alfred the Great, King of England, translated it into English in the latter part of the ninth century.

In translating Orosius' work, Alfred rearranged the geographical introduction and added to it the reports of two of his thanes, Ohthere and Wulfstan. While Wulfstan's description of the lands surrounding the Baltic is important, Ohthere's description—based on his own travels—of northernmost Norway and of the northern coast of

---

From Reinhold Pauli, *The Life of Alfred the Great,* with Alfred's Anglo-Saxon version of Orosius, trans. B. Thorpe (London, 1853), pp. 249 and 251.

2. C. R. Beazley, *The Dawn of Modern Geography* (London: J. Murray, 1897–1906), I, 326–327.

Scandinavia as far as the White Sea, makes this a truly significant document, the first English report on a region that was to remain unexplored by Western Europeans until the sixteenth century.

Ohthere told his lord King Ælfred, that he dwelt northmost of all the Northmen. He said that he dwelt in the land to the northward, along the West-Sea; he said, however, that that land is very long north from thence, but it is all waste, except in a few places, where the Fins here and there dwell, for hunting in the winter, and in the summer for fishing in that sea. He said that he was desirous to try, once on a time, how far that country extended due north, or whether any one lived to the north of the waste. He then went due north along the country, leaving all the way the waste land on the right, and the wide sea on the left, for three days: he was as far north as the whale-hunters go at the farthest. Then he proceeded in his course due north, as far as he could sail within another three days; then the land there inclined due east, or the sea into the land, he knew not which, but he knew that he there waited for a west wind, or a little north, and sailed thence eastward along that land as far as he could sail in four days; then he had to wait for a due north wind, because the land there inclined due south, or the sea in on that land, he knew not which; he then sailed thence along the coast due south, as far as he could sail in five days. There lay a great river up in that land; they then turned up in that river, because they durst not sail on by that river, on account of hostility, because all that country was inhabited on the other side of that river; he had not before met with any land that was inhabited since he came from his own home; but all the way he had waste land on his right, except fishermen, fowlers, and hunters, all of whom were Fins, and he had constantly a wide sea to the left. The Beormas had well cultivated their country, but they did not dare to enter it; and the Terfinna land was all waste, except where hunters, fishers, or fowlers had taken up their quarters.

The Beormas told him many particulars both of their own land, and of the other lands lying around them; but he knew not what was true, because he did not see it himself; it seemed to him that the Fins and the Beormas spoke nearly one language. He went thither chiefly, in addition to seeing the country, on account of the walrusses, because they have very noble bones in their teeth, some of those teeth they brought to the king: and their hides are good for ship-ropes. This whale is much less than other whales, it being not longer than seven ells; but in his own country is the best whale-hunting, there they are eight-and-forty ells long, and most of them fifty ells long; of these he said that he and five others had killed sixty in two days. He was a very wealthy man in those possessions in which their wealth consists, that is in wild deer. He had at the time he came to the king, six wild deer. These deer they call rein-deer, of which there were six decoy rein-deer, which are very valuable amongst the Fins, because they catch the wild rein-deer with them.

He was one of the first men in that country, yet he had not more than twenty horned cattle, and twenty sheep, and twenty swine, and the little that he ploughed he ploughed with horses. But their wealth consists for the most part in the rent paid them by the Fins. That rent is in skins of animals, and birds' feathers, and whalebone, and in ship-ropes made of whales' hides, and of seals'. Every one pays according to his birth; the best-born, it is said, pay the skins of fifteen martens, and five rein-deer's, and one bear's skin, ten ambers of feathers, a bear's or otter's skin kyrtle, and two ship-ropes, each sixty ells long, made either of whale-hide or of seal's.

He said that the Northmen's land was very long and very narrow; all that his man could either pasture or plough lies by the sea, though that is in some parts very rocky; and to the east are wild mountains, parallel to the cultivated land. The Fins inhabit these mountains, and the cultivated land is broadest to the eastward, and continually narrower the more north. To the east it may be sixty miles broad, or a little broader, and towards the middle thirty, or broader; and northward, he said, where it is narrowest, that it might be three miles broad to the mountain, and the mountain then is in some parts so broad that a man may pass over in two weeks, and in some parts so broad that a man may pass over in six days. Then along this land southwards, on the other side of the mountain, is Sweden.

Geography in the Byzantine Empire

For a thousand years after the Roman Empire was split into two parts, the eastern portion, better known as the Byzantine Empire, survived, though its size was gradually reduced over time. It is to the Byzantines that we owe the survival of Ptolemy's work, but few examples of actual Byzantine geographical writing exist. In this segment two such writings are quoted: the first, the work of Procopius, a historian of the age of Justinian; the second, that of Constantine VII, an emperor-statesman who ruled the Byzantine Empire during the tenth century.

47. Procopius on Byzantium and the waterway leading to it

Procopius of Caesarea (c. 495–c. 562), historian, lawyer, and civil servant, is best known for his official *Histories of the Byzantine Empire*, dealing with events occurring during his lifetime; his "unofficial" chronicle, an attack on the Emperor Justinian and the Empress Theodora; and his book *De aedificiis*, a description of important public buildings erected during Justinian's reign. The following excerpts, describing the site of Byzantium and the maritime approaches to it, are taken from *De aedificiis*.

The prosperity of Byzantium is increased by the sea which enfolds it, contracting itself into straits, and connecting itself with the ocean, thus rendering the city remarkably beautiful, and affording a safe protection

---

From Procopius, *Of The Buildings of Justinian*, trans. Aubrey Stewart (London, 1896), pp. 24–25, 128–129.

in its harbours to seafarers, so as to cause it to be well supplied with provisions and abounding with all necessaries; for the two seas which are on either side of it, that is to say the Ægean and that which is called the Euxine, which meet at the east part of the city and dash together as they mingle their waves, separate the continent by their currents, and add to the beauty of the city while they surround it. It is, therefore, encompassed by three straits connected with one another, arranged so as to minister both to its elegance and its convenience, all of them most charming for sailing on, lovely to look at, and exceedingly safe for anchorage. The middle one of them, which leads from the Euxine Sea, makes straight for the city as though to adorn it. Upon either side of it lie the several continents, between whose shores it is confined, and seems to foam proudly with its waves because it passes over both Asia and Europe in order to reach the city; you would think that you beheld a river flowing towards you with a gentle current. That which is on the left hand of it rests on either side upon widely extended shores, and displays the groves, the lovely meadows, and all the other charms of the opposite continent in full view of the city. As it makes it way onward towards the south, receding as far as possible from Asia, it becomes wider; but even then its waves continue to encircle the city as far as the setting of the sun. The third arm of the sea joins the first one upon the right hand, starting from the place called Sycae, and washes the greater part of the northern shore of the city, ending in a bay. Thus the sea encircles the city like a crown, the interval consisting of the land lying between it in sufficient quantity to form a clasp for the crown of waters. This gulf is always calm, and never crested into waves, as though a barrier were placed there to the billows, and all storms were shut out from thence, through reverence for the city. Whenever strong winds and gales fall upon these seas and this strait, ships, when they once reach the entrance of this gulf, run the rest of their voyage unguided, and make the shore at random; for the gulf extends for a distance of more than forty stadia in circumference, and the whole of it is a harbour, so that when a ship is moored there the stern rests on the sea and the bows on the land, as though the two elements contended with one another to see which of them could be of the greatest service to the city.

The sea as far as the Hellespont is contained in a narrow strait; for the two continents there approach nearest to one another, forming the beginning of the strait near Sestos and Abydos. Ships bound for Constantinople consequently anchor when they reach this place, because they are unable to proceed any further unless the wind blows from the south. When, therefore, the fleet of corn-ships sail thither from Alexandria, if it meets with a favourable wind, the merchants in a very short time moor their ships in the harbours of Byzantium, and as soon as they have unloaded them, depart at once, in order that they may all make this voyage for a second or even a third time before the winter, while those of them who choose take in some other merchandise for the return voyage. If, however, the wind blows against them at the Hellespont, both the corn and the ships become injured by delay. Reflecting upon these things, the Emperor Justinian has clearly proved that nothing is impossible for man, even when he has to contend with the greatest difficulties; for he built granaries on the island of Tenedos, which is close to the strait, of a sufficient size to contain the freight of the whole fleet, being in width no less than ninety feet, in length two hundred and eighty, and of great height. After the Emperor had constructed these, when those who were conveying the public supply of corn were detained by contrary winds at this point, they used to unload their cargo into the granary, and, disregarding the northerly and westerly winds and all the other winds which were unfavourable for them, would prepare for another voyage. They therefore at once sailed home, while afterwards, whenever it became convenient to sail from Tenedos to Byzantium, the corn was conveyed from Tenedos thither in other ships by persons appointed to perform this duty.

48. Constantine VII describes the great water road of Russia, the trade routes of the Byzantine Empire, and the city of Venice

Constantine VII, surnamed Porphyrogennetos ("born to the purple"), ruled the Byzantine Empire from 911 to 959. Gibbon saw him first and foremost as a geographer, and C. R. Beazley considered him to be the greatest of the Byzantine geographers. His treatise On the Administration of the Empire offers a detailed survey of the subjects, vassals, and neighbors of the empire during the tenth century, and is our chief source for the historical geography of Eastern Europe at that time. Reading it, however, one must keep in mind that the terms used by Constantine VII were those of the Byzantines, and that for "Russian" one must substitute Norse; for "Slavic," Russian and Ukrainian; and for "Turk," Hungarian.

Of the coming of the Russians in "monoxyla" from Russia to Constantinople

The 'monoxyla' which come down from outer Russia to Constantinople are from Novgorod, where Sviatoslav, son of Igor, prince of Russia, had his seat, and others from the city of Smolensk and from Teliutza and

---

From Constantine VII (Porphyrogennetos), De administrando imperio, Greek text ed. Gy. Moravcsik, trans. R. J. H. Jenkins, new rev. ed. (Washington, D. C.: Dumbarton Oaks Center for Byzantine Studies, 1967), pp. 57–63, 119–121, 183–189.

Chernigov and from Vyshegrad. All these come down the river Dnieper, and are collected together at the city of Kiev, also called Sambatas. Their Slav tributaries, the so-called Krivichians and the Lenzanenes and the rest of the Slavonic regions, cut the "monoxyla" on their mountains in time of winter, and when they have prepared them, as spring approaches, and the ice melts, they bring them on to the neighbouring lakes. And since these *lakes* debouch into the river Dnieper, they enter thence on to this same river, and come down to Kiev, and draw *the ships* along to be finished and sell them to the Russians. The Russians buy these bottoms only, furnishing them with oars and rowlocks and other tackle from their old "monoxyla," which they dismantle; *and so* they fit them out. And in the month of June they move off down the river Dnieper and come to Vitichev, which is a tributary city of the Russians, and there they gather during two or three days; and when all the "monoxyla" are collected together, then they set out, and come down the said Dnieper river. And first they come to the first barrage, called Essoupi, which means in Russian and Slavonic "Do not sleep!"; the barrage itself is as narrow as the width of the Polo-ground; in the middle of it are rooted high rocks, which stand out like islands. Against these, then, comes the water and wells up and dashes down over the other side, with a mighty and terrific din. Therefore the Russians do not venture to pass between them, but put in to the bank hard by, disembarking the men on to dry land leaving the rest of the goods on board the "monoxyla"; they then strip and, feeling with their feet to avoid striking on a rock ... This they do, some at the prow, some amidships, while others again, in the stern, punt with poles; and with all this careful procedure they pass this first barrage, edging round under the river-bank. When they have passed this barrage, they re-embark the others from the dry land and sail away, and come down to the second barrage, called in Russian Oulvorsi, and in Slavonic Ostrovouniprach, which means "the Island of the Barrage." This one is like the first, awkward and not to be passed through. Once again they disembark the men and convey the "monoxyla" past, as on the first occasion. Similarly they pass the third barrage also, called Gelandri, which means in Slavonic "Noise of the Barrage," and then the fourth barrage, the big one, called in Russian Aeifor, and in Slavonic Neasit, because the pelicans nest in the stones of the barrage. At this barrage all put into land prow foremost, and those who are deputed to keep the watch with them get out, and off they go, these men, and keep vigilant watch for the Pechenegs. The remainder, taking up the goods which they have on board the "monoxyla," conduct the slaves in their chains past by land, six miles, until they are through the barrage. Then, partly dragging their "monoxyla," partly portaging them on their shoulders, they convey them to the far side of the barrage; and then, putting them on the river and loading up their baggage, they embark themselves, and again sail off in them. When they come to the fifth barrage, called in Russian Varouforos, and in Slavonic Voulniprach, because it forms a large lake, they again convey their "monoxyla" through at the edges of the river, as at the first and second barrages, and arrive at the sixth barrage, called in Russian Leanti, and in Slavonic Veroutzi, that is "the Boiling of the Water," and this too they pass similarly. And thence they sail away to the seventh barrage, called in Russian Stroukoun, and in Slavonic Naprezi, which means "Little Barrage." This they pass at the so-called ford of *Vrar,* where the Chersonites cross over from Russia and the Pechenegs to Cherson; which ford is as wide as the Hippodrome, and, measured upstream from the bottom as far as the rocks break surface, a bow-shot in length. It is at this point, therefore, that the Pechenegs come down and attack the Russians. After traversing this place, they reach the island called St. Gregory, on which island they perform their sacrifices because a gigantic oak-tree stands there; and they sacrifice live cocks. Arrows, too, they peg in round about, and others bread and meat, or something of whatever each may have, as is their custom. They also throw lots regarding the cocks, whether to slaughter them, or to eat them as well, or to leave them alive. From this island onwards the Russians do not fear the Pecheneg until they reach the river Selinas. So then they start off thence and sail for four days, until they reach the lake which forms the mouth of the river, on which is the island of St. Aitherios. Arrived at this island, they rest themselves there for two or three days. And they re-equip their "monoxyla" with such tackle as is needed, sails and masts and rudders, which they bring with them. Since this lake is the mouth of this river, as has been said, and carries on down to the sea, and the island of St. Aitherios lies on the sea, they come thence to the Dniester river, and having got safely there they rest again. But when the weather is propitious, they put to sea and come to the river called Aspros, and after resting there too in like manner, they again set out and come to the Selinas, to the so-called branch of the Danube river. And until they are past the river Selinas, the Pechenegs keep pace with them. And if it happens that the sea casts a "monoxylon" on shore, they all put in to land, in order to present a united opposition to the Pechenegs. But after the Selinas they fear nobody, but, entering the territory of Bulgaria, they come to the mouth of the Danube. From the Danube they proceed to the Konopas, and from the Konopas to Constantia, *and from Constantia* to the river of Varna, and from Varna they come to the river Ditzina, all of which are Bulgarian territory. From the Ditzina they reach the district of Mesembria, and there at last their voyage, fraught with such travail and terror, such difficulty and danger, is at an end. The severe manner of life of these same Russians in winter-time is as follows. When the month of November begins, their chiefs together with all the Russians at once leave Kiev and go off on the "poliudia," which means "rounds," that is, to the Slavonic regions of the Vervians and Drugovichians and Krivichians and Severians and the rest of the Slavs who are tributaries of the Russians. There they are maintained throughout the winter, but then once more, starting from the month of April, when the ice of the Dnieper river melts, they come back to Kiev. They then pick up their "monoxyla," as has been said above, and fit them out, and come down to Romania.

Geographical description from Thessalonica to the Danube river and the city of Belgrade; of Turkey and Patzinacia to the Chazar city of Sarkel and Russia and to the Nekropyla, that are in the sea of Pontus, near the Dnieper river; and to Cherson together with Bosporus, between which are the cities of the Regions; then to the lake of Maeotis, which for its size is also called a sea, and to the city called Tamatarcha; and of Zichia, moreover, and of Papagia and of Kasachia and of Alania and of Abasgia and to the city of Sotirioupolis.

From Thessalonica to the river Danube where stands the city called Belgrade, is a journey of eight days, if one is not travelling in haste but by easy stages. The Turks live beyond the Danube river, in the land of Moravia, but also on this side of it, between the Danube and the Save river. From the lower reaches of the Danube river, opposite to Distra, Patzinacia stretches along, and its inhabitants control the territory as far as Sarkel, the city of the Chazars, in which garrisons of 300 men are posted and annually relieved. Sarkel among them means "white house," and it was built by the spatharocandidate Petronas, surnamed Camaterus, when the Chazars requested the emperor Theophilus that this city should be built for them. For the then chagan and the pech of Chazaria sent envoys to this same emperor Theophilus and begged that the city of Sarkel might be built for them, and the emperor acceded to their request and sent to them the aforesaid spatharocandidate Petronas with ships of war of the imperial navy, and *sent* also ships of war of the captain-general of Paphlagonia. This same Petronas arrived at Cherson and left the ships of war at Cherson, and, having embarked his men on ships of burden, went off to that place on the Tanaïs river where he was to build the city. And since the place had no stones suitable for the building of the city, he made some ovens and baked bricks in them and with these he carried out the building of the city, making mortar out of tiny shells from the river. Now this aforesaid spatharocandidate Petronas, after building the city of Sarkel, went to the emperor Theophilus and said to him: "If you wish complete mastery and dominion over the city of Cherson and of the places in Cherson, and not that they should slip out of your hand, appoint your own military governor and do not trust to their primates and nobles." For up till the time of Theophilus the emperor, there was no military governor sent from here, but all administration was in the hands of the so-called primate, with those who were called the fathers of the city. The emperor Theophilus took counsel in this matter, whether to send as military governor so-and-so or such-an-one, and at last made up his mind that the aforesaid spatharocandidate Petronas should be sent, as one who had acquired local *experience* and was not unskilled in affairs, and so he promoted him to be protospatharius and appointed him military governor and sent him out to Cherson, with orders that the then primate and everyone else were to obey him; and from that time until this day it has been the rule for military governors in Cherson to be appointed from here. So much, then, for the building of the city of Sarkel. From the Danube river to the aforesaid city of Sarkel is a journey of 60 days. In this land between are many rivers: the two biggest of them are the Dniester and the Dnieper. But there are other rivers, that which is called the Syngoul and the Hybyl *and* the Almatai and the Kouphis and the Bogou and many others. On the higher reaches of the Dnieper river live the Russians, and down this river they sail and arrive at the Romans. Patzinacia possesses all the land *as far as* Russia and Bosporus and as far as Cherson and up to Sarat, Bourat and the 30 places. The distance along the sea-coast from the Danube river to the Dniester river is 120 miles. From the Dniester river to *the* river Dnieper is 80 miles, the so-called "gold-coast." After the mouth of the river Dnieper comes Adara, and there is a great gulf, called Nekropyla, where it is utterly impossible for a man to pass through. From the Dnieper river to Cherson is 300 miles, and between are marshes and harbours, in which the Chersonites work the salt. Between Cherson and Bosporus are the cities of the Regions, and the distance is 300 miles. After Bosporus comes the mouth of the Maeotic lake, which for its size everybody calls a sea. Into this same Maeotic sea run rivers many and great; on its northern side runs the Dnieper river, from which the Russians come through to Black Bulgaria and Chazaria and Syria. This same gulf of Maeotis comes opposite to, and within about four miles of, the Nekropyla that are near the Dnieper river, and joins them where the ancients dug a ditch and carried the sea through, enclosing within all the land of Cherson and of the Regions and the land of Bosporus, which cover up to 1,000 miles or even rather more. In the course of many years this same ditch has silted up and become a great forest, and there are in it but two roads, along which the Pechenegs pass through to Cherson and Bosporus and the Regions. Into the eastern side of the Maeotic lake debouch many rivers, the Tanaïs river that comes down from the city of Sarkel, and the Charakoul, in which they fish for sturgeon, and there are other rivers, the Bal and the Bourlik, the Chadir and other rivers very numerous. From the Maeotic lake debouches a mouth called Bourlik and flows down into the sea of Pontus where Bosporus is, and opposite to Bosporus is the city called Tamatarcha; the width of the strait of this mouth is 18 miles. In the middle of these 18 miles is a large, low island, called Atech. After Tamatarcha, some 18 or 20 miles from it, is a river called Oukrouch, which divides Zichia and Tamatarcha, and from the Oukrouch to the Nikopsis river, on which stands a city with the same name as the river, is the country of Zichia; the distance is 300 miles. Beyond Zichia is the country called Papagia, and beyond the country of Papagia is the country called Kasachia, and beyond Kasachia are the Caucasian mountains, and beyond the mountains is the country of Alania. Off the seaboard of Zichia lie islands, the great island and the three islands; and, closer to shore than these, are yet other islands, which have been used for pasturage and built upon by the Zichians, Tourganirch and Tzarbaganin and another island; and in the harbour of Spalaton another island; and at Pteleai another, where the Zichians take refuge during Alan incursions. The coastal area from the limit of Zichia, that is, from the Nikopsis river, is the country of Abasgia, as far as the city of Sotirioupolis; it is 300 miles.

Story of the settlement of what is now called Venice

Of old, Venice was a desert place, uninhabited and swampy. Those who are now called Venetians were Franks from Aquileia and from the other places in Francia, and they used to dwell on the mainland opposite Venice. But when Attila, the king of the Avars, came and utterly devastated and depopulated all the parts of Francia, all the Franks from Aquileia and from the other cities of Francia began to take to flight, and to go to the uninhabited islands of Venice and to built huts there, out of their dread of king Attila. Now when this king Attila had devastated all the country of the mainland and had advanced as far as Rome and Calabria and had left Venice far behind, those who had fled for refuge to the islands of Venice, having obtained a breathing-space and, as it were, shaken off their faintness of heart, took counsel jointly to settle there, which they did, and have been settled there till this day. But again, many years after the withdrawal of Attila, king Pippin arrived, who at that time was ruling over Papia and other kingdoms. For this Pippin had three brothers, and they were ruling over all the Frank and Slavonic regions. Now when king Pippin came against the Venetians with power and a large army, he blockaded them along the mainland, on the far side of the crossing between it and the islands of Venice, at a place called Aeibolas. Well, when the Venetians saw king Pippin coming against them with his power and preparing to take ship with the horses to the island of Madamaucon (for this is an island near the mainland), they laid down spars and fenced off the whole crossing. The army of king Pippin, being brought to a stand (for it was not possible for them to cross at any other point), blockaded them along the mainland six months, fighting with them daily. The Venetians would man their ships and take up position behind the spars they had laid down, and king Pippin would take up position with his army along the shore. The Venetians assailed them with arrows and javelins, and stopped them from crossing over to the island. So then king Pippin, at a loss, said to the Venetians: "You are beneath my hand and my providence, since you are of my country and domain." But the Venetians answered him: "We want to be servants of the emperor of the Romans, and not of you." When, however, they had for long been straitened by the trouble that had come upon them, the Venetians made a treaty of peace with king Pippin, agreeing to pay him a very considerable tribute. But since that time the tribute has gone on diminishing year by year, though it is paid even to this day. For the Venetians pay to him who rules over the kingdom of Italy, that is, Papia, a twopenny fee of 36 pounds of uncoined silver annually. So ended the war between Franks and Venetians. When the folk began to flee away to Venice and to collect there in numbers, they proclaimed as their doge him who surpassed the rest in nobility. The first doge among them had been appointed before king Pippin came against them. At that time the doge's residence was at a place called Civitanova, which means "new city." But because this island aforesaid is close to the mainland, by common consent they moved the doge's residence to another island, where it now is at this present, because it is at a distance from the mainland, as far off as one may see a man on horseback.

The Norse Contribution

The inhabitants of northern Europe remained remote and almost completely unknown until well into the early Middle Ages; they are seldom mentioned in classical and early Christian texts. Among the early reports based on firsthand experience, the description by the Arab diplomat, Ibn Fadhlan, stands out in terms of its detailed and accurate portrayal of people and places. The first western chronicler of Northern Europe is the historian Adam of Bremen, who wrote in the eleventh century.

In the ninth, tenth, and eleventh centuries the Norse succeeded in surrounding Europe. They sailed out of their Scandinavian domain to England, northern Germany, France, and all the way to Sicily; they crossed the Baltic, penetrating deep into Russia, and traded regularly in these regions as well as serving as mercenaries in the Byzantine Empire. The Norse also accomplished exceptional feats of seamanship in the North Atlantic, sailing from Norway first to the Orkneys, Shetlands, Hebrides, and Scotland; then later to Iceland, Greenland, and northeastern North America. The feats of these Atlantic pioneers are related in the great sagas, excerpts of which are presented here. *The King's Mirror,* written in the thirteenth century, sums up the Norse seafaring experience.

49. An Arab ambassador among the Norsemen: the report of Ibn Fadhlan

In 921 the Caliph of Baghdad sent Ibn Fadhlan, one of his courtiers, on an embassy to the ruler of the Bulgars, who controlled what is now

From *Eaters of the Dead,* by Michael Crichton. Copyright © 1976 by Michael Crichton. Reprinted by permission of Alfred A. Knopf, Inc., pp. 51, 54, 68–70.

the middle and lower valley of the Volga River. Ibn Fadhlan spent three years away from Baghdad, most of it in the company of Norsemen who, at that time, ruled the great river routes between the Baltic and Byzantium. His report is one of the few contemporary descriptions of the land and customs of the Norsemen, an eyewitness account of those people who were the outstanding seamen and warriors of their era.

These Northmen are by their own accounting the best sailors in the world, and I saw much love of the oceans and waters in their demeanor. Of the ship there is this: it was as long as twenty-five paces, and as broad as eight and a little more than that, and of excellent construction, of oak wood. Its color was black at every place. It was fitted with a square sail of cloth and trimmed with sealskin ropes. The helmsman stood upon a small platform near the stern and worked a rudder attached to the side of the vessel in the Roman fashion. The ship was fitted with benches for oars, but never were the oars employed; rather we progressed by sailing alone. At the head of the ship was the wooden carving of a fierce sea monster, such as appears on some Northman vessels; also there was a tail at the stern. In water this ship was stable and quite pleasant for travelling, and the confidence of the warriors elevated my spirit.

Ibn Fadhlan, coming from a much more southerly region, was struck by the short northern nights and the phenomenon of the aurora.

These North people never travel at night. Nor do they often sail at night, but prefer every evening to beach their ship and await the light of dawn before continuing farther. Yet this was the occurrence: during our travels, the period of the night became so short you could not cook a pot of meat in the time of it. Verily it seemed that as soon as I lay down to sleep I was awakened by the Northmen who said, "Come, it is day, we must continue the journey." Nor was the sleep refreshing in these cold places.

Also, (Herger) explained to me that in this North Country the day is long in the summer, and night is long in the winter, and rarely are they equal. Then he said to me I should watch in the night for the sky curtain; and upon one evening I did, and I saw in the sky shimmering pale lights, of green and yellow and sometimes blue, which hung as a curtain in the night air. I was much amazed by the sight of this sky curtain but the Northmen count it nothing strange.

In his travels with the Norsemen, Ibn Fadhlan visited the fortified "ring" town of Trelleborg, on the island of Sjaelland, west of present-day Copenhagen. Recent excavations, leading to a complete reconstruction of that Viking fortress, confirmed Ibn Fadhlan's description of it in every detail.

For the space of two days we sailed along a flat coast among many islands that are called the land of the land of Dans, coming finally to a region of marsh with a crisscross of narrow rivers that pour onto the sea. These rivers have no names themselves but are each one called "wyk," and the peoples of the narrow rivers are called "wykings," which means to the Northmen warriors who sail their ships up the rivers and attack settlements in such fashion.

Now in this marshy region we stopped at a place called Trelburg, which was a wonder to me. Here is no town but rather a military camp, and its people are warriors, with few women or children among them. The defenses of this camp of Trelburg are constructed with great care and skill of workmanship in the Roman fashion.

Trelburg lies at the joining point of two wyks, which then run to the sea. The main part of the town is encircled by a round earthwork wall, as tall as five men standing one atop the other. Above this earthen ring there stands a wooden fence for greater protection. Outside the earthen ring there is a ditch filled with water, the depth I do not know.

These earthworks are excellently made, of a symmetry and quality to rival anything we know. And there is this further: on the landward side of the town, a second semicircle of high wall, and a second ditch beyond.

Inside the town stand sixteen wooden dwellings, all the same: they are long houses, for so the Northmen call them, with walls that curve so that they resemble overturned boats with the ends cut flat front and back. In length they are thirty paces, and wider in the middle portion than either end. They are arranged thus: four long houses precisely set, so as to form a square. Four squares are arranged to make sixteen houses in all.

The society of Trelburg is mostly men, and women are all slaves. There are no wives among the women, and all women are taken freely as the men desire. The people of Trelburg live on fish, and some little bread; they do no agriculture or farming, although the marshlands surrounding the town contain areas suitable for growing. I asked... why there was no agriculture and was told, "These are warriors. They do not till the soil."

50. Adam of Bremen on "the northern islands"

Adam of Bremen (fl. 1069–1076) was the chronicler of the important archbishopric of Hamburg and Bremen, the main center of evangelical activity among the Norse. In his description of "the northern islands" he is the first man to mention Vinland, one of the distant places reached by the Norse that has been definitely identified with northeastern North America.

In going beyond the islands of the Danes there opens up another world in the direction of Sweden and Norway, which are the two most extensive kingdoms of the north and until now nearly unknown to our parts. About these kingdoms the very well-informed king of the Danes told me that Norway can hardly be crossed in the course of a month, and Sweden is not easily traversed in two months. "I myself found this out," he said, "when a while ago I fought for twelve years in those regions under King James. Both these countries are shut in by exceedingly high mountains—higher ones, however, in Norway which encircles Sweden with its alps." About Sweden, too, the ancient writers, Solinus and Orosius are not silent. They say that the Swedes hold a very large part of Germany and, besides, that their highland regions extend up to the Rhiphaean Mountains. There, also, is the Elbe River to which Lucan appears to have referred. It rises in the alps before mentioned and courses through the midst of the Gothic peoples into the ocean; hence, also, it is called the Götaälv. The Swedish country is extremely fertile; the land is rich in fruits and honey besides excelling all others in cattle raising, exceedingly happy in streams and woods, the whole region everywhere full of merchandise from foreign parts. Thus you may say that the Swedes are lacking in none of the riches, except the pride that we love or rather adore. For they regard as nothing every means of vainglory; that is gold, silver, stately chargers, beaver and marten pelts, which make us lose our minds admiring them. Only in their sexual relations with women do they know no bounds; a man according to his means has two or three or more wives at one time, rich men and princes an unlimited number. And they also consider the sons born of such unions legitimate. But if a man knows another man's wife, or by violence ravishes a virgin, or spoils another of his goods, or does him an injury, capital punishment is inflicted on him. Although all the Hyperboreans are noted for their hospitality, our Swedes are so in particular. To deny wayfarers entertainment is to them the basest of all shameful deeds, so much so that there

---

From Adam of Bremen, *History of the Archbishops of Hamburg-Bremen,* trans. Francis J. Tschan (New York: Columbia University Press, 1959), pp. 202–203, 210–211, 218–220. By permission of the publisher.

is strife and contention among them over who is worthy to receive a guest. They show him every courtesy for as many days as he wishes to stay, vying with one another to take him to their friends in their several houses. These good traits they have in their customs. But they also cherish with great affection preachers of the truth, if they are chaste and prudent and capable, so much so that they do not deny bishops attendance at the common assembly of the people that they call the *Warh*. There they often hear, not unwillingly, about Christ and the Christian religion. And perhaps they might readily be persuaded of our faith by preaching but for bad teachers who, in seeking "their own; not the things that are Jesus Christ's," give scandal to those whom they could save . . .

As Nortmannia is the farthest country of the world, so we properly place consideration of it in the last part of the book. By moderns it is called Norway. Of its location and extent we made some mention earlier in connection with Sweden, but this in particular must now be said, that in its length that land extends into the farthest northern zone, whence also it takes its name. It begins with towering crags at the sea commonly called the Baltic; then with its main ridge bent toward the north, after following the course of the shore line of a raging ocean, it finally has its bounds in the Rhiphaean Mountains, where the tired world also comes to an end. On account of the roughness of its mountains and the immoderate cold, Norway is the most unproductive of all countries, suited only for herds. They browse their cattle, like the Arabs, far off in the solitudes. In this way do the people make a living from their livestock by using the milk of the flocks or herds for food and the wool for clothing. Consequently, there are produced very valiant fighters who, not softened by any overindulgence in fruits, more often attack others than others trouble them. Even though they are sometimes assailed—not with impunity—by the Danes, who are just as poor, the Norwegians live in terms of amity with their neighbors, the Swedes. Poverty has forced them thus to go all over the world and from piratical raids they bring home in great abundance the riches of the lands. In this way they bear up under the unfruitfulness of their own country. . .

In the ocean there are very many other islands of which not the least is Greenland, situated far out in the ocean opposite the mountains of Sweden and the Rhiphaean range. To this island they say it is from five to seven days' sail from the coast of Norway, the same as to Iceland. The people there are greenish from the salt water, whence, too, that region gets its name. The people live in the same manner as the Icelanders except that they are fiercer and trouble seafarers by their piratical attacks. Report has it that Christianity of late has also winged its way to them.

The third island is Helgeland, nearer to Norway but in extent not unequal to the rest. That island sees the sun upon the land for fourteen days continuously at the solstice in summer and, similarly, it lacks the sun for the same number of days in the winter. To the barbarians, who do not know that the difference in the length of days is due to the accession and recession of the sun, this is astounding and mysterious. For on account of the rotundity of the earth, the sun in its course necessarily brings day as it approaches a place, and leaves night as it recedes from it. While the sun is approaching the summer solstice, the day is lengthened for those who live in the northern parts and the nights shortened; when it descends to the winter solstice, that is for the same reason the experience of south-erners. Not knowing this, the pagans call that land holy and blessed which affords mortals such a wonder. The king of the Danes and many others have attested the occurrence of this phenomenon there, as in Sweden and in Norway and on the rest of the islands in those parts.

He spoke also of yet another island of the many found in that ocean. It is called Vinland because vines producing excellent wine grow wild there. That unsown crops also abound on that island we have ascertained not from fabulous reports but from the trustworthy relation of the Danes. Beyond that island, he said, no habitable land is found in that ocean, but every place beyond it is full of impenetrable ice and intense darkness. Of this fact Martianus makes mention as follows: "Beyond Thule," he says, "the sea is congealed after one day's navigation." The very well-informed prince of the Norwegians, Harold, lately attempted this sea. After he had explored the expanse of the Northern Ocean in his ships, there lay before their eyes at length the darksome bounds of a failing world, and by retracing his steps he barely escaped in safety the vast pit of the abyss.

51. The sagas: Norse discoveries in North America

It is assumed that the achievements of the Norse—their landings and settling in Iceland, Greenland, and North America—were first preserved in oral tradition. Composed at the time the events took place, in the late tenth and early eleventh century, these oral sagas were put into written form in Iceland, two or more centuries later. The following passages describe, first, Bjarni Herjulfsson's voyage, in which he saw the coastline of North America; and second, Leif Eiriksson's voyage to Helluland (Baffin Land), Markland (Labrador), and Vinland (probably Newfoundland).

Bjarni sails to Greenland (from The Groenlandinga Saga)
Bjarni arrived in Iceland at Eyrar in the summer of the year that his father

The first two of these excerpts are from "The Groenlendinga Saga" and the third is from "Eirik's Saga"; all are taken from *The Vinland Sagas*, trans. Magnus Magnusson and Hermann Pálsson (Harmondsworth: Penguin Classics, 1965), pp. 52–54, 55–56, 94–96. © Magnus Magnusson and Hermann Pálsson, 1965. Reprinted by permission of Penguin Books Ltd.

had left for Greenland. The news came as a shock to Bjarni, and he refused to have his ship unloaded. His crew asked him what he had in mind; he replied that he intended to keep his custom of enjoying his father's hospitality over the winter—"so I want to sail my ship to Greenland, if you are willing to come with me."

They all replied that they would do what he thought best. Then Bjarni said, "This voyage of ours will be considered foolhardy, for not one of us has ever sailed the Greenland Sea."

However, they put to sea as soon as they were ready and sailed for three days until land was lost to sight below the horizon. Then the fair wind failed and northerly winds and fog set in, and for many days they had no idea what their course was. After that they saw the sun again and were able to get their bearings; they hoisted sail and after a day's sailing they sighted land.

They discussed amongst themselves what country this might be. Bjarni said he thought it could not be Greenland. The crew asked him if he wanted to land there or not; Bjarni replied, "I think we should sail in close."

They did so, and soon they could see that the country was not mountainous, but was well wooded and with low hills. So they put to sea again, leaving the land on the port quarter; and after sailing for two days they sighted land once more.

Bjarni's men asked him if he thought this was Greenland yet; he said he did not think this was Greenland, any more than the previous one—"for there are said to be huge glaciers in Greenland."

They closed the land quickly and saw that it was flat and wooded. Then the wind failed and the crew all said they thought it advisable to land there, but Bjarni refused. They claimed they needed both firewood and water; but Bjarni said, "You have no shortage of either." He was criticized for this by his men.

He ordered them to hoist sail, and they did so. They turned the prow out to sea and sailed before a south-west wind for three days before they sighted a third land. This one was high and mountainous, and topped by a glacier. Again they asked Bjarni if he wished to land there, but he replied, "No, for this country seems to me to be worthless."

They did not lower sail this time, but followed the coastline and saw that it was an island. Once again they put the land astern and sailed out to sea before the same fair wind. But now it began to blow a gale, and Bjarni ordered his men to shorten sail and not to go harder than ship and rigging could stand. They sailed now for four days, until they sighted a fourth land.

The men asked Bjarni if he thought this would be Greenland or not.

"This tallies most closely with what I have been told about Greenland," replied Bjarni. "And here we shall go in to land."

They did so, and made land as dusk was falling at a promontory which had a boat hauled up on it. This was where Bjarni's father, Herjolf, lived, and it has been called Herjolfsness for that reason ever since.

Bjarni now gave up trading and stayed with his father, and carried on farming there after his father's death.

Leif explores Vinland (from The Groenlandinga Saga)

They made their ship ready and put out to sea. The first landfall they made was the country that Bjarni had sighted last. They sailed right up to the shore and cast anchor, then lowered a boat and landed. There was no grass to be seen, and the hinterland was covered with great glaciers, and between glaciers and shore the land was like one great slab of rock. It seemed to them a worthless country.

Then Leif said, "Now we have done better than Bjarni where this country is concerned—we at least have set foot on it. I shall give this country a name and call it *Helluland.*"

They returned to their ship and put to sea, and sighted a second land. Once again they sailed right up to it and cast anchor, lowered a boat and went ashore. This country was flat and wooded, with white sandy beaches wherever they went; and the land sloped gently down to the sea.

Leif said, "This country shall be named after its natural resources: it shall be called *Markland.*"

They hurried back to their ship as quickly as possible and sailed away to sea in a north-east wind for two days until they sighted land again. They sailed towards it and came to an island which lay to the north of it.

They went ashore and looked about them. The weather was fine. There was dew on the grass, and the first thing they did was to get some of it on their hands and put it to their lips, and to them it seemed the sweetest thing they had ever tasted. Then they went back to their ship and sailed into the sound that lay between the island and the headland jutting out to the north.

They steered a westerly course round the headland. There were extensive shallows there and at low tide their ship was left high and dry, with the sea almost out of sight. But they were so impatient to land that they could not bear to wait for the rising tide to float the ship; they ran ashore to a place where a river flowed out of a lake. As soon as the tide had refloated the ship they took a boat and rowed out to it and brought it up the river into the lake, where they anchored it. They carried their hammocks ashore and put up booths. Then they decided to winter there, and built some large houses.

There was no lack of salmon in the river or the lake, bigger salmon than they had ever seen. The country seemed to them so kind that no winter fodder would be needed for livestock: there was never any frost all winter and the grass hardly withered at all.

In this country, night and day were of more even length than in either

Greenland or Iceland: on the shortest day of the year, the sun was already up by 9 A.M. and did not set until after 3 P.M.

Karlsefni explores Vinland (from The Eiriks Saga)

They sailed first up to the Western Settlement, and then to the Bjarn Isles. From there they sailed before a northerly wind and after two days at sea they sighted land and rowed ashore in boats to explore it. They found there many slabs of stone so huge that two men could stretch out on them sole to sole. There were numerous foxes there. They gave this country a name and called it *Helluland*.

From there they sailed for two days before a northerly wind and sighted land ahead; this was a heavily-wooded country abounding with animals. There was an island to the south-east, where they found bears, and so they named it *Bjarn Isle;* they named the wooded mainland itself *Markland*.

After two days they sighted land again and held in towards it; it was a promontory they were approaching. They tacked along the coast, with the land to starboard.

It was open and harbourless, with long beaches and extensive sands. They went ashore in boats and found a ship's keel on the headland, and so they called the place *Kjalarness*. They called this stretch of coast *Furdustrands* because it took so long to sail past it. Then the coastline became indented with bays and they steered into one of them.

When Leif Eirikson had been with King Olof Tryggvason and had been asked to preach Christianity in Greenland, the king had given him a Scottish couple, a man called Haki and a woman called Hekja. The king told Leif to use them if he ever needed speed, for they could run faster than deer. Leif and Eirik had turned them over to Karlsefni for this expedition.

When the ships had passed Furdustrands the two Scots were put ashore and told to run southwards to explore the country's resources, and to return within three days. They each wore a garment called a *bjafal*, which had a hood at the top and was open at the sides; it had no sleeves and was fastened between the legs with a loop and button. That was all they wore.

The ships cast anchor there and waited, and after three days the Scots came running down to the shore; one of them was carrying some grapes, and the other some wild wheat. They told Karlsefni that they thought they had found good land.

They were taken on board, and the expedition sailed on until they reached a fjord. They steered their ships into it. At its mouth lay an island around which there flowed very strong currents, and so they named it *Straum Island*. There were so many birds on it that one could scarcely set foot between their eggs.

They steered into the fjord, which they named *Straumfjord;* here they unloaded their ships and settled down. They had brought with them livestock of all kinds and they looked around for natural produce. There were mountains there and the country was beautiful to look at, and they paid no attention to anything except exploring it. There was tall grass everywhere.

They stayed there that winter, which turned out to be a very severe one; they had made no provision for it during the summer, and now they ran short of food and the hunting failed. They moved out to the island in the hope of finding game, or stranded whales, but there was little food to be found there, although their livestock throve. Then they prayed to God to send them something to eat, but the response was not as prompt as they would have liked.

Meanwhile Thorhall the Hunter disappeared and they went out to search for him. They searched for three days; and on the fourth day Karlsefni and Bjarni found him on top of a cliff. He was staring up at the sky with eyes and mouth and nostrils agape, scratching himself and pinching himself and mumbling. They asked him what he was doing there; he replied that it was no concern of theirs, and told them not to be surprised and that he was old enough not to need them to look after him. They urged him to come back home with them, and he did.

A little later a whale was washed up and they rushed to cut it up. No one recognized what kind of a whale it was, not even Karlsefni, who was an expert on whales. The cooks boiled the meat, but when it was eaten it made them all ill.

Then Thornhall the Hunter walked over and said, "Has not Redbeard turned out to be more successful than your Christ? This was my reward for the poem I composed in honour of my patron, Thor; he has seldom failed me."

When the others realized this they refused to use the whalemeat and threw it over a cliff, and committed themselves to God's mercy. Then a break came in the weather to allow them to go out fishing, and after that there was no scarcity of provisions.

In the spring they went back to Straumfjord and gathered supplies, game on the mainland, eggs on the island, and fish from the sea.

52. The King's Mirror: a medieval handbook on the northern lands

The *Konungs Skuggsja,* or *King's Mirror,* was called by George Sarton, a distinguished historian of science, the most important geographical work of Christendom in the first half of the thirteenth century. Writ-

From *The King's Mirror,* trans. Lawrence M. Larson (New York: Twayne, 1917), pp. 141–144, 149–150, 156–160.

ten in the form of a dialogue between learned father and aspiring/ inquiring son, the *King's Mirror* is a compendium of information on the Norse discoveries and on Norse seafaring knowledge and practices.

––––––––

The animal life of Greenland and the character of the land in those regions

Son. These things must seem wonderful to all who may hear of them,— both what is told about the fishes and that about the monsters which are said to exist in those waters. Now I understand that this ocean must be more tempestuous than all other seas; and therefore I think it strange that it is covered with ice both in winter and in summer, more than all other seas are. I am also curious to know why men should be so eager to fare thither, where there are such great perils to beware of, and what one can look for in that country which can be turned to use or pleasure. With your permission I also wish to ask what the people who inhabit those lands live upon; what the character of the country is, whether it is ice-clad like the ocean or free from ice even though the sea be frozen; and whether corn grows in that country as in other lands. I should also like to know whether you regard it as mainland or as an island, and whether there are any beasts or such other things in that country as there are in other lands.

Father. The answer to your query as to what people go to seek in that country and why they fare thither through such great perils is to be sought in man's three-fold nature. One motive is fame and rivalry, for it is in the nature of man to seek places where great dangers may be met, and thus to win fame. A second motive is curiosity, for it is also in man's nature to wish to see and experience the things that he has heard about, and thus to learn whether the facts are as told or not. The third is desire for gain; for men seek wealth wherever they have heard that gain is to be gotten, though, on the other hand, there may be great dangers too. But in Greenland it is this way, as you probably know, that whatever comes from other lands is high in price, for this land lies so distant from other countries that men seldom visit it. And everything that is needed to improve the land must be purchased abroad, both iron and all the timber used in building houses. In return for their wares the merchants bring back the following products: buckskin, or hides, sealskins, and rope of the kind that we talked about earlier which is called "leather rope" and is cut from the fish called walrus, and also the teeth of the walrus.

As to whether any sort of grain can grow there, my belief is that the country draws but little profit from that source. And yet there are men among those who are counted the wealthiest and most prominent who have tried to sow grain as an experiment; but the great majority in that country do not know what bread is, having never seen it. You have also asked about the extent of the land and whether it is mainland or an island; but I believe that few know the size of the land, though all believe that it is continental and connected with some mainland, inasmuch as it evidently contains a number of such animals as are known to live on the mainland but rarely on islands. Hares and wolves are very plentiful and there are multitudes of reindeer. It seems to be generally held, however, that these animals do not inhabit islands, except where men have brought them in; and everybody seems to feel sure that no one has brought them to Greenland, but that they must have run thither from other mainlands. There are bears, too, in that region; they are white, and people think they are native to the country, for they differ very much in their habits from the black bears that roam the forests. These kill horses, cattle, and other beasts to feed upon; but the white bear of Greenland wanders most of the time about on the ice in the sea, hunting seals and whales and feeding upon them. It is also as skillful a swimmer as any seal or whale.

In reply to your question whether the land thaws out or remains icebound like the sea, I can state definitely that only a small part of the land thaws out, while all the rest remains under the ice. But nobody knows whether the land is large or small, because all the mountain ranges and all the valleys are covered with ice, and no opening has been found anywhere. But it is quite evident that there are such openings, either along the shore or in the valleys that lie between the mountains, through which beasts can find a way; for they could not run thither from other lands, unless they should find open roads through the ice and the soil thawed out. Men have often tried to go up into the country and climb the highest mountains in various places to look about and learn whether any land could be found that was free from ice and habitable. But nowhere have they found such a place, except what is now occupied, which is a little strip along the water's edge.

There is much marble in those parts that are inhabited; it is variously colored, both red and blue and streaked with green. There are also many large hawks in the land, which in other countries would be counted very precious,—white falcons, and they are more numerous there than in any other country; but the natives do not know how to make any use of them . . .

You asked whether the sun shines in Greenland and whether there ever happens to be fair weather there as in other countries; and you shall know of a truth that the land has beautiful sunshine and is said to have a rather pleasant climate. The sun's course varies greatly, however; when winter is on, the night is almost continuous; but when it is summer, there is almost constant day. When the sun rises highest, it has abundant power to shine and give light, but very little to give warmth and heat;

still, it has sufficient strength, where the ground is free from ice, to warm the soil so that the earth yields good and fragrant grass. Consequently, people may easily till the land where the frost leaves, but that is a very small part.

But as to that matter which you have often inquired about, what those lights can be which the Greenlanders call the northern lights, I have no clear knowledge. I have often met men who have spent a long time in Greenland, but they do not seem to know definitely what those lights are. However, it is true of that subject as of many others of which we have no sure knowledge, that thoughtful men will form opinions and conjectures about it and will make such guesses as seem reasonable and likely to be true. But these northern lights have this peculiar nature, that the darker the night is, the brighter they seem; and they always appear at night but never by day,—most frequently in the densest darkness and rarely by moonlight. In appearance they resemble a vast flame of fire viewed from a great distance. It also looks as if sharp points were shot from this flame up into the sky; these are of uneven height and in constant motion, now one, now another darting highest; and the light appears to blaze like a living flame. While these rays are at their highest and brightest, they give forth so much light that people out of doors can easily find their way about and can even go hunting, if need be. Where people sit in houses that have windows, it is so light inside that all within the room can see each other's faces. The light is very changeable. Sometimes it appears to grow dim, as if a black smoke or a dark fog were blown up among the rays; and then it looks very much as if the light were overcome by this smoke and about to be quenched. But as soon as the smoke begins to grow thinner, the light begins to brighten again; and it happens at times that people think they see large sparks shooting out of it as from glowing iron which has just been taken from the forge. But as night declines and day approaches, the light begins to fade; and when daylight appears, it seems to vanish entirely.

The winds with respect to navigation

Son. I should indeed consider it highly informing, if I could remember all the things that you have now told me. I gather from your remarks, however, that you seem to think that I have asked about too many things in these our talks. But if you are not wearied with my questions, there still remains a little matter which, with your permission, I should like to ask about, one that also seems to belong to the knowledge of seafarers.

In a talk some time ago you said that whoever wishes to be a merchant ought to be prepared early in spring, and be careful not to remain out at sea too late in the autumn; but you did not indicate the earliest time in the spring when you think one may risk a journey overseas to other countries, nor how late you consider it safe to sail the seas in autumn. You told how the ocean manages to quiet its storms, but you did not show under what circumstances it begins to grow restless. Therefore I would fain ask you again to answer this question, even if it does annoy you, for I think that a time may come when it will seem both needful to know this and instructive to understand it.

Father. The matters to which you are now referring can scarcely be grouped under one head; for the seas are not all alike, nor are they all of equal extent. Small seas have no great perils, and one may risk crossing them at almost any time; for one has to make sure of fair winds to last a day or two only, which is not difficult for men who understand the weather. And there are many lands where harbors are plentiful as soon as the shore is reached. If the circumstances are such that a man can wait for winds in a good haven or may confidently expect to find good harbors as soon as he has crossed, or if the sea is so narrow that he needs to provide for a journey of only a day or two, then he may venture to sail over such waters almost whenever he wishes. But where travel is beset with greater perils, whether because the sea is wide and full of dangerous currents, or because the prow points toward shores where the harbors are rendered insecure by rocks, breakers, shallows, or sand bars,—wherever the situation is such, one needs to use great caution; and no one should venture to travel over such waters when the season is late.

Now as to the time that you asked about, it seems to me most correct to say that one should hardly venture over-seas later than the beginning of October. For at that time the sea begins to grow very restless, and the tempests always increase in violence as autumn passes and winter approaches. And about the time when we date the sixteenth of October, the east wind begins to look sorrowful and thinks himself disgraced, now that his headgear, the golden crown, is taken away. He puts a cloud-covered hat on his head and breathes heavily and violently, as if mourning a recent loss. But when the southeast wind sees how vexed his neighbor is, he is stricken with a double grief: the one sorrow is that he fears the same deprivation as the east wind has suffered; the other is grief over the misfortunes of his good and estimable neighbor. Stirred by the distress of a resentful mind, he knits his brows under the hiding clouds and blows the froth violently about him. When the South wind sees the wrath of his near neighbors, he wraps himself in a cloud-lined mantle in which he conceals his treasures and his wealth of warm rays and blows vigorously as if in terrifying defence. And when the southwest wind observes how friendship has cooled, now that the truce is broken, he sobs forth his soul's grief in heavy showers, rolls his eyes above his tear-moistened beard, puffs his cheeks under the cloudy helmet, blows the chilling scud violently forward, leads forth huge billows, wide-breasted waves, and breakers that yearn for ships, and orders all the tempests to dash forward in angry contest.

But when the west wind observes that a wrathful blast and a sorrowful sighing are coming across to him from the east, whence formerly he was accustomed to receive shining beams with festive gifts, he understands clearly that the covenant is broken and that all treaties are renounced. Deeply grieved and pained because of the unpeace, he puts on a black robe of mourning over which he pulls a cloud-gray cloak, and, sitting with wrinkled nose and pouting lips, he breathes heavily with regretful care. And when the ill-tempered northwest wind observes how sorrowful his neighbors look, and sees how he himself has suffered the loss of the evening beauty which he was formerly accustomed to display, he shows at once his temper in stern wrath: he knits his brows fiercely, throws rattling hail violently about, and sends forth the rolling thunder with terrifying gleams of lightning, thus displaying on his part a fearful and merciless anger. But when the north wind misses the friendliness and the kind gifts which he was wont to get from the south wind, he seeks out his hidden treasures and displays the wealth that he has most of: he brings out a dim sheen which glitters with frost, places an ice-cold helmet on his head above his frozen beard, and blows hard against the hail-bearing cloud-heaps. But the chill northeast wind sits wrathful with snowy beard and breathes coldly through his wind-swollen nostrils. Glaring fiercely under his rimy brows, he wrinkles his cheeks beneath his cold and cloudy temples, puffs his jowl with his icy tongue, and blows the piercing drift-snow vigorously forth.

But since peace has been broken among these eight chiefs and the winds are stirred to stormy violence, it is no longer advisable for men to travel over-seas from shore to shore because of great perils: the days shorten; the nights grow darker; the sea becomes restless; the waves grow stronger and the surf is colder; showers increase and storms arise; the breakers swell and the shores refuse good harbors; the sailors become exhausted, the lading is lost, and there is great and constant destruction of life due to a too great venturesomeness; souls are placed in perils of judgment because of recklessness and sudden death. Therefore all sensible men should beware and not venture upon the sea too late in the season; for there are many dangers to look out for and not one alone, if a man dares too much at such times. Consequently, the better plan is to sail while summer is at its best; for one is not likely to meet misfortune if there has been careful and wise forethought. But it would surely pass all expectations if that were to succeed which was foolishly advised and planned at the beginning, though sometimes the outcome may be favorable. I consider it a more sensible plan for a man to remain quiet as long as much danger may be looked for, and to enjoy during the winter in proper style and in restful leisure what he labored to win during the summer, than to risk in a little while through his own obstinate contriving the loss of all the profit which he strove to gain in the summer. But first of all a man must have care for his own person; for he can have no further profit, if it fares so ill that he himself goes under.

Moslem Geography

George Sarton called the achievements of Islam in the sciences "an Arabic miracle. The creation of a new civilization of international and encylopedic magnitude within less than two centuries is something we can describe, but not completely explain."[1]

Two "pillars" of the Moslem faith help explain Islam's intense interest in, and concern with, geography. Moslems are admonished to pray at certain times of the day, every day, turning toward the holy city of Mecca. For this reason, a special niche exists in every mosque to indicate to the faithful the direction of Mecca. And during his lifetime every Moslem is expected to accomplish the *hajj,* or pilgrimage to Mecca. The Moslems, then, were concerned with the establishment of exact reference points and cardinal directions. While accomplishing the pilgrimage to Mecca, Moslems traveled across much of the Old World, relying on written geographical information and, eventually, adding greatly to it.

The Moslem contribution to geography took many forms: the translation and absorption of learning from both West and East—from Greek, Persian, and Sanskrit sources, and possibly from Chinese ones as well; the preservation of writings that were essential, later on, to the revival of geography in the Christian world (in particular the writings of Ptolemy); the compilation of geographical compendia on the world of Islam, which extended from the Atlantic to the Pacific; and original contributions in mathematical geography and surveying.

The following excerpts represent Moslem geographical writing

---

1. George Sarton, "Early History of Science," in *The Practical Cogitator* (Boston: Houghton Mifflin, 1945) p. 225.

from the ninth to the fourteenth century of the Christian era; they include views on the structure of the cosmos and on the principal features of the earth; descriptions of regions both within and outside the domain of Islam; passages concerned with geodesy; excerpts from the writings of Moslem travelers; and a summing up of Islamic geography by the Moslem polymath of the fourteenth century, Ibn Khaldūn.

53. Al-Muqaddasi: a geographer's experiences in pursuit of knowledge

Al-Muqaddasi (c. 945–c. 1000) was a native of Jerusalem and this name, by which he is best known, is derived from the Arabic name of that city, Al-Quds. His work, *The Best Divisions for the Classification of Regions,* is one of the most important geographical treatises of the Golden Age of Arabic literature. The following excerpt describes his method of collecting information, and his experiences as a traveler through much of the world of Islam.

———

You should know that a number of scholars and ministers of state have written about this subject, and then rather confusedly; in fact, most of them—let me say all of them—have relied on hearsay. But for our part, no region remains that we did not enter, and even the slightest matters we have brought within the purview of our knowledge. At the same time, we have not omitted research, or making enquiries, or speculating. Our book, then, comprises three elements: firstly, what we ourselves have experienced; secondly, what we have heard from reliable persons; and, thirdly, what we found in books devoted to this subject or to others. There is no king's library that I did not assiduously examine, no sect but I studied its writings, no juridical schools but I became acquainted with them, no ascetics but I associated with them, no preachers in any country but I attended their assemblies. In this way I found a correct knowledge of what I was seeking in this subject.

I have been given thirty-six names by which I have been called and spoken to: Muqaddasi [Jerusalemite], Filastini [Palestinian], Misri [Egyptian] Maghribi [North African], Khurasani [Khorasanian], Salami [Salamian], Muqri' [teacher in reading the Quran], Faqih [jurist], Sufi [mystic], Wali [saint], Abid [devout man], Zahid [ascetic], Sayya [pilgrim], Warraq [copyist], Mujallid [bookbinder], Tajir [merchant], Mudhakkir

From Basil Anthony Collins, *Al-Muqaddasi, The Man and His Work,* Michigan Geographical Publication no. 10 (Ann Arbor, 1974). © Basil Anthony Collins. By permission of the author.

[discourser], Imam [leader in prayer], Mu'addin [summoner to prayer], Khatib [preacher], Gharib [foreigner], Iraqi, Baghdadi, Shami [Syrian], Hanifi [follower of the jurist Abu Hanifa], Muta'addib [scholar], Kari [lodger at charitable foundation], Mutafaqqih [jurist], Muta'allim [savant], Fara'idhi [notary], Ustadh [master], Danishumand [sage], Sheikh, Nishastah [learned one], Rakib [courier], Rasul [messenger]. All these names were given me through the number of countries in which I resided, and the many places I visited. There is nothing that befalls travelers of which I did not have my share, barring begging and grievous sin.

I studied law and letters, practiced asceticism and piety, taught theology and letters, preached on the pulpits, called to prayer on the minarets, led the prayers in the mosques, preached in the congregational mosques, frequented the schools. I said the prayers in convocation, spoke in the councils. I have eaten their *harisa* [porridge] with the mystics; supped *tharid* [broth] with the monks, and *'asida* [pudding] with the sailors. I have been ejected from mosques at night, have traveled in the solitudes, gone astray in the deserts. At times I have been scrupulously pious; at times I have openly eaten forbidden food. I have become acquainted with the devout men on the mountain of Lubnan [Mount Lebanon], and also been on intimate terms with governors. I have owned slaves, but have had to carry baskets on my own head.

A number of times I was close to drowning; our caravans were waylaid on the highway. I have served the judges and the great ones, have spoken with rulers and ministers. I have accompanied the licentious on the way, sold goods in the markets, been confined in jail, been accused as a spy. I have witnessed war with the Romaeans in battleships, and the ringing of church bells at night. I have been a bookbinder to earn money, have bought water at a high price. I have ridden in sedans and on horseback, have walked in the sandstorms and snows. I have been in the courtyards of the kings, standing among the nobles; I have lived among the ignorant in the workshops of weavers. What glory and honor I have been given! Yet my death has been plotted more than once. I have performed the Pilgrimage, and visited Medina; have been on forays, and in frontier posts. I have drunk *sawiq* [a potage of grains] from the public fountains in Mecca, eaten bread and chickpeas in a monastery.

I have depended on the hospice of Abraham, the Friend of God, and on the free sycamore figs at Ascalon. I have worn the robes of honor of the kings, and they have ordered gifts to be given me. Many times I have been naked and destitute. Yet rulers have corresponded with me. Nobles have reproved me, yet I have been asked to take charge of religious endowments. I have had to submit to blackguards; heresy has been imputed to me; I have been accused of greed; princes and judges have placed me in their confidence. I have been entered in wills, made an administrator; have had good experience of pickpockets, and seen the artifices of scoundrels. Contemptible people have hounded me, and the envious have opposed me; I have been slandered to the authorities. I have entered the baths of Tabariyya [Tiberias], and the fortresses of Fars [South Iran]; I have witnessed the festival of al-Fawwara [the fountain], the Feast of Saint Barbara, the Well of Budha'a, the castle of Ya'qub [Jacob], and his villages.

Experiences of this kind are many, but I have mentioned such a number of them that the reader of our book may know that we did not compile it haphazardly, nor arrange it in any random fashion. The reader will thus be able to distinguish between it and the others; for, after all, what a difference there is between one who has undergone all these experiences, and one who has compiled his book in perfect ease, basing it on the reports of others.

In my travels I spent more than ten thousand dirhams; I have, moreover, been lax in matters of the *shar'ia* [sacred law]. For there was no license of any juridical school, but I took advantage of it; doing, for washing the feet, a simple rubbing; getting by with praying using the one word "mudhammatani" [the dark-green gardens; the pious Muslim includes at least one verse of the Quran in his prayers, and Sura LV v.64 consists of this one word]; I have departed [from Mina, at Mecca] before sunset; have said the prescribed prayers on the back of an animal, having detestable filth on my clothes. I omitted the *tasbih* [prayer of glorification] in my *ruku* [bowings] and *sujud* [prostrations], and made the prostration of negligence before the prayer of salutation. I have grouped prayers together, and curtailed pious journeys freely undertaken for God's sake. Just the same, I did not depart from the prescriptions of the leading jurists, nor did I ever, at any time, defer saying my prayers.

In fine, I have never travelled the highroad when there was a distance of ten *farsakhs* or less between me and a town but that I quit the caravan and made my way to the city, that I might see it before the others. Sometimes I employed some men to accompany me, making my journey at night, so that I could return to my companions—at a great cost of money and labor.

54. Ibn Hauqal on the world of Islam and the lands beyond it

Ibn Hauqal, also known as Mohammed Abul Kassem (fl. 943–969), was born in Baghdad, the leading center of Islamic learning of his day.

From Ibn Hauqal, *Configuration de la Terre*, trans. J. H. Kramers & G. Wiet (Beyrouth: Commission internationale pour la traduction des chefs d'oeuvre, 1964), pp. 9–12; Editor's translation.

His description of the Islamic *oikumene*, the *Dar-ul-Islam*, or world of Islam, is far more extensive than that of his Christian contemporaries and includes lands extending across the whole of the Old World. Ibn Hauqal's account is based on factual information and the only allusion to the fabled unknown is the statement referring to the people of "Gog and Magog," an allusion found also in medieval Christian texts.

This therefore is the entire earth with its inhabited districts and uncultivated regions. It is divided according to the empires. The most important empires of the earth are four in number. The most prosperous, the richest in wealth, where the conduct of government, the conservation of resources and the abundance of revenue are in the best conditions, is the empire of Iranshahr, with its center in the region of Babil, that is the Persian empire. The frontiers of this empire were well defined in the times of the Persians, but after the introduction of Islam this empire acquired territories from others: from the Byzantine empire it annexed Syria, Egypt, Maghreb and Spain; from the empire of China it annexed Transoxiana. These important lands became part of the empire of Iranshahr.

The Byzantine empire extends to the borders of the lands of the Slavs and of their neighbours, Russians, Sarir, Alanians, Armenians and all those who profess the Christian religion.

The Chinese empire comprises also all of the land of the Turks, one part of Tibet, and heathen peoples.

The Indian empire includes Sind, Kashmir, a part of Tibet and peoples of the same religion.

The earth is divided into two parts, north and south. Starting from the East, from the gulf of the Ocean that surrounds China, to the gulf of the same ocean in the West, between the lands of Spain and Tangiers, the earth is divided into two parts. The dividing line extends from the sea of China, across India, the central portions of the Empire of Islam and westward, having crossed Egypt. To the north of this dividing line people are of a white colour and the whiteness of their skin becomes more intense as you move north toward the glacial regions. To the south of this line people are black and their blackness is even more pronounced as you move southward. The most temperate regions of these empires are along or near the dividing line.

The empire of Islam, is limited to the east by India and the sea of Persia; to the west, by the land of the Blacks, who live along the shores of the Ocean, in the general area of the plains and deserts of Audaghust, facing Ulil; to the north is the Byzantine empire as well as the neighbouring peoples, the Armenians, Alanians . . . Kazars, Russians, Bulgarians, Slavs and certain Turkish tribes; also to the north is part of the Chinese empire and its neighbouring Turkish regions; to the south is the sea of Persia.

The Byzantine empire is limited to the east by the empire of Islam, to the west and south by the Ocean, to the north by the frontiers of China, for I have included here in the Byzantine lands the Slavs located between the lands of the Turks and the Byzantine territories, as well as all the other people who live near the lands of the Byzantines.

The Chinese empire is limited to the north and east by the Ocean, to the south by the Islamic empire and India, to the west by the Ocean again because one has to include within his empire Gog and Magog and the neighbouring people whose land extends to the shores of the ocean.

The territory of India is limited to the east by the sea of Persia, to the west and to the south by the region of Khorasan and to the north by the empire of China.

These are the limits of the empires that I have described...

As to the size of the earth, from the extreme north to its southern limits, it begins from the shores of the ocean to join the lands of Gog and Magog, passes across the lands of the Slavs, crosses the lands of the interior Bulgarians and the Slavs, continues into the lands of the Byzantines, the land of Egypt and Nubia, extends to the deserts between the land of the Blacks and that of the Zendj to terminate at the ocean; this is the line which extends from the north to the south of the earth.

The Caspian sea does not receive any waters from the two seas in any way. They tell all sorts of stories about the sea referring to important authors. I have read in more than one version of the Geography according to Ptolemy that this sea is fed by the Mediterranean. The Lord may keep us from admitting that a scholar like Ptolemy could have uttered absurdities or affirmed something that was contrary to the existing state of affairs.

This sea is located in a depression of the earth and is fed by sweet water. The rivers which run in its direction and end in it are: the Volga, the greatest of its tributaries, which is the river of the Russians, the Kura, the Arax, and the rivers coming from the provinces of Ghilan, Dailam, Tabaristan, and Ghuz. All of these rivers carry sweet water, but the soil is filled with disease, causes fevers in such a way that the water is spoiled. Thus whoever travels along the shores of this sea, starting from the lands of the Kazars passing across the territory of Azerbaidjan, Dailam, of Tabaristan, of Jurjan and the desert bordering the mountain of Siyah-Kuh, returning thus to his point of departure, does not re-encounter any salt water but only those rivers carrying sweet water that we have described.

55. Ibn Hauqal on Spain, the Byzantine lands, and Sicily

In his *Description of the Earth* Ibn Hauqal included sketches of places such as Spain and Italy, which were under Moslem rule at the time and were well known to Islamic writers from the reports of government administrators. He also included descriptions of other places known through trade and travelers's reports such as "the lands of the Romans"—a Moslem appellation for the Byzantine Empire.

_____

Spain

Spain is one of the most magnificent peninsulas; she occupies an important position because of all the things that are within her and I shall describe most of it. I traveled there at the beginning of the year 337 (948).

Spain is about a months' travel in length and about twenty or so days travel in width. One finds there uncultivated land but the greatest part of it is cultivated and densely populated. There is running water everywhere, forests and orchards, and rivers with sweet water. Abundance and ease are characteristic of life; the enjoyment of it and the means of acquiring wealth are equally accessible to the great men and to the little ones, and these blessings are even extended to workers and to artisans thanks to low taxes, to the excellent condition of the land, and because the prince does not impose heavy demands in taxes upon his people.

Lands of the Romans

It takes eight days by land following the post roads from Antalia to Constantinople and fifteen days by sea with favourable winds. The lands between the two cities are fertile, densely populated and traffic flows, uninterrupted, along the entire road starting with the suburbs of Antalia and the rural districts surrounding it, all of which are flourishing and very productive, to the strait of Constantinople. Upon that strait a chain is stretched, to stop ships coming from the open sea who have not received permission, certified by an official document, to go beyond. At the end of this chain there is a customs office.

Beyond this strait westward one reaches the lands of Athens and Rome. Each of these metropolises serves as a capital for districts, cities, and towns that depend upon it and support it, cultivated fields, strongholds and

_____

The first two of these excerpts are from Ibn Hauqal, *Configuration de la Terre,* trans. J. H. Kramers and G. Wiet (Beyrouth: Commission internationale pour la traduction des chefs d'oeuvre, 1964), pp. 107, 196–197. The third is from Ibn Hauqal, *Description de Palerme au 10e siècle,* trans. M. Amari (Paris, 1845). Editor's translation.

fortresses, the whole governed by a large number of princes. The cities of Rome and Athens are meeting places for Chistians; they are both close to the sea. Athens was the very center of Greek philosophy, the place where their science and their theories have been preserved. Rome is one of the most important cities of the kingdom of the Christians; it is a Christian episcopal see, like the see of Antioch, the see of Alexandria, and the most recent see of Jerusalem, which did not exist in the time of the Apostles and was established later to raise the prestige of Jerusalem.

Sicily

The length of this island is seven days of walking, the width four days; it is mountainous with many castles and fortresses; it is inhabited and cultivated everywhere. Palermo, the most populous and the most famous city of this island, is also its capital. Situated on the northern coast, Palermo is divided into five districts, each one distinct from the others, even though they lie very near each other. The first is the city itself, called Palermo, surrounded by a stone wall which is high and awe-inspiring. This district is the merchants' place. Here is the great Mosque of Friday, in earlier times a Christian church, where you find a large chapel. I heard about this chapel from a dialectician who said that it is rumoured that a wise man of ancient Greece, that is, Aristotle, is suspended in a coffin in this very chapel, which the Moslems have converted into a mosque. The Christians, they say, venerated this personage and addressed their prayers to him to have rain, because of the extraordinary talent and many merits that the ancient Greeks have recognised in him. They add that the suspension between heaven and earth was done to get his protection to get rain, or to heal the sick, and for any other serious circumstance which leads men to pray to God (may he be praised!) and to make him offerings in times of misery, disease or civil war. As a matter of fact, I did see in this place a large box which undoubtedly held within it a coffin.

The largest markets, such as those of the oil vendors, are located between the mosque of Ibn-Aqlab and the El-Jadid district. The money changers and the drug sellers are outside the walls. The tailors, the armorers, the coppersmiths, and the wheat markets are all outside the wall and so are the other artisans distributed according to their various crafts.

Within the city there are one hundred and fifty butcher shops and perhaps even more, where they sell meat. Yet there is but a small number of butchers and this circumstance makes us understand their number and their importance. The size of their mosque shows the profits of their business. In fact, one day when this mosque was full of the people who usually worship there, I figured that the crowd consisted of more than seven thousand individuals; for there were thirty-two rows of people praying, and within each row there were about two hundred individuals.

Palermo is surrounded by several rivers which run from west to east
and which have enough water in them to provide power for two mills.
Along the banks of these rivers, all the way from their sources to the mouth
of the sea shore, there are a number of marshes where the Persian reeds
grow, yet neither the marshes nor the dry places have any malaria.

In the center of the island there is a valley covered largely by papyrus,
the reed of which is used to make scrolls for writing. As far as I know, the
only place upon earth where there is papyrus equal in quality to that of
Egypt is in Sicily. The largest part of the papyrus production is twisted
into cordage for ships; the rest of it is used to make paper for the Sultan,
and they only make as much as he needs for his own personal use.

56. Al-Masudi on the earth and its inhabitable portion; on Syria, Egypt, and Iraq

Al-Masudi (c. 895–956/957) was the author of The Meadows of Gold,
one of the most popular Moslem geographical works. Like many
Moslems, he had traveled widely and considered geography to be essential to the understanding of history. He believed that the environment influences the character and structure of plant and animal life.
His writings include numerous allusions to Greek authorities; one of
the most interesting of these allusions occurs in the first excerpt and
presumably refers to the now-lost maps that illustrated the Geography
of Ptolemy. Al-Masudi, however, was not above disputing the views
of Ptolemy and maintained that the Indian Ocean was an open rather
than a closed sea, a statement he based on reports of Moslem seamen.

The first excerpt from Al-Masudi's book presents his views on the
earth and the seas, and includes a Ptolemaic model of the cosmos.
The second excerpt offers descriptions of three regions in the heart of
the world of Islam: Syria, Egypt, and Iraq.

Men of learning divide the earth between the four cardinal points, east,
west, north, and south. They also divide it into two parts, that which is
inhabited and that which is desert, that which is cultivated and that which
is uncultivated. The earth, they say, is round, its center passed by the axis
of the sphere, air surrounds it on all sides and compared to the sphere of
the zodiac it is as small as a mathematical point. Its inhabited part extends
from a group of six islands called the Eternal Islands [Fortunate Islands],
which are situated in the western ocean, to the farthest part of China. Since
this distance corresponds to twelve hours, the daily journey of the sun,

---

From Al-Masudi, Les Prairies d'Or, trans. C. B. Meynard and P. de Courteille
(Paris, 1861–1877), pp. 179–187, 193–95. Editor's translation.

they have established that the sun rises for the Eternal Islands in the Western Ocean when it is setting for the farthest parts of China, and it is rising for that remote part of the earth when it is setting in those islands. This extent is half of the terrestrial circumference and it is that much that they have observed. If we attempt to evaluate in miles the measure of the earth we find that it is 13,500 miles. The studies of the learned men upon the width of the earth have shown that the inhabited part extends from the equator northwards to the island of Thule in Great Britain, where the longest day lasts twenty hours. According to them, the equator passes between east and west across an island situated between India and Abyssinia, somewhat to the south of these two countries. This midpoint between north and south is also intersected by the midpoint between the Eternal Islands and the farthest part of China; this is what we call the cupola of the earth. It is approximately sixty degrees of latitude from the equator to the island of Thule, that is, a sixth of the circumference of the earth. In multiplying this sixth, which is the extent of latitude, by a half, which represents longitude, you find that the inhabited part of the northern hemisphere represents one-twelfth of the surface of the earth.

The philosopher [Ptolemy] in his book entitled *Geography* describes the earth, the cities, the mountains, the seas, the islands, the rivers and the sources that it includes; he speaks of inhabited cities and cultivated lands; states the numbers of the cities to be 4,530 in his time, and lists them by order of the climates. He also distinguishes in the same work the mountains of the earth by their color, be they red, yellow, green, or other, and numbers them at more than 200; he also describes their height and the mines and precious stones that they contain. This philosopher counts five seas around the earth and speaks of cultivated and uncultivated, known and unknown islands that are situated there.

In the *Geography* [of Ptolemy] these seas are illuminated in various colours and differ by their extent and their appearance. Some have the shape of a short jacket, others that of a harness or that of a gun; still others are triangular, but all their names are shown in Greek in this book, and therefore cannot be understood.

The spheres or heavens number nine; the first, which is the smallest and the closest to the earth, is the sphere of the moon; the second, of Mercury; the third, of Venus; the fourth, of the Sun; the fifth, of Mars; the sixth, of Jupiter; the seventh, of Saturn; the eighth, of the fixed stars, and the ninth, of the zodiac. All of these spheres are shaped like globes, one included in the other. That of the zodiac is called the Universal Sphere, and its revolution creates day and night, for within one day and one night it carries the sun, the moon, and all of the stars from east to west around two immobile poles, of which one, situated in the north, is called the North Pole, and the other, the South Pole or Canopus.

The line which cuts this globe into two halves from east to west is called the equinoxial line, because when the sun is on this line, day and night are of equal length in all lands. That part of the sphere extending from north to south is called latitude, that part extending from east to west is called longitude.

The general shape of the seas has been the subject of much discussion. Most of the ancient philosophers of India and the wise men of Greece, except those who have adopted the revelation [Christians], maintain that the sea follows the spherical movement of the earth and prove it by numerous arguments. Thus, when one is sailing away, it is first the earth, then the mountains, which disappear and at the very last you lose sight of their summit; on the contrary when one approaches the coast it is the mountains which appear first, and when you are very close to the shore, you can distinguish the earth, the land, and trees.

This is the case of Mount Demavend in Persia. One can see the summit of this mountain, lost in the clouds, from a distance of 500 kilometers; a thick smoke rises from it and it is covered with eternal snow. On its base a great river rises, whose sulphurous waters are yellow like gold. To climb to the top of the mountain you have to spend three days and three nights. Once at the top you find a level area of about a thousand square feet, even though when you see it from below it appears to have a conical shape. This plateau is covered with red sand, into which one's feet tend to sink. Wild animals and birds cannot even reach this high peak because of its elevation, the wind, and the cold that prevails there. One also finds there thirty or so fissures, whence rise thick, sulphurous fumes, and a roar which resembles the loudest thunder. This roar comes from fire within. He who risks his life to climb this summit will be able to gather at the opening of these caverns pieces of sulphur, yellow like gold, which are used in alchemy and other arts. Seen from this height the highest surrounding mountains resemble but small hills. The Demavend is roughly twenty parasangs from the sea of Tabaristan [the Caspian Sea]. Ships that move out to sea lose all sight of it but at a distance of a hundred parasangs approaching the mountains of Tabaristan, they see, first, part of the peak of Demavend, which becomes more and more visible as they approach the shore. This fact, they say, proves the thesis of the sphericity of the sea.

Syria, Egypt, and Iraq

According to the traditional writers, after the Lord had given to the Muslims Iraq, Syria, Egypt, and a great many other countries, Omar, son of Alkhattab, wrote thus to a scholar of the time: "We are Arabs and the Lord has given us the conquest of the world; we now wish to settle down and dwell in the principal cities. Therefore describe for me the different lands of the earth, their climate, their position and the influence that soil and temperature exert upon their inhabitants."

Here is the response of that learned man: "Prince of the believers, know that the Lord has divided the countries of this earth between north, west, east, and south. All that is in the far east and in those regions where the sun rises is dangerous to dwell in because of the burning hot temperature and excessive heat which devours those who dare to venture there. The lands are most remote toward the west and are even more dangerous than those to the far east because they are less exposed to the rays of the sun. In those regions that are toward the north, cold, frost, and snow offer nothing to man but danger and sufferings of all sorts. Finally in the southern regions the difficulties of climate and the wild animals who dwell there make living there also very unpleasant for man. There remains therefore but a small portion of this earth which is inhabitable due to its temperate climate and other advantages that Providence has given it. It is this inhabitable part of the earth of which I shall give you a description.

O Prince of the believers... Syria is a green and fertile land watered by abundant streams, covered with trees and criss-crossed by rivers, with well-cultivated soil, and prosperous. It is the land of the prophets, the district where stands Jerusalem the chosen city, the land where had lived the most noble creatures of the Lord, the most austere holy men, and its mountains had given asylum to the most holy anchorites.

Egypt is a flat and depressed country; it is the old home of the Pharaohs and the dwelling place of tyrants. In spite of the advantages that it owes to the Nile it deserves more condemnation than praise; its climate is heavy and characterised by excessive heat, though not particularly bothersome. Its people are of darkish colour and slow of mind. It has mines of gold and precious stones, of emeralds and wealth of all sorts, a land of rich harvests and many palm trees. But the climate wears out the body and darkens the skin. People live there to an advanced age. Egyptians are smart, tricky, and given to low tricks of bad faith and to dishonesty. In a word, it is a land where one can become rich but where one does not want to dwell because of troubles and disorders that continue to depress one.

Iraq is the torch of the east, the center and heart of the earth criss-crossed by rivers that spread fertility in it; this land enjoys an always temperate climate. One notices among the people who dwell there a healthy constitution, a keen mind, an analytical spirit, a strong character, which just the same tends to tricks, good powers of reasoning, and good judgement. Iraq is the heart of the earth, the key to the east, and the road of the light. Its inhabitants surpass all others by the clarity of their complexion, the beauty of their race, the vigor of their temperament; they have outstanding qualities and possess the roots of all that is good. Thus Iraq offers great advantages: clear skies, fresh air, fertile soils, abundant waters, and a pleasant and easy life."

A certain number of biographers, men conversant with traditions tell the story of Omar, son of el-Hattab, who finding out that the Persian army was massing at Nehawend decided to go to Iraq, but before so doing questioned Kaab el-Akbar on that land; "Prince of the believers," replied Kaab, "when the Lord created the universe, each creature sought out another creature. 'I shall attach myself to Iraq,' said Reason; 'I shall follow thee,' said Science. 'Me,' said Wealth, 'I shall attach myself to Syria.' Discord cried out, 'I shall go with thee.' 'As for me,' said Health, 'I shall follow the nomads of the desert.' Sincerity said, 'I shall accompany thee.' "

57. Al-Biruni on the determination of longitude

Al-Biruni, Abu Rayhan Mohammed Ibn Ahmad (973– c. 1050), was a native of Khwarazm, present-day Soviet Central Asia. He was known to his contemporaries as "The Master," a title bestowed in recognition of his achievements in geography and geodesy, his translations of Indian treatises, and his linguistic skills. He is described by his contemporaries as being fluent in Persian, Arabic, Greek, Syriac, and Hebrew. His book on the determination of geographical coordinates of places, *Tahdid-al-Amakin,* is the source of the following excerpt.

We mentioned before that the people of this profession have accepted, by common consent, the rule to diminish the distance by a sixth in such problems, because the geographical distance follows the flight of the arrow. However, it does not state the thing which determines this amount, nor the circumstance that leads to it; because distances vary with the even flatness and the steepness (of the road), and also with the plurality of curves and depressions or with their paucity. If the decrease is due to these factors, then the amount of decrease should vary with their variations, according to the impressions and estimates of eyewitnesses of those variations, on the roads under consideration. But, even when the roads are free from ascents and descents, distances would be increased by such an increment, if they run between mountains and through valleys, because of the windings, and the interception of rivers the crossings of which lengthen the road, or of muddy swamps which have to be rounded by long circuitous roads, and by the compulsory deviation of the horse from the straight road towards the source of water and the safe abode, both of which are indispensable for travel in stages, and there are other similar causes.

From Al-Biruni, *The Determination of the Co-ordinates of Positions for the Correction of Distances between Cities,* trans. Jamil Ali (Beirut: American University of Beirut, 1967), pp. 202–204. By permission of the publisher.

Let A represent the position of Baghdad on the surface of the earth or the zenith of its inhabitants on the celestial sphere, and let AZ be part of its parallel of latitude, E the north pole, and EDA its meridian. Then EA is its colatitude. Let B represent Rayy, BD part of its parallel of latitude, and EBZ its meridian. Then EB is its colatitude; AD is the latitudinal difference, and AB, part of a great circle, is the distance between them. Though the distance between Baghdad and Hulwān and that between Hamadān and Rayy covers mountain passages, it requires a decrease of less than one sixth, but that between Hulwān and Hamadān requires a decrease of one sixth or more.

The number of farsakhs between Baghdad and Rayy is 158, and by dropping its sixth, (there remains) approximately 132, which is obtained by multiplying it (the 158) by five and dividing the product by six. If the quotient is multiplied by three, it is converted into 397 miles, and if this is divided by 56;40, according to the well-known opinion of the moderns, which is not far from my own estimate mentioned above (223:14), then the distance in degrees would be 7;0,21°.

Because the trapezoid whose sides are the chords AD, DB, BZ, and ZA is cyclic, and because chord AD is equal to chord BZ, and chord AZ is parallel to chord BD, the diagonals AB and ZD are equal. The square of AB, the magnitude of the distance, is equal to the square of the chord AD plus the product of chord AZ times chord DB. But the ration of chord AZ to chord DB is equal to the ratio of the radius of the parallel AZ, which is the sine of EA or the cosine of the latitude of Baghdad, to the radius of the parallel DB, which is the sine of EB, the colatitude of Rayy.

As to the latitude of Baghdad, different observers have found that it is neither less than 33;20° nor greater than 33;30°, and the approved one is 33;25°, because it is also the mean between those two. As to the latitude of Rayy, it was observed by Abū Maḥmūd al-Khujandī who found it to be 35;34,39°, and this is in agreement with what Abū al-Faḍl al-Hirawī had found in the days of Rukn al-Dowla. Hence AD, the latitudinal difference between Rayy and Baghdad is 2;9,39°. Its chord is 2;15,45, and the square of this is 5;7,8,3,45. The chord of the displacement AB is 7;19,54, and the square of this is 53;45,12,0,36. The difference between the two squares is 48;38,3,56,51. We multiplied this difference by the cosine of the latitude of Rayy, which is 48;47, 59, and obtained the product 2373;20,48,0,12,51,9. We divided this product by the cosine of the latitude of Baghdad, which is 50;4,52, and obtained the quotient 47;23,24, 12,8. We extracted the (square) root of this quotient and obtained 6;53,2. We multiplied this root by the total sine and obtained the product 413;2,0; then we divided this product by the cosine of the latitude of Rayy and obtained the quotient 8;27,50, which is the measure of a chord whose arc is 8;5,20°, the longitudinal difference between Baghdad and Rayy.

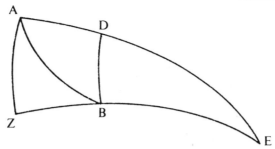

58. Al-Biruni reflects on the geography of earlier times

Al-Biruni (973–c. 1050) included in his treatise on geodesy some descriptive passages that are of particular interest because of his remarks on the physical geography of the past, on the Creation, on water, and on the relations between the distribution of population and the shape of the earth. Also worthy of notice are the references to both the Pentateuch and the Koran in his account of the Creation.

This steppe of Arabia was at one time sea which was later upturned, and the traces of that sea become evident on digging for wells and springs, because the desert then unfolds various strata of earth, sand, and soft pebbles, intermingled with pieces of pottery, glass, and bones, which could not have been buried there intentionally. Again, a variety of stones is excavated which reveals, on breaking up, definite sea products: shells, cowrie shells, and what are called «fish ears». These products will be found, either fully preserved, or in a state of complete decay, and in the latter case they will have left their figures completely imprinted in cavities in the stones. Such remains can also be found in (the city of) Bāb al-Abwāb on the coast of the Caspian Sea. The durations and dates of such transformations are completely unknown.

The Arabs have lived there since the days of their forefather Yuqṭān, but it is possible that they lived in the mountains of the Yaman, when the Arabian desert was a sea. Those were the ancient pure Arabs who dwelt in the Yaman, behind a dam between two mountains, where the water rose to the top of the mountains, and enabled them to cultivate two rows of gardens: one on the right, and one on the left of their dwelling places. But a torrential flood destroyed the dam; their habitation became desolate,

From Al-Biruni, *The Determination of the Co-ordinates of Positions for the Correction of Distances between Cities,* trans, Jamil Ali (Beirut: American University of Beirut, 1967), pp. 18–21, 23–25. By permission of the publisher.

and their two rows of gardens were transformed to two other gardens, "Producing bitter fruit and tamarisks, and some few Lote trees" (Qur'ān, Sura 34:16).

We find the like of these stones, with "fish ears" in their middle, in the sandy desert between Jurjān and Khwārizm. This desert was like a lake in the past, because the course of the Jayhūn (the Oxus), I mean the river of Balkh, ran through that desert to a town called Balkhān, on the Caspian Sea. Ptolemy relates in his book, the Geography, that the mouth of the Oxus is at the sea of Hyrcania, i.e. Jurjān. The time that has elapsed since those days of Ptolemy is about eight hundred years, and in those days the Oxus ran through the region, which is now a complete desert, between the towns of Zamm and Āmūya up to Balkhān, giving rise to prosperous towns and villages, and finally emptying its water into the sea, between Jurjān and the land of the Khazars ...

Ibn al-ᶜAmīd related in his book, *Fī Binā' Al-Mudun,* (on the Construction of Cities), that an earthquake took place in Rūyān, not long ago, which made two mountains collide and tumble down, and that the debris of the collision blocked the courses of the rivers which ran between them, and that the waters of the rivers receded and formed a lake. This is what usually happens when the water has no outlet, like the Dead Sea which is formed by the water of the Jordan River.

He also quoted from the histories of the Syrians that in the year three hundred and thirty-eight of Alexander, the second year of the Caesar Justinian, an earthquake took place in Antioch, and was accompanied by a fault in the earth's crust, and that a mountain above Qalūdhia (Clodia) had crumbled, and its debris fell into the Euphrates. The river was blocked; its water rose and flooded the area, causing a great deal of destruction; then the water swerved round its barrier, found an outlet for itself, and returned to its normal flow.

Aristotle related in his book, al-Āthār al-ᶜUlwiya, the Meteorologica, that the land of Egypt, in ancient times, was completely covered by the Nile which made it look like a sea; then the Nile water receded; the emerging land hardened gradually and was inhabited, and many towns sprang up which were full of people. But the Egyptians of today are ignorant of the art of construction. In ancient times Egypt was called Thiba (Thebes) after the name of one of its upper cities which was early constructed and inhabited; it is not the great city of Minf which is now called Mimfiyas (Memphis). The poet Homer, who is comparatively more recent than the early Egyptians, also calls it Thebes.

When the land of Egypt was a sea, the kings of Persia, during their partial domination over Egypt, wisely decided to excavate between it and the Red Sea, and to construct a canal which would connect the two seas; so that a ship could be navigated from the sea in the west to that in the east, and all this was done in good faith, to promote the country's welfare.

The first to render this service was the king Sāsṭrāṭis, (Sesostris) and the second was Darius. They both ordered the excavation of a long canal which still remains, and the water of the Red Sea enters it with the tide and exits with its ebb. However, when they measured the depth of the water in the canal, they decided to drop the project because the level of the Red Sea was found to be higher than that of the river of Egypt, and the risk that the sea water might spoil that of the river. The canal was completed by Ptolemy the Third, and the risk mentioned was eliminated by the expert advice of Archimedes. But the canal was later filled up with earth by order of one of the Roman kings, to prevent the Persians from coming to Egypt through it . . .

It is well known that civilization demands water, and that it shifts in pursuit of it, because the former is dependent on the latter. Aristotle related in his book, the Meteorologica, that some ancient people held the view that the earth was moist, and that the sun and the moon evaporated the moisture from some regions until they dried up, and the vapor produced winds which circulated in the air, and that the moisture that remained has formed the sea which will diminish, steadily decrease, and finally become dry land. This view, though systematic, apparently contradicts observed natural phenomena, but if some interpretation is provided, it can explain the realities of Nature. For it has been established in fundamental astronomy that the earth is spherical, amid a spherical universe, and that the plumb line naturally moves towards the center from any initial position or direction. This makes it clear that the surface of the water must be spherical, except for irregularities of surface produced by the waves, because of lack of cohesion between the particles of water.

It is also known from observation that the natural position of dry land is below that of the water, for soil precipitates in water, and that the penetration of water, falling vertically into soil or earth, is due to the rarefaction of the air, and the natural tendency of the water to settle below the air which permeates the space between the agglomerated particles of the soil.

It is also known that if parts of the earth are disrupted by violent force, then they move about the center and that if a disruption takes place, then water surrounds a disrupted part equally on all sides.

What follows is the state of things which was prevailing at the beginning of creation, as related in the Torah; I mean the movement of the breath of God on the surface of the water, when the earth was void and desolate. The same state was reported in the Revelation, where God Almighty says: "And His Throne was over the waters" (Qur'ān, Sura 11:7). When God Almighty intended the creation of mankind, He purposely designed the creation of the earth at first, and gave it the consolidating force to evolve its natural shape, I mean that which is truly spherical. He also elevated parts of the earth above the water, and this made the water run down into parts of the earth which were sunk, because of the elevation of others. He called the water that gathered in a depression a sea, and gave it the taste of salinity. This salinity, according to Thābit bin Qurra, prevents the water from getting foul, and eliminates putrefaction which would be disastrous to His intended creatures. The sea was also intended to be a reservoir of water for man's special needs, and as the lives of both man and animal, which is put in man's service, are dependent on fresh water, and as his habitation is far away from reservoirs, so God Almighty has designed the continuous motion of the sun and the moon, and commanded them both to produce motion in the water, to evaporate it, and to lift its vapor upwards. Further, the elevation of parts of the earth above the water was designed as a connecting link between the earth and the air, and water was made to possess the properties of permeation and diffusion, but all that is hardly possible without the influence of heat . . .

If we notice the distribution of people and civilization where we live, in the inhabited parts of the world which are on the same circle of latitude, where climatic conditions are identical and means of civilization are available, where causes of diseases are eliminated and atmospheric conditions are the same, then we can not attribute the absence of people in some places and their concentration in others to anything other than freedom of choice, or to agreement with others, or to the possibility that no human beings ever set foot on those uninhabited places. But the elevation of land above the water in the southern quarter, diametrically opposite the northern quarter, is possible if the spherical shape of the earth has been distorted to a cylindrical shape; so that the empirical sphere shall envelop both the spheres of earth and of water, and the center of gravity of the Whole shall be at the middle of the axis of the cylinder, in order to preserve the equilibrium of the weight. However, a portion of the spherical surface can be removed, and cavities then produced will be partly filled with water from the surrounding sea, while the tops of those cavities will remain above the water, and so the water will be all over the land, except that portion from which the mountains are made.

59. Ibn Khordadbeh describes Byzantium, some trade routes, and the divisions of the inhabitable world

Ibn Khordadbeh (c. 820–c. 912), a native of Persia, was postmaster general of the Abbasid caliphate and author of *The Book of Roads and Provinces*. This is one of the earliest examples of a popular genre of

From Ibn Khordadbeh, *Description du Maghreb et de l'Europe au IIIe-IXe siècle*, trans. Hadj Sadok (Alger: Editions Carbonel, 1949), pp. 15–25 passim. Editor's translation.

Moslem geographical writing and consists of a compendium of information on places and regions both within and beyond the boundaries of Islam.

The first excerpt from this work is a description of Byzantium, the city that the Moslems called Rūm (Rome); the second describes the range of the travels of Jewish merchants within and beyond the Moslem world. The third passage refers to the trade between early Russia and the Byzantines; the final one is typical of the author's rather practical approach to the more generalized part of geography.

Rome (Byzantium) and its marvels

Rome touches on the sea to the east, the south, and the west; only in the north does it have any contact with the mainland. From the eastern to the western gate the distance is twenty miles. The city is surrounded by two stone walls separated by a space of sixty feet. The inside wall is twelve feet high and seventy-two feet wide; the outside wall eight feet wide and forty-two in height. A river runs between them covered by large copper plates each one of which is forty-six feet long. This river is called Fistilatus.

From the Golden Gate to the gate of the King, the distance is twelve miles. There is a bread market that is one parasang long; the great market extends from the east to the western gate and is formed by three rows of columns and two arcades. The central arcade is surrounded by columns made of yellow Roman copper, their base and their capitals are all made of this metal. It is in this market that the merchants' shops are located. The columns are thirty feet high. In front of these columns and these shops a canal, paved with yellow copper (brass), runs from east to west, it is fed by a branch of the sea in such a way that the ships can enter with their cargo and come all the way to the merchants' shops.

The itinerary of the Jewish merchants

These merchants speak Arabic, Persian, Roman (Greek and Latin), Frank, Spanish, and Slavic, and travel from east to west by land and by sea. They bring to the West slaves of both sexes, eunuchs, brocades, beaver pelts, the pelts of mink and other animals, and sables. They leave from the land of the Franks on the western sea to al-Farama, and from there they carry their merchandise on camelback to al-Qulzum. There they get onto the eastern sea to get from al-Qulzum to al-Jar, the port of Medina, then to Jiddah and thence to Sind, to Hind, and to China. They bring back from China musk, aloe wood, camphor, cinnamon, and other products of those lands. They come back to al-Qulzum whence they take their merchandise to al-Farama and start out on the Western Sea.

Sometimes they go to Constantinople to sell their products to the Byzantines. If they so desire, they go from France on the Western Sea to Antioch, from thence in three stages to al-Jabiya, then they sail down the Euphrates to Baghdad, down the Tigris to Oman, Sind, Hind, and China, lands that are next to each other.

Itinerary of the Russian merchants

These are Slavs from the most distant regions of their land, the land of the Slavs; they bring beaver, black fox pelts and sables, and come to the Black Sea. The sovereign of the Byzantines taxes them for one-tenth of the value of their merchandise. When they go down the Don, the river of the Slavs passing by the capital of the Khazars, the sovereign again taxes them for one tenth of their merchandise. Thence they go to the Caspian and land at any particular place on the coast that they decide upon. This sea is sixty parasangs in diameter. Sometimes they transport their products on camelback from the Caspian to Baghdad. There, Slav servants serve them as interpreters. They call themselves Christians and pay special tax.

Divisions of the inhabited world

The inhabited world is divided into four parts;

1. Urufa (Europe), which includes Spain, the lands of the Slavs, the lands of the Rūm (Byzantines and Romans), the lands of the Franks and the lands of Tangiers to the frontier of Egypt.

2. Lubiya (Libya), which includes Egypt, al-Qulzum, Abyssinia, the land of the Berbers, and the adjacent lands all the way to the southern sea. These lands do not have any wild boar, or deer, or wild horses or ibex.

3. Ityufiya (Ethiopia), which includes Tihama, the Yemen, Sind, Hind, and China.

4. Isqutiya (Scythie), including Armenia, Khorasan, Turkestan and the land of the Khazars.

60. Al-Muqaddasi on Tiberias, Traq, and Kairouan

In his travels, Al-Muqaddasi (c. 945–c. 1000) acquired firsthand knowledge of places within his native land, such as Tiberias, and of others situated both in the eastern part of Islam (Iraq), and in the western part (Kairouan in Tunisia).

From Al Muqaddasi, *"The Best Divisions for the Classification of Regions,"* an unpublished translation by Professor Basil A. Collins; used with his permission.

Tiberias

Tiberias is the capital of Jordan and a city of the Valley of Canaan, situated between the mountain and the lake. It is narrow, with suffocating heat in summer, and unhealthy. Its length is about a parasang, but it has virtually no width. Its market extends from one gate to the other, with its cemetery on the hill. There are eight bath houses there that need not be heated, along with numerous basins of hot water. A large fine mosque stands in the marketplace. The mosque is on pillars that differ from one another, and its floor is paved with small stones.

It is said of the people of Tiberias that for two months they dance, for two months they stuff themselves; for two months they flail about, and for two months they go naked; for two months they play the flute, and for two months they wallow. The meaning of this is as follows: they dance from the number of fleas, then fill themselves with the lotus fruit; they slap about at the hornets with fly-swatters, to drive them from their meat and fruits. They go naked from the intense heat, and they suck the sugar-cane: then they must wallow in their mud. At the lower end of the lake is a great rampart over which goes the road to Damascus. The drinking water is from the lake. All around the lake are villages and palm trees, and on it, ships come and go. Water from the baths and the mills flows into the lake so that it is distasteful to strangers. Still the water is light to drink, and teems with fish. The towering mountain overhangs the town.

Iraq

The climate in this region varies; for Baghdad and Wasit and the area between has a gentle climate, but it is quick to change. Sometimes it is burning and irritating in the summer here, while in Kufa it is just the opposite. In Basra there is dreadful heat, except when the north wind blows: then it is pleasant there. I read in the records of Basra: "Our life in Basra is delightful; when the north wind blows it is fair and fresh." However, when the south wind blows, it is terribly depressing; for I saw them in a depressed mood with the south wind blowing, and one man would meet another and say, "Do you not see what we are in now?" and the other would answer: "We hope to God for relief." Sometimes at night something like a syrup falls there.

Hulwan has an equable climate; but God preserve us from the climate of al-Batai'h (the Swamps). Anyone who has experienced them in the summer has been amazed to see how they sleep under veils, because there is a flea there with a sting like a needle, and it will kill.

The province of Medina has many jurisprudents, Koran readers, litterateurs, imams, and kings, especially Baghdad and Basra; preachers there are held in low estimation. Snow is brought thither from afar. It is cold in the winter: sometimes the water freezes in Basra, but always in Baghdad. The people of Kufa and Basra are tan in color.

Kairouan

Kairouan is the metropolis of its region; it is a large, delightful place. The bread is very good, the meat excellent. There is every kind of fruit there: around it is plain, and mountain, and sea. It is both prosperous and learned, and prices are remarkably low. For a dirham, one can buy five *minns* of meat or ten of figs. You need not ask about its raisins, dates, grapes, and olive oil. It is the point of departure of travelers going west, and it is the center of commerce for the two seas. No area has more towns, more agreeable people. They are none other than Hanifites and Malikites there, and there is remarkable harmony among them—no discord, no factionalism. Without doubt, they follow the light of their Lord, putting up with discomfort, and rejecting revenge from their hearts. This area is the glory of the Maghrib, the seat, and one of the pillars of power. It is more gracious than Nisbur, bigger than Damascus, more glorious than Isfahan.

Yet their water is poor, there is little culture or elegance there. The water has to be stored in cisterns. Taxes are heavy on storekeepers, so they live in Sabra, so the markets of the metropolis are abandoned. The populace is left like a flock of unattended sheep. They do not perform the night prayers during Ramadhan: in fact, neither of the two sects has much influence there.

This city is a little less than three miles from the coast, and is unwalled. Their drinking water is from cisterns and tanks in which they collect the rainwater. Al-Mu'izz made a *qanat* for them from the mountains which fills the cisterns after going through his palace in Sabra. Their buildings are of clay and brick. There are several cisterns there for olive oil. The mosque is in a place called al-Simat al-Kibir, in the middle of the markets, in the center of the city. It is bigger than the mosque of Ibn Tailun; it has columns of marble, and is paved with marble tiles. The drain pipes of the mosque are of lead. It has the following gates: [nine names], a special gate for the date market, [two other gates].

61. Idrisi on the cities and countries of the Christian and Moslem worlds

Abu Abdallah Mohammed Idrisi (1100–1166) was a native of Morocco, the *Maghrib,* or West, of the Islamic world. His fame rests

From Idrisi, *Geographie,* trans. P. Amédée Jaubert (Paris, 1836–1840), vol. VI passim. Editor's translation.

both on his *Geography,* the source of the following excerpt, and on his achievements as a cartographer. Idrisi resided for many years at the court of King Roger II of Sicily—the *Geography* is also called the *Book of Roger*—and his writing reflects the wealth and variety of information available at that Norman court, a meeting place for West the East. His most famous map, called the "Silver Map" because it is reported to have been engraved for King Roger on a large slab of silver, is lost; but other, smaller maps of remarkable detail and accuracy have survived.

The city of Seville

This is a large and well-populated city. The buildings are high, the walls solid, the markets busy, and frequented by a considerable amount of people. The principal trade of this city is in oils, which they send to the East and to the West, by land and by sea; this oil comes from a district called al-Sharf, which is forty miles long and completely planted with olives and fig trees, it extends from Seville to the city of Niebla, and is approximately twelve miles wide. It is said that there are 8,000 flourishing villages there, a large number of baths, and country residences. The district is approximately three miles from Seville. It is called al-Sharf, because in effect it is the highest part of the Seville area, it extends from the north to the south in the shape of a hill. The olive plantations extend as far as the port of Niebla. Seville stands on the banks of the Wadi al-Kebir [Guadalquivir] that is, the river of Cordoba.

The city of Lisbon

Lisbon is built on the banks of the river called Tagus, or the river of Toledo. It is six miles wide near Lisbon and the tides can be observed in their full force. It is this beautiful city that extends along the river; it is enclosed by walls and protected by a castle. In the center of the city there is a spring emitting warm water in winter and in summer. Situated next to the Sea of Darkness [the ocean], there is across the river from this city, and near the mouth of the river, the Fort of the Mine, so called because the sea leaves pure gold pebbles on the shoreline. During the winter the people of this area go out to search for this metal, and manage to find a good deal of it during the cold season. This is a curious fact which we have observed ourselves.

The city of Antioch

Antioch is a rather smallish city, but is built on the most pleasant site in the midst of a fertile district. There is not another city except for Damascus that has such pleasant surroundings and aspect. There are a great many streams which cool the bazaars, the streets, and even the buildings; there is a high wall surrounding the city and gardens. This wall, which is surprisingly strong and built of stones, surrounds the city as well as the mountain on which Antioch stands; within its limits there are mills, gardens, orchards, and charming walks. The markets of the city are flourishing, its buildings are magnificent, its industry active, its trade prosperous, its resources and products are well known. They make beautiful cloth of a single color as well as some of the most beautiful silk cloth, brocades called destouri, isfahani, and others.

The city of Samarkand

Samarkand is a large and handsome city, situated to the south of the Soghd River; it is the capital of the province of the same name. The streets and the squares are large, the buildings very high, and so are the bazaars and the baths. It is surrounded by an earthen wall, and a deep ditch. The territory surrounding the city is very fertile and produces large amounts of fruit. The city has four gates. The water necessary for the people of the city is brought in from the south through the main gate. There stands on the banks of this river an aquaduct, which in some places stands very high above the ground; it is by this aquaduct that the water is carried into the city and distributed among the buildings. There are guardians of it who take very good care that the water is kept clean as it enters the city. The citadel is large and very handsome. The great mosque, situated above the citadel, is separated from it by a wide road. There are, in Samarkand, a number of houses and palaces, and there are hardly any buildings of any importance that do not have gardens, orchards, or running water.

The seat of the government in the olden days was Samarkand, but it has been transferred to Bukhara. At the time at which we are speaking the larger part of the city is in ruins, and the people have moved to Bukhara. According to tradition, Samarkand was founded by the king of Arabia Felix, and owes its progress to Alexander the Great.

England

England is a very large island, filled with settlements, castles, villages, cultivated fields, rivers and streams, mountains, valleys, and inhabited places. The southern part of this island is included in the section of which we are now speaking and of which we describe the most important cities. This part of England includes a number of inhabited places, that is, Chichester (?), Wareham, Shoreham, Hastings, Dover, Dartmouth, Ipswich (?), Lynn Regis, London, Wallingford, Winchester, and a number of places that belong in this section.

Norway

Norway is a large island, but most of it is a desert. This island has two capes, of which one in the west is opposite the island of Denmark, a half-day's sail away, while the other is opposite the coast of Finmark.

It is said that there are three flourishing cities in Norway, two are opposite Finmark and the third opposite Denmark. These cities look all the same, they are not very populous, and rather poor, for it rains a great deal in this country and fogs prevail a good deal of the time. The Norwegians, having sown their grain, harvest it while it is still green and take it to their dwellings to dry it by the fire, for there is but little sunshine. There are a great many trees in this country, with immense trunks, that are very heavy. They also state that there lives there a race of savages, whose heads are next to their shoulders so that they do not have any neck at all. They live in the midst of dark forests where they build their homes and live entirely on nuts and chestnuts.

62. Al-Dimashqi on the divisions of the world and on the stone called emery

Al-Dimashqi, Shems Ed Din Abu Abdallah Mohammed (1256–1327), was a native of Damascus and spent his life in Syria. His approach to geography closely resembles that of Christian geographers of the Middle Ages, relying as it does on the writings of ancient authorities, both historical and Biblical; his book, however, also includes important firsthand information gathered from traders.

Ardeshir ben Babek divided the Earth into four parts: one belongs to the Turks, the second to the Arabs, the third to the Persians, the fourth to the Negroes. King Feridun believed that the Earth is shaped like a bird, with China forming its head, India, its right wing, the lands of the Khazars and the Turks its left wing; Hejaz, Yemen, Iraq, Syria and Egypt its breast; the West its tail, with the tail feathers extending towards the Sudan.

Alexander the Great divided the peoples of the Earth: he called one part Europe, including Andalusia, the lands of the Slavs, France, Tangiers, and Italy; the second, Africa, including Egypt, the Persian Gulf, Abyssinia, eastern Africa, and the southern ocean; the third, Esqamunia, including Armenia, the lands of the Khazar, the Turks, and the district of Khorasan;

From Al-Dimashqi, *Description of the most remarkable things in our time, the marvels of the earth and of the sea,* trans. A. F. M. van Mehren (Amsterdam, 1874), pp. 22–25, 82. Editor's translation from the French.

and the fourth part, Benushia, comprising Arabia, Yemen, the Indies, and China.

Hermes and his early successors on the throne of Persia divided the Earth into seven climates or circles; three occupy the middle, two lie above, one on the right, the other on the left; and two lie below, arranged in the same manner. The first of the climates in the middle includes Syria and the Maghreb [Tunisia, Algeria, and Morocco]; the second in the middle, called Iranshehr, comprises Khorasan, Persia and Iraq; the third, on the left, Tibet and China. Of those situated above, the climate on the right includes the Arabian Peninsula and Yemen, that on the left the Indies and Sind. The circle on the right, situated above, comprises the lands of Rum [the Byzantine Empire], of the Slavs, and of the nations to the north and the west; the circle on the left includes the land of the Khazar, of the Turks with their several tribes, and the nations of Gog and Magog towards the east and the north. No account was taken in this division of Abyssinia, of the Sudan, of the Berbers, or of Egypt, either because these kingdoms did not exist at the time, or because they were included in the divisions cited above.

Noah, too, divided the Earth among his three sons: Japhet and his descendants received the north; Ham and his descendants were given the south and the west; and the center was given to Shem, including the Arabs, the Persians, and the Greeks. The descendants of Japhet are the Turks, the Slavs, Gog and Magog; the descendants of Ham, the Copts, the Berbers, and the Negroes. According to Said the Andalusian, the Negroes, and the Berbers form a single race, with the Copts and the Franks living to the north of them; the Indians and the peoples of East Africa form another race, with the peoples of Arabia, Syria, Iraq, and Persia to the north of them. The Chinese constitute a race that includes, to the north, the Kithai, the Turks, and the people of Gog and Magog; the Greeks and Romans form a race that includes the Russians; finally, there are the Slavs.

The Greeks and the Romans occupied the center of the Earth; that is why they were so distinguished in the sciences, like Hippocrates and Galen in medicine and natural history, Aristotle and Plato in philosophy and theology, Euclid and Pythagoras in mathematics and the exact sciences, Euclimon and Hylaos in physiognomy.

Of stones

Emery, a ferrous, rough, and hard stone, has the property of cutting all stones and minerals except emeralds and pearls . . . It is yellow or black in color, somewhat mixed with yellow; it is mined in China, in the Indies, on the isle of Ceylon, and in eastern Africa; the best quality is the Nubian, found in the district of Aswan. Thoroughly ground and mixed with lacquer, melted in fire, until the emery is fully saturated, it serves for the preparation of round chunks, used by polishers and jewelers to polish and shape all sorts of gemstones.

63. Ibn Battuta: his travels

Ibn Battuta (1304–c. 1368) was the greatest Moslem traveler of all time. In his book, *Rehla* ("Travels"), he describes what he has seen from the westernmost reaches of the world of Islam to the lands of India and China.

The Egyptian Nile

The Egyptian Nile surpasses all rivers of the earth in sweetness of taste, breadth of channel and magnitude of utility. Cities and villages succeed one another along its banks without interruption and have no equal in the inhabited world, nor is any river known whose basin is so intensively cultivated as that of the Nile. There is no river on earth but it which is called a sea; God Most High has said "If thou fearest for him, cast him into the *yamm*," thus calling it *yamm*, which means "sea" (*bahr*). It is related in an unimpeachable Tradition that the Prophet of God (God's blessing and peace upon him) reached on the night of his Ascension the Lote-Tree of the Extremity, and lo, at its base were four streams, two outer streams and two inner streams. He asked Gabriel (peace be upon him) what streams these were, and he replied "The two inner streams flow through Paradise, and as for the two outer streams they are the Nile and Euphrates." It is also related in the Traditions of the Prophet that the Nile, Euphrates, Saihān and Jaihān are, each one, rivers of Paradise. The course of the Nile is from south to north, contrary to all the [great] rivers. One extraordinary thing about it is that it begins to rise in the extreme hot weather, at the time when rivers generally diminish and dry up, and begins to subside at the time when rivers increase in volume and overflow. The river of Sind [Indus] resembles it in this respect, and will be mentioned later. The first beginning of the Nile flood is in Hazīrān, that is June; and when its rise amounts to sixteen cubits, the land-tax due to the Sultan is payable in full. If it rises another cubit, there is plenty in that year, and complete well-being. But if it reaches eighteen cubits it does damage to the cultivated lands and causes an outbreak of plague. If it falls short of sixteen by a cubit, the Sultan's land-tax is diminished, and if it is two cubits short the people make solemn prayers for rain and there is the greatest misery.

The Nile is one of the five great rivers of the world, which are the Nile, Euphrates, Tigris, Saihūn [Syr Darya] and Jaihūn [Amu Darya]; five other rivers rival these, the river of Sind, which is called Panj Ab [i.e. Five

The first two selections are taken from *The Travels of Ibn Baṭṭuṭa, A.D. 1325–1354,* H. A. R. Gibb, ed. (London: Hakluyt Society, 2d ser., no. 110, 1958), I, 48–50, © The Hakluyt Society, London. The third selection is from Ibn Battuta, *Travels in Asia and Africa,* trans. H. A. R. Gibb (New York, R. M. McBride & Company, 1929), pp. 242–243, 282–283, 284–285.

Rivers], the river of Hindustān which is called the Kank [or Gang, i.e. Ganges]—to it the Hindus go on pilgrimage, and when they burn their dead they throw the ashes of them into it, and they say that it comes from Paradise—the river Jūn, also in Hindustān, the river Itil [Volga] in the Qifjaq [Kipchak] steppe, on the shore of which is the city of al-Sarā, and the river Sarū in the land of al-Khitā [Cathay], on the banks of which is the city of Khān-Bāliq [Peking], whence it descends to the city of al-Khansā [Hang-Chow] and from there to the city of al-Zaitūn [Zayton] in the land of China. We shall speak of all these in their proper places, if God will. Some distance below Cairo the Nile divides into three sections, and none of these streams can be crossed except by boat, winter or summer. The inhabitants of every township have canals led off the Nile; when it is in flood it fills these and they inundate the cultivated fields.

The islands in the Indian Ocean

The people of the Maldive Islands are upright and pious, sound in belief and sincere in thought; their bodies are weak, they are unused to fighting, and their armour is prayer. Once when I ordered a thief's hand to be cut off, a number of those in the room fainted. The Indian pirates do not raid or molest them, as they have learned from experience that anyone who seizes anything from them speedily meets misfortune. In each island of theirs there are beautiful mosques, and most of their buildings are made of wood. They are very cleanly and avoid filth; most of them bathe twice a day to cleanse themselves, because of the extreme heat there and their profuse perspiration. They make plentiful use of perfumed oils, such as oil of sandal-wood. Their garments are simply aprons; one they tie round their waists in place of trousers, and on their backs they place other cloths resembling the pilgrim garments. Some wear a turban, others a small kerchief instead. When any of them meets the qádí or preacher, he removes his cloth from his shoulders, uncovering his back, and accompanies him thus to his house. All, high or low, are bare-footed; their lanes are kept swept and clean and are shaded by trees, so that to walk in them is like walking in an orchard. In spite of that every person entering a house must wash his feet with water from a jar kept in a chamber in the vestibule, and wipe them with a rough towel of palm matting which he finds there. The same practice is followed on entering a mosque.

From these islands there are exported the fish we have mentioned, coconuts, cloths, and cotton turbans, as well as brass utensils, of which they have a great many, cowrie shells, and *qanbar*. This is the hairy integument of the coconut, which they tan in pits on the shore, and afterwards beat out with bars; the women then spin it and it is made into cords for sewing [the planks of] ships together. These cords are exported to India, China, and Yemen, and are better than hemp. The Indian and Yemenite ships are sewn together with them, for the Indian Ocean is full of reefs, and if a ship is nailed with iron nails it breaks up on striking the rocks, whereas if it is sewn together with cords, it is given a certain resilience and does not fall to pieces. The inhabitants of these islands use cowrie shells as money. This is an animal which they gather in the sea and place in pits, where its flesh disappears, leaving its white shell. They are used for buying and selling at the rate of four hundred thousand shells for a gold dinar, but they often fall in value to twelve hundred thousand for a dinar. They sell them in exchange for rice to the people of Bengal, who also use them as money, as well as to the Yemenites, who use them instead of sand [as ballast] in their ships. These shells are used also by the negroes in their lands; I saw them being sold at Málli and Gawgaw at the rate of 1,150 for a gold dinar.

China

The land of China is of vast extent, and abounding in produce, fruits, grain, gold and silver. In this respect there is no country in the world that can rival it. It is traversed by the river called the "Water of Life," which rises in some mountains, called the "Mountain of Apes," near the city of Khán-Báliq [Peking] and flows through the centre of China for the space of six months' journey, until finally it reaches Sín as-Sín [Canton]. It is bordered by villages, fields, fruit gardens, and bazaars, just like the Egyptian Nile, only that [the country through which runs] this river is even more richly cultivated and populous, and there are many waterwheels on it. In the land of China there is abundant sugar-cane, equal, nay superior, in quality to that of Egypt, as well as grapes and plums. I used to think that the 'Othmání plums of Damascus had no equal, until I saw the plums in China. It has wonderful melons too, like those of Khwárizm and Isfahán. All the fruits which we have in our country are to be found there, either much the same or of better quality. Wheat is very abundant in China, indeed better wheat I have never seen, and the same may be said of their lentils and chick-peas.

The Chinese pottery [porcelain] is manufactured only in the towns of Zaytún and Sín-kalán. It is made of the soil of some mountains in that district, which takes fire like charcoal, as we shall relate subsequently. They mix this with some stones which they have, burn the whole for three days, then pour water over it. This gives a kind of clay which they cause to ferment. The best quality of [porcelain is made from] clay that has fermented for a complete month, but no more, the poorer quality [from clay] that has fermented for ten days. The price of this porcelain there is the same as, or even less than, that of ordinary pottery in our country. It is exported to India and other countries, even reaching as far as our own lands in the West, and it is the finest of all makes of pottery.

The hens and cocks in China are very big indeed, bigger than geese in our country, and hens' eggs there are bigger than our goose eggs. On the other hand their geese are not at all large. We bought a hen once and set about cooking it, but it was too big for one pot, so we put it in two. Cocks over there are about the size of ostriches; often a cock will shed its feathers and [nothing but] a great red body remains. The first time I saw a Chinese cock was in the city of Kawlam. I took it for an ostrich and was amazed at it, but its owner told me that in China there were some even bigger than that, and when I got to China I saw for myself the truth of what he had told me about them.

The Chinese themselves are infidels, who worship idols and burn their dead like the Hindus. The king of China is a Tatar, one of the descendants of Tinkiz [Chingiz] Khán. In every Chinese city there is a quarter for Muslims in which they live by themselves, and in which they have mosques both for the Friday services and for other religious purposes. The Muslims are honoured and respected. The Chinese infidels eat the flesh of swine and dogs, and sell it in their markets. They are wealthy folk and well-to-do, but they make no display either in their food or their clothes. You will see one of their principal merchants, a man so rich that his wealth cannot be counted, wearing a coarse cotton tunic. But there is one thing that the Chinese take a pride in, that is, gold and silver plate. Every one of them carries a stick, on which they lean in walking, and which they call "the third leg." Silk is very plentiful among them, because the silk-worm attaches itself to fruits and feeds on them without requiring much care. For that reason it is so common to be worn by even the very poorest there. Were it not for the merchants it would have no value at all, for a single piece of cotton cloth is sold in their country for the price of many pieces of silk. It is customary amongst them for a merchant to cast what gold and silver he has into ingots, each weighing a hundredweight or more or less, and to put those ingots above the door of his house.

The Chinese use neither [gold] dinars nor [silver] dirhams in their commerce. All the gold and silver that comes into their country is cast by them into ingots, as we have described. Their buying and selling is carried on exclusively by means of pieces of paper, each of the size of the palm of the hand, and stamped with the sultan's seal. Twenty-five of these pieces of paper are called a *bálisht,* which takes the place of the dinar with us [as the unit of currency]. When these notes become torn by handling, one takes them to an office corresponding to our mint, and receives their equivalent in new notes on delivering up the old ones. This transaction is made without charge and involves no expense, for those who have the duty of making the notes receive regular salaries from the sultan. Indeed the direction of that office is given to one of their principal amírs. If anyone goes to the bazaar with a silver dirham or a dinar, intending to buy something, no one will accept it from him or pay any attention to him until he changes it for *bálisht,* and with that he may buy what he will.

All the inhabitant of China and of Cathay use in place of charcoal a kind of lumpy earth found in their country. It resembles our fuller's earth, and its colour too is the colour of fuller's earth. Elephants [are used to] carry loads of it. They break it up into pieces about the size of pieces of charcoal with us, and set it on fire and it burns like charcoal, only giving out more heat than a charcoal fire. When it is reduced to cinders, they knead it with water, dry it, and use it again for cooking, and so on over and over again until it is entirely consumed. It is from this clay that they make the Chinese porcelain ware, after adding to it some other stones, as we have related.

64. Ibn Khaldūn on geography

The work of Ibn Khaldūn (1332–1406) has been described as "a summing up of Muslim medieval civilization."[1] His book the *Muqaddimah* is principally concerned with the philosophy of history, but it also includes a brilliant statement on geography. Ibn Khaldūn relied on both Moslem and non-Moslem authorities; Ptolemy is one of the latter and Idrisi, author of the *Book of Roger,* one of the former.

In the books of philosophers who speculated about the condition of the world, it has been explained that the earth has a spherical shape and is enveloped by the element of water. It may be compared to a grape floating upon water.

The water withdrew from certain parts of (the earth), because God wanted to create living beings upon it and settle it with the human species that rules as (God's) representative over all other beings. One might from this get the impression that the water is below the earth. This is not correct. The natural "below" of the earth is the core and middle of its sphere, the center to which everything is attracted by its gravity. All the sides of the earth beyond that and the water surrounding the earth are "above." When some part of the earth is said to be "below," it is said to be so with reference to some other region (of the earth).

The part of the earth from which the water has withdrawn is one-half the surface of the sphere of the earth. It has a circular form and is surrounded on all sides by the element of water which forms a sea called "the Surrounding Sea" (*al-Baḥr al-Muḥīṭ*). It is also called *lablâyah,* with thick-

From Ibn Khaldūn, *The Muqaddimah: An Introduction to History,* trans. Franz Rosenthal. Bollingen Series XLIII. © 1958 and 1967 by Princeton University Press. Selections from vol. 1, pp. 94–97, 101–102, and 167–173 reprinted by permission of Princeton University Press.

1. Franz Rosenthal, "Ibn Khaldūn," *Dictionary of Scientific Biography* (New York: Charles Scribner's Sons, 1973), VII, 322.

ening of the second *l*, or *oceanos*. Both are non-Arabic words. It is also called "the Green Sea" and "the Black Sea."

The part of the earth that is free from water (and thus suitable) for human civilization has more waste and empty areas than cultivated (habitable) areas. The empty area in the south is larger than that in the north. The cultivated part of the earth extends more toward the north. In the shape of a circular plane it extends in the south to the equator and in the north to a circular line, behind which there are mountains separating (the cultivated part of the earth) from the elemental water. Enclosed between (these mountains) is the Dam of Gog and Magog. These mountains extend toward the east. In the east and the west, they also reach the elemental water, at two sections (points) of the circular (line) that surrounds (the cultivated part of the earth).

The part of the earth that is free from water is said to cover one-half or less of the sphere (of the earth). The cultivated part covers one-fourth of it. It is divided into seven zones.

The equator divides the earth into two halves from west to east. It represents the length of the earth. It is the longest line on the sphere of (the earth), just as the ecliptic and the equinoctial line are the longest lines on the firmament. The ecliptic is divided into 360 degrees. The geographical degree is twenty-five parasangs, the parasang being 12,000 cubits or three miles, since one mile has 4,000 cubits. The cubit is twenty-four fingers, and the finger is six grains of barley placed closely together in one row. The distance of the equinoctial line, parallel to the equator of the earth and dividing the firmament into two parts, is ninety degrees from each of the two poles. However, the cultivated area north of the equator is (only) sixty-four degrees. The rest is empty and uncultivated because of the bitter cold and frost, exactly as the southern part is altogether empty because of the heat. We shall explain it all, if God wills.

Information about the cultivated part and its boundaries and about the cities, towns, mountains, rivers, waste areas, and sandy deserts it contains, has been given by men such as Ptolemy in the *Geography* and, after him, by the author of the *Book of Roger* [Idrisi]. These men divided the cultivated area into seven parts which they called the seven zones. The borders of the seven zones are imaginary. They extend from east to west. In width (latitudinal extension) they are identical, in length (longitudinal extension) different. The first zone is longer than the second. The same applies to the second zone, and so on. The seventh zone is the shortest. This is required by the circular shape that resulted from the withdrawal of the water from the sphere of the earth.

According to these scholars, each of the seven zones is divided from west to east into ten contiguous sections. Information about general conditions and civilization is given for each section . . .

The Nile begins at a large mountain, sixteen degrees beyond the equator at the boundary of the fourth section of the first zone. This mountain is called the Mountain of the Qumr. No higher mountain is known on earth. Many springs issue from the mountain, some of them flowing into one lake there, and some of them into another lake. From these two lakes, several rivers branch off, and all of them flow into a lake at the equator which is at the distance of a ten days' journey from the mountain. From that lake, two rivers issue. One of them flows due north, passing through the country of the Nûbah and then through Egypt. Having traversed Egypt, it divides into many branches lying close to each other. Each of these is called a "channel." All flow into the Mediterranean at Alexandria. This river is called the Egyptian Nile. It is bordered by Upper Egypt on the east, and by the oases on the west. The other river turns westward, flowing due west until it flows into the Surrounding Sea. This river is the Sudanese Nile. All the Negro nations live along its borders . . .

We have explained that the cultivated region of that part of the earth which is not covered by water has its center toward the north, because of the excessive heat in the south and the excessive cold in the north. The north and the south represent opposite extremes of cold and heat. It necessarily follows that there must be a gradual decrease from the extremes toward the center, which, thus, is moderate. The fourth zone is the most temperate cultivated region. The bordering third and fifth zones are rather close to being temperate. The sixth and second zones which are adjacent to them are far from temperate, and the first and seventh zones still less so. Therefore, the sciences, the crafts, the buildings, the clothing, the foodstuffs, the fruits, even the animals, and everything that comes into being in the three middle zones are distinguished by their temperate (well-proportioned character). The human inhabitants of these zones are more temperate (well-proportioned) in their bodies, color, character qualities, and (general) conditions. They are found to be extremely moderate in their dwellings, clothing, foodstuffs, and crafts. They use houses that are well constructed of stone and embellished by craftsmanship. They rival each other in production of the very best tools and implements. Among them, one finds the natural minerals, such as gold, silver, iron, copper, lead, and tin. In their business dealings they use the two precious metals (gold and silver). They avoid intemperance quite generally in all their conditions. Such are the inhabitants of the Maghrib, of Syria, the two 'Irâqs, Western India (as-Sind), and China, as well as of Spain; also the European Christians nearby, the Galicians, and all those who live together with these peoples or near them in the three temperate zones. The 'Irâq and Syria are directly in the middle and therefore are the most temperate of all these countries.

The inhabitants of the zones that are far from temperate, such as the first, second, sixth, and seventh zones, are also farther removed from being temperate in all their conditions. Their buildings are of clay and reeds. Their foodstuffs are durra and herbs. Their clothing is the leaves of trees, which they sew together to cover themselves, or animal skins. Most of them go naked. The fruits and seasonings of their countries are strange and inclined to be intemperate. In their business dealings, they do not use the two noble metals, but copper, iron, or skins, upon which they set a value for the purpose of business dealings. Their qualities of character, moreover, are close to those of dumb animals. It has even been reported that most of the Negroes of the first zone dwell in caves and thickets, eat herbs, live in savage isolation and do not congregate, and eat each other. The same applies to the Slavs. The reason for this is that their remoteness from being temperate produces in them a disposition and character similar to those of the dumb animals, and they become correspondingly remote from humanity. The same also applies to their religious conditions. They are ignorant of prophecy and do not have a religious law, except for the small minority that lives near the temperate regions. (The minority includes,) for instance, the Abyssinians, who are neighbors of the Yemenites and have been Christians from pre-Islamic and Islamic times down to the present; and the Mâlî, the Gawgaw, and the Takrûr who live close to the Maghrib and, at this time, are Muslims. They are said to have adopted Islam in the seventh [thirteenth] century. Or, in the north, there are those Slav, European Christian, and Turkish nations that have adopted Christianity. All the other inhabitants of the intemperate zones in the south and in the north are ignorant of all religion. (Religious) scholarship is lacking among them. All their conditions are remote from those of human beings and close to those of wild animals. "And He creates what you do not know."

The (foregoing statement) is not contradicted by the existence of the Yemen, the Hadramawt, al-Ahqâf, the Hijâz, the Yamâmah, and adjacent regions of the Arabian Peninsula in the first and second zones. As we have mentioned, the Arabian Peninsula is surrounded by the sea on three sides. The humidity of (the sea) influences the humidity in the air of (the Arabian Peninsula). This diminishes the dryness and intemperance that (otherwise) the heat would cause. Because of the humidity from the sea, the Arabian Peninsula is to some degree temperate.

Genealogists who had no knowledge of the true nature of things imagined that Negroes are the children of Ham, the son of Noah, and that they were singled out to be black as the result of Noah's curse, which produced Ham's color and the slavery God inflicted upon his descendants. It is mentioned in the Torah that Noah cursed his son Ham. No reference is made there to blackness. The curse included no more than that Ham's descendants should be the slaves of his brothers' descendants. To attribute the blackness of the Negroes to Ham, reveals disregard of the true nature of heat and cold and of the influence they exercise upon the air (climate) and upon the creatures that come into being in it. The black color (of skin) common to the inhabitants of the first and second zones is the result of the composition of the air in which they live, and which comes about under the influence of the

greatly increased heat in the south. The sun is at the zenith there twice a
year at short intervals. In (almost) all seasons, the sun is in culmination for
a long time. The light of the sun, therefore, is plentiful. People there have
(to undergo) a very severe summer, and their skins turn black because of the
excessive heat. Something similar happens in the two corresponding zones
to the north, the seventh and sixth zones. There, a white color (of skin) is
common among the inhabitants, likewise the result of the composition of
the air in which they live, and which comes about under the influence of
the excessive cold in the north. The sun is always on the horizon within the
visual field (of the human observer), or close to it. It never ascends to the
zenith, nor even (gets) close to it. The heat, therefore, is weak in this region,
and the cold severe in (almost) all seasons. In consequence, the color of the
inhabitants is white, and they tend to have little body hair. Further consequences of the excessive cold are blue eyes, freckled skin, and blond hair.

The fifth, fourth, and third zones occupy an intermediate position. They
have an abundant share of temperance, which is the golden mean. The
fourth zone, being the one most nearly in the center, is as temperate as can
be. We have mentioned that before. The physique and character of its inhabitants are temperate to the (high) degree necessitated by the composition of the air in which they live. The third and fifth zones lie on either side
of the fourth, but they are less centrally located. They are closer to the hot
south beyond the third zone and the cold north beyond the fifth zone. However, they do not become intemperate.

The four other zones are intemperate, and the physique and character
of their inhabitants show it. The first and second zones are excessively hot
and black, and the sixth and seventh zones cold and white. The inhabitants
of the first and second zones in the south are called the Abyssinians, the
Zanj, and the Sudanese (Negroes). These are synonyms used to designate
the (particular) nation that has turned black. The name "Abyssinians,"
however, is restricted to those Negroes who live opposite Mecca and the
Yemen, and the name "Zanj" is restricted to those who live along the
Indian Sea. These names are not given to them because of an (alleged)
descent from a black human being, be it Ham or any one else. Negroes
from the south who settle in the temperate fourth zone or in the seventh
zone that tends toward whiteness, are found to produce descendants whose
color gradually turns white in the course of time. Vice versa, inhabitants
from the north or from the fourth zone who settle in the south produce
descendants whose color turns black. This shows that color is conditioned
by the composition of the air. In his *rajaz* poem on medicine, Avicenna said:

> Where the Zanj live is a heat that changes their bodies
> Until their skins are covered all over with black.
> The Slavs acquire whiteness
> Until their skins turn soft.

The inhabitants of the north are not called by their color, because the people who established the conventional meanings of words were themselves white. Thus, whiteness was something usual and common (to them), and they did not see anything sufficiently remarkable in it to cause them to use it as a specific term. Therefore, the inhabitants of the north, the Turks, the Slavs, the Tughuzghuz, the Khazars, the Alans, most of the European Christians, the Gog and Magog are found to be separate nations and numerous races called by a variety of names.

The inhabitants of the middle zones are temperate in their physique and character and in their ways of life. They have all the natural conditions necessary for a civilized life, such as ways of making a living, dwellings, crafts, sciences, political leadership, and royal authority. They thus have had (various manifestations of) prophecy, religious groups, dynasties, religious laws, sciences, countries, cities, buildings, horticulture, splendid crafts, and everything else that is temperate.

Now, among the inhabitants of these zones about whom we have historical information are, for instance, the Arabs, the Byzantines (Rûm), the Persians, the Israelites, the Greeks, the Indians, and the Chinese. When genealogists noted differences between these nations, their distinguishing marks and characteristics, they considered these to be due to their (different) descents. They declared all the Negro inhabitants of the south to be descendants of Ham. They had misgivings about their color and therefore undertook to report the afore-mentioned silly story. They declared all or most of the inhabitants of the north to be the descendants of Japheth, and they declared most of the temperate nations, who inhabit the central regions, who cultivate the sciences and crafts, and who possess religious groups and religious laws as well as political leadership and royal authority, to be the descendants of Shem. Even if the genealogical construction were correct, it would be the result of mere guesswork, not of cogent, logical argumentation. It would merely be a statement of fact. It would not imply that the inhabitants of the south are called "Abyssinians" and "Negroes" because they are descended from "black" Ham. The genealogists were led into this error by their belief that the only reason for differences between nations is in their descent. This is not so. Distinctions between races or nations are in some cases due to a different descent, as in the case of the Arabs, the Israelites, and the Persians. In other cases, they are caused by geographical location and (physical) marks, as in the case of the Zanj (Negroes), the Abyssinians, the Slavs, and the black (Sudanese) Negroes. Again, in other cases, they are caused by custom and distinguishing characteristics, as well as by descent, as in the case of the Arabs. Or, they may be caused by anything else among the conditions, qualities, and features peculiar to the different nations. But to generalize and say that the inhabitants of a specific geographical location in the south or in the north are the descendants of such-and-such a well-known person because they have a common color, trait, or (physical) mark which that (alleged) forefather had, is one of those errors which are caused by disregard, (both) of the true nature of created beings and of geographical facts. (There also is disregard of the fact that the physical circumstances and environment) are subject to changes that affect later generations; they do not necessarily remain unchanged.

This is how God proceeds with His servants.——And verily, you will not be able to change God's way.

Revival of Geography in the West

The revival of geography in western Christendom is represented here by excerpts from the work of two popular writers on geographical subjects: Robert Grosseteste (c. 1168–1253) and John of Holywood, better known by his scholastic name of Sacrobosco (1195?–?1244). Grosseteste, a Franciscan, taught at Oxford. Sacrobosco, a member of the Augustinian order and an Oxonian by training, taught mathematics at the University of Paris. Sacrobosco's *On the Sphere* became one of the most popular treatises on geography and astronomy; it was also the second book about astronomy ever printed (Ferrara, 1472).

65. Robert Grosseteste on the heat of the sun

As our main purpose is to discuss whatever may be the principle of generation of the heat of the sun, we may ask universally: How many principles of generation are there? Since there are three principles from which heat is generated, namely a hot body, motion and a concentration of rays, we should realise that the heat in these is of a single nature (*univocum*); from this single nature an effect of a single nature is produced in them. And since they have an effect of a single nature, it follows that in all these principles there is a cause of a single nature: for, of every effect of a single nature there is a cause of a single nature. That the heat in all of them is of a single nature is clear, because, from whichever of them it is generated,

---

Robert Grosseteste, "On the Heat of the Sun," from A. C. Crombie, "Grosseteste's Position in the History of Science," in *Robert Grosseteste, Scholar and Bishop*, ed. D. A. Callus (Oxford: Clarendon Press, 1955), pp. 116–120. By permission of the Oxford University Press.

it has the same powers and produces the same effects. Therefore it is univocally, not equivocally, named.

Let us, then, look for this univocal cause. In all of them the proximate cause of heat is scattering (*disgregatio*). Whence, since a hot body generates heat, this is by the scattering of materials. But how this explanation fits motion and a concentration of rays it is difficult to see.

Local motion, from which heat is generated, is divided into natural and violent motion, natural motion again into rectilinear and circular motion. First let us discuss violent motion, or a heavy body violently moved. A heavy body can be moved violently in three ways, up, down, or down but not directly towards the centre of the earth. In all cases it is clear that in the violent motion there is scattering because of the motion. For in violent motion there is a two-fold motive power, one part natural, the other violent, and these move every part of the moving body in different directions. As a result of this tendency to go in different directions, scattering takes place. Because of the violent motion, the moving body must be scattered part from part, and so heat results. Since, in the first way of moving violently, there is the greatest amount of opposition between the tendencies of the moving powers (they are moving in completely opposite directions), this produces the most scattering and the most heat; the second and third ways produce only moderate heat. This is in the highest degree clear from both theory (*ratione*) and from observation (*experimento*).

The same thing is shown in the case of natural motion. For heat is generated during motion in anything moving naturally downwards. Acting on every part there is actually a two-fold motive power, partly natural, partly violent. It is obvious that a natural power is operating here. That a violent power is also operating I show as follows:

Everything that is heavy and is moved downwards not directly towards the centre of the earth, is moved violently. But all heavy parts are moved downwards not directly towards the centre. Therefore all heavy parts are moved violently. The minor premiss I prove thus: The parts of a heavy body always keep the same distance apart in the whole. Therefore, since they are moved downwards with the motion of the whole, they move along lines which remain a constant distance apart. But lines which remain a constant distance apart when extended to infinity in either direction never meet. Therefore the parts of a naturally moved heavy body are moved downwards along non-intersecting lines. Therefore they are not moved directly towards the centre of the earth, because, if they were, they would be moved along lines running directly together there. So the principle is evident, namely that acting on every part of a body moved naturally down there is a two-fold power tending in different directions. But the opposition between these tendencies is weaker than the opposition between the tendencies of the parts in violent motion; so, of all agencies generating heat, natural motion generates the least natural heat in motion. Thus it is plain that from natural rectilinear motion, and from violent motion, heat is generated, and from a hot body heat is generated by a univocal cause.

The same can be shown, by similar reasoning, of the third principle of generation of heat. That some heat may be generated, from a univocal cause of heat, by the concentration of rays is shown by *On Mirrors;* rays from a concave mirror directed towards the sun produce combustion, and this is on account of scattering. A ray in a denser transparent medium is more incorporated than in a less dense one (and we are not speaking of total incorporation, like heat, but of a certain partial incorporation). Because of this incorporation, the incorporated parts of the air fly apart when the rays are concentrated at one point, each part, at the point itself, going along its own straight line. As a result there will be, round this point, the greatest dispersion of the air into different directions, and so there will be a scattering and heat will result. Thus it is evident that in these three genera heat is present from a cause of a single nature.

If, then, the sun generates heat, it will do so either as a hot body does, or as motion does, or as a concentration of rays does. That the sun does not generate heat in the manner in which a hot body does is evident from the following: It is proved in the seventh book of the *Physics* that an agent producing a change of quality, and the subject undergoing change, must be in immediate contact. Hence, if there were a medium between the original agent of change and the ultimate changed subject, that medium would first have to be changed by the heat of the hot sun, rather than the ultimate subject; otherwise the original agent and the first subject to be changed would not be in immediate contact. Therefore, since there are several media between the sun and the air, and next to the sun (which produces change according to the heat it possesses) is the fifth element or part of the fifth element, it must follow that the heat of the sun must first bring about a change of heat in the fifth element, rather than in the air. But this is impossible, because if the fifth element can undergo change of quality, it is corruptible. Therefore the first premiss is impossible, namely that the sun generates heat in the manner in which a hot body does. Perhaps some would say that heat is present in the sun virtually, as it is in pepper. But this is not to the point, because, in so far as heat is present in pepper virtually and not actually, it cannot produce movement unless it is moved by something else, nor qualitative change unless it is changed by something else. And similarly for the sun. But this is impossible; so therefore is the first premiss.

That heat is not generated from the motion of the sun is shown as follows. Now motion does not generate heat, unless there are, in every part of the moved thing, different tendencies moving the part in different directions. But in everything that is moved circularly and not violently, any part has the same tendency as the whole; there is no difference: the tendency of every part is to move in a circle. Therefore no heat is generated by circular motion. You might perhaps say that although there is no intrinsic cause of heat in anything moving in a circle, nevertheless there is an extrinsic cause, as there is with inferior bodies from the resistance of the medium. But this is false for two reasons: one reason is that in these inferior bodies the resistance of the medium is not the source of the heat produced by motion. If it were, since the medium resists equally things moving with natural and with violent motion, the same amount of heat should be generated by violent and by natural motion. But this is false, as observation shows; therefore so is the first premiss. The other reason why it is false is that the sun and stars have no resistance to their motion, because they are not moved by motors of their own. But since they are fixed in their spheres, they are moved by the motion of their spheres, like a ship in a river, which is moved by the motion of the river, as the Philosopher [Aristotle] shows in the second book of the *De Caelo et Mundo*.

There remains therefore only the theory that the sun generates heat by the concentration of rays. This is shown as follows: The sun's rays in the transparent medium of the air are, through the nature of the dense body, to some extent incorporated in it. Rays falling downwards on to the plane, concave, or convex surface of the earth are reflected at equal angles, as shown by the last of the principles taught in *On Mirrors*. Therefore, if they fall perpendicularly, they are reflected perpendicularly; and so the incident and the reflected rays go along the same line in totally opposite directions, and there is a maximum of scattering. This is the case on the equator, when the sun is at the zenith of this region, and at any place south or north of the equator at a distance from it less than that of the tropic of cancer or of the tropic of capricorn. In these regions the sun's rays must fall perpendicularly twice a year. But at a place on the tropic of cancer or the tropic of capricorn, the sun can reach the zenith only once, and only once send rays perpendicularly to this place. When this happens there is a maximum of scattering and of heat in these places. This is a violent scattering, such as is brought about by a concentration of rays refracted through a spherical body or reflected from a concave mirror, though in it the rays go in totally opposite directions, whereas in the case of the spherical refracting body and the concave mirror they do not.

In regions at greater distances from the equator than the tropic of cancer, since the sun does not come north far enough to reach the zenith, the rays fall at an angle less than a right angle, and are reflected at the same angle and so not in a totally opposite direction. The further the place from the equator, the more obtuse the angles at which the solar rays fall and are reflected, and the less opposed the directions in which the incident and reflected rays go, the less the scattering, and the less the heat generated. This is shown by observation.

If it is asked why heat is not generated from the rays of the sun in the fifth element, two replies can be given: first, that reflected rays do not intersect there; secondly, even if they did intersect after being reflected in a totally opposite direction [to the incident rays], heat would not be generated. For, since in this transparent medium there is no dense nature, the solar rays are not in any way incorporated in it, and so cannot scatter the parts of matter. In the uppermost layer of air, where the air is thinnest, the least amount of heat is generated, as observation shows. For there are more clouds on the summit of mountains, where the solar rays are brightest, than in valleys, though, nevertheless, rays are reflected there, just as they are in a valley; but, because of the thinness of the air there, the density of the air is small, so is the incorporation of light with it, and hence so also is the scattering of the parts of the air in the concentration of rays. In a valley there is a greater incorporation of rays and therefore more scattering and more heat.

66. John of Holywood (Sacrobosco) on the sphere

On spheres and the shape of the world

Sphere defined.—A sphere is thus described by Euclid: A sphere is the transit of the circumference of a half-circle upon a fixed diameter until it revolves back to its original position. That is, a sphere is such a round and solid body as is described by the revolution of a semicircular arc.

By Theodosius a sphere is described thus: A sphere is a solid body contained within a single surface, in the middle of which there is a point from which all straight lines drawn to the circumference are equal, and that point is called the "center of the sphere." Moreover, a straight line passing through the center of the sphere, with its ends touching the circumference in opposite directions, is called the "axis of the sphere." And the two ends of the axis are called the "poles of the world."

Sphere divided.—The sphere is divided in two ways, by substance and by accident. By substance it is divided into the ninth sphere, which is called the "first moved" or the *primum mobile;* and the sphere of the fixed stars, which is named the "firmament"; and the seven spheres of the seven planets, of which some are larger, some smaller, according as they the more approach, or recede from, the firmament. Wherefore, among them the sphere of Saturn is the largest, the sphere of the moon the smallest.

By accident the sphere is divided into the sphere right and the sphere oblique. For those are said to have the sphere right who dwell at the equator, if anyone can live there. And it is called "right" because neither pole is elevated more for them than the other, or because their horizon intersects

From *The Sphere of Sacrobosco and its Commentators* by Lynn Thorndike (Chicago: University of Chicago Press, 1949), pp. 118–125, 138–140. By permission of the University of Chicago Press. © 1949 The University of Chicago.

the equinoctial circle and is intersected by it at spherical right angles. Those are said to have the sphere oblique who live this side of the equator or beyond it. For to them one pole is always raised above the horizon, and the other is always depressed below it. Or it is because their artificial horizon intersects the equinoctial at oblique and unequal angles.

The four elements.—The machine of the universe is divided into two, the ethereal and the elementary region. The elementary region, existing subject to continual alteration, is divided into four. For there is earth, placed, as it were, as the center in the middle of all, about which is water, about water air, about air fire, which is pure and not turbid there and reaches to the sphere of the moon, as Aristotle says in his book of *Meteorology*. For so God, the glorious and sublime, disposed. And these are called the "four elements" which are in turn by themselves altered, corrupted and regenerated. The elements are also simple bodies which cannot be subdivided into parts of diverse forms and from whose commixture are produced various species of generated things. Three of them, in turn, surround the earth on all sides spherically, except in so far as the dry land stays the sea's tide to protect the life of animate beings. All, too, are mobile except earth, which, as the center of the world, by its weight in every direction equally avoiding the great motion of the extremes, as a round body occupies the middle of the sphere.

The heavens.—Around the elementary region revolves with continuous circular motion the ethereal, which is lucid and immune from all variation in its immutable essence. And it is called "Fifth Essence" by the philosophers. Of which there are nine spheres, as we have just said: namely, of the moon, Mercury, Venus, the sun, Mars, Jupiter, Saturn, the fixed stars, and the last heaven. Each of these spheres incloses its inferior spherically.

Their movements.—And of these there are two movements. One is of the last heaven on the two extremities of its axis, the Arctic and Antarctic poles, from east through west to east again, which the equinoctial circle divides through the middle. Then there is another movement, oblique to this and in the opposite direction, of the inferior spheres on their axes, distant from the former by 23 degrees. But the first movement carries all the others with it in its rush about the earth once within a day and night, although they strive against it, as in the case of the eighth sphere one degree in a hundred years. This second movement is divided through the middle by the zodiac, under which each of the seven planets has its own sphere, in which it is borne by its own motion, contrary to the movement of the sky, and completes it in varying spaces of time—in the case of Saturn in thirty years, Jupiter in twelve years, Mars in two, the sun in three hundred and sixty-five days and six hours, Venus and Mercury about the same, the moon in twenty-seven days and eight hours.

Revolution of the heavens from east to west.—That the sky revolves from east to west is signified by the fact that the stars, which rise in the east, mount gradually and successively until they reach mid-sky and are always at the same distance apart, and, thus maintaining their relative positions, they move toward their setting continuously and uniformly. Another indication is that the stars near the North Pole, which never set for us, move continuously and uniformly, describing their circles about the pole, and are always equally near or far from one another. Wherefore, from those two continuous movements of the stars, both those that set and those which do not, it is clear that the firmament is moved from east to west.

The heavens spherical.—There are three reasons why the sky is round: likeness, convenience, and necessity. Likeness, because the sensible world is made in the likeness of the archetype, in which there is neither end nor beginning; wherefore, in likeness to it the sensible world has a round shape, in which beginning or end cannot be distinguished. Convenience, because of all isoperimetric bodies the sphere is the largest and of all shapes the round is most capacious. Since largest and round, therefore the most capacious. Wherefore, since the world is all-containing, this shape was useful and convenient for it. Necessity, because if the world were of other form than round—say, trilateral, quadrilateral, or many-sided—it would follow that some space would be vacant and some body without a place, both of which are false, as is clear in the case of angles projecting and revolved.

A further proof.—Also, as Alfraganus says, if the sky were flat, one part of it would be nearer to us than another, namely, that which is directly overhead. So when a star was there, it would be closer to us than when rising or setting. But those things which are closer to us seem larger. So the sun when in mid-sky should look larger than when rising or setting, whereas the opposite is the case; for the sun or another star looks bigger in the east or west than in mid-sky. But, since this is not really so, the reason for its seeming so is that in winter and the rainy season vapors rise between us and the sun or other star. And, since those vapors are diaphanous, they scatter our visual rays so that they do not apprehend the object in its true size, just as is the case with a penny dropped into a depth of limpid water, which appears larger than it actually is because of a like diffusion of rays.

The earth a sphere.—That the earth, too, is round is shown thus. The signs and stars do not rise and set the same for all men everywhere but rise and set sooner for those in the east than for those in the west; and of this there is no other cause than the bulge of the earth. Moreover, celestial phenomena evidence that they rise sooner for orientals than for westerners. For one and the same eclipse of the moon which appears to us in the first hour of the night appears to orientals about the third hour of the night, which proves that they had night and sunset before we did, of which setting the bulge of the earth is the cause.

Further proofs of this.—That the earth also has a bulge from north to south and vice versa is shown thus: To those living toward the north, certain stars are always visible, namely, those near the North Pole, while others which are near the South Pole are always concealed from them. If, then, anyone should proceed from the north southward, he might go so far that the stars which formerly were always visible to him now would tend toward their setting. And the farther south he went, the more they would be moved toward their setting. Again, that same man now could see stars which formerly had always been hidden from him. And the reverse would happen to anyone going from the south northward. The cause of this is simply the bulge of the earth. Again, if the earth were flat from east to west, the stars would rise as soon for westerners as for orientals, which is false. Also, if the earth were flat from north to south and vice versa, the stars which were always visible to anyone would continue to be so wherever he went, which is false. But it seems flat to human sight because it is so extensive.

Surface of the sea spherical.—That the water has a bulge and is approximately round is shown thus: Let a signal be set up on the seacoast and a ship leave port and sail away so far that the eye of a person standing at the foot of the mast can no longer discern the signal. Yet if the ship is stopped, the eye of the same person, if he has climbed to the top of the mast, will see the signal clearly. Yet the eye of a person at the bottom of the mast ought to see the signal better than he who is at the top, as is shown by drawing straight lines from both to the signal. And there is no other explanation of this thing than the bulge of the water. For all other impediments are excluded, such as clouds and rising vapors.

Also, since water is a homogeneous body, the whole will act the same as its parts. But parts of water, as happens in the case of little drops and dew on herbs, naturally seek a round shape. Therefore, the whole, of which they are parts, will do so.

The earth central.—That the earth is in the middle of the firmament is shown thus. To persons on the earth's surface the stars appear of the same size whether they are in mid-sky or just rising or about to set, and this is because the earth is equally distant from them. For if the earth were nearer to the firmament in one direction than in another, a person at that point of the earth's surface which was nearer to the firmament would not see half of the heavens. But this is contrary to Ptolemy and all the philosophers, who say that, wherever man lives, six signs rise and six signs set, and half of the heavens is always visible and half hid from him.

And a mere point in the universe.—That same consideration is a sign that the earth is as a center and point with respect to the firmament, since, if the earth were of any size compared with the firmament, it would not be possible to see half the heavens. Also, suppose a plane passed through the center of the earth, dividing it and the firmament into equal halves. An eye at the earth's center would see half the sky, and one on the earth's surface would see the same half. From which it is inferred that the magnitude of the earth from surface to center is inappreciable and, consequently, that the magnitude of the entire earth is inappreciable compared to the firmament. Also Alfraganus says that the least of the fixed stars which we can see is larger than the whole earth. But that star, compared with the firmament, is a mere point. Much more so is the earth, which is smaller than it.

The earth immobile.—That the earth is held immobile in the midst of all, although it is the heaviest, seems explicable thus. Every heavy thing tends toward the center. Now the center is a point in the middle of the firmament. Therefore, the earth, since it is heaviest, naturally tends toward that point. Also, whatever is moved from the middle toward the circumference ascends. Therefore, if the earth were moved from the middle toward the circumference, it would be ascending, which is impossible.

Measuring the earth's circumference.—The total girth of the earth by the authority of the philosophers Ambrose, Theodosius, and Eratosthenes in defined as comprising 252,000 stades, which is allowing 700 stades for each of the 360 parts of the zodiac (sic). For let one take an astrolabe on a clear starry night and, sighting the pole through both apertures in the indicator, note the number of degrees where it is. Then let our measurer of the cosmos proceed directly north until on another clear night, observing the pole as before, the indicator stands a degree higher. After this let the extent of his travel be measured, and it will be found to be 700 stades. Then, allowing this many stades for each of 360 degrees, the girth of the earth is found.

And diameter.—From these data the diameter of the earth can be found thus by the rule for the circle and diameter. Subtract the twenty-second part from the circuit of the whole earth, and a third of the remainder— that is, 80,181 stades and a half and third part of one stade—will be the diameter or thickness of the terrestrial ball.

Of the circles and their names

Celestial circles.—Of these circles some are larger, some smaller, as sense shows. For a great circle in the sphere is one which, described on the surface of the sphere about its center, divides the sphere into two equal parts, while a small circle is one which, described on the surface of the sphere, divides it not into two equal but into two unequal portions.

The equinoctial.—Of the great circles we must first mention the equinoctial. The equinoctial is a circle dividing the sphere into two equal parts and equidistant at its every point from either pole. And it is called "equinoctial" because, when the sun crosses it, which happens twice a year, namely, in the beginning of Aries and in the beginning of Libra, there is equinox the world over. Wherefore it is termed the "equator of day and night," because it makes the artificial day equal to the night. And 'tis called the "belt of the first movement."

The two movements again.—Be it understood that the "first movement" means the movement of the *primum mobile,* that is, of the ninth sphere or last heaven, which movement is from east through west back to east again, which is also called "rational motion" from resemblance to the rational motion in the microcosm, that is, in man, when thought goes from the Creator through creatures to the Creator and there rests.

The second movement is of the firmament and planets contrary to this, from west through east back to west again, which movement is called "irrational" or "sensual" from resemblance to the movement of the microcosm from things corruptible to the Creator and back again to things corruptible.

The north and south poles.—'Tis called the "belt of the first movement" because it divides the *primum mobile* or ninth sphere into two equal parts and is itself equally distant from the poles of the world. It is to be noted that the pole which always is visible to us is called "septentrional," "arctic," or "boreal." "Septentrional" is from *septentrio,* that is, from Ursa Minor, which is derived from *septem* and *trion,* meaning "ox," because the seven stars in Ursa move slowly, since they are near the pole. Or those seven stars are called *septentriones* as if *septem teriones,* because they tread the parts about the pole. "Arctic" is derived from *arthos,* which is Ursa Maior, for 'tis near Ursa Maior. It is called "boreal" because it is where the wind Boreas comes from. The opposite pole is called "Antarctic" as opposed to "Arctic." It also is called "meridional" because it is to the south, and it is called "austral" because it is where the wind Auster comes from. The two fixed points in the firmament are called the "poles of the world" because they terminate the axis of the sphere and the world revolves on them. One of these poles is always visible to us, the other always hidden. Whence Virgil:

> This vertex is ever above us, but that
> Dark Styx and deep Manes hold beneath our feet.

The zodiac.—There is another circle in the sphere which intersects the equinoctial and is intersected by it into two equal parts. One half of it tips toward the north, the other toward the south. That circle is called "zodiac" from *zoe,* meaning "life," because all life in inferior things depends on the movement of the planets beneath it. Or it is derived from *zodias,* which means "animal," because, since it is divided into twelve equal parts, each part is called a sign and has its particular name from the name of some animal, because of some property characteristic of it and of the animal, or because of the arrangement of the fixed stars there in the outline of that kind of animal. That circle in Latin is called *signifer* because it bears the "signs" or because it is divided into them. By Aristotle in *On Generation and Corruption* it is called the "oblique circle," where he says that, according to the access and recess of the sun in the oblique circle, are produced generations and corruptions in things below.

The twelve signs.—The names, order, and number of the signs are set forth in these lines:

> There are Aries, Taurus, Gemini, Cancer, Leo, Virgo,
> Libra and Scorpio, Architenens, Caper, Amphora, Pisces.

Moreover, each sign is divided into 30 degrees, whence it is clear that in the entire zodiac there are 360 degrees. Also, according to astronomers, each degree is divided into 60 minutes, each minute into 60 seconds, each second into 60 thirds, and so on. And as the zodiac is divided by astronomers, so each circle in the sphere, whether great or small, is divided into similar parts.

While every circle in the sphere except the zodiac is understood to be a line or circumference, the zodiac alone is understood to be a surface, 12 degrees wide of degrees such as we have just mentioned. Wherefore, it is clear that certain persons in astrology lie who say that the signs are squares, unless they misuse this term and consider square and quadrangle the same. For each sign is 30 degrees in longitude, 12 in latitude.

The ecliptic.—The line dividing the zodiac in its circuit, so that on one side it leaves 6 degrees and on the other side another 6, is called the "ecliptic," since when sun and moon are on that line there occurs an eclipse of sun or moon. The sun always moves beneath the ecliptic, but all the other planets decline toward north or south; sometimes, however, they are beneath the ecliptic. The part of the zodiac which slants away from the equinoctial to the north is called "northern" or "boreal" or "Arctic," and those six signs which extend from the beginning of Aries to the end of Virgo are called "northern." The other part of the zodiac which tips from the equinoctial toward the south is called "meridional" or "austral," and the six signs from the beginning of Libra to the end of Pisces are called "meridional" or "austral" . . .

The seven climes.—Let a circle be imagined on the earth's surface directly under the equinoctial. And suppose another circle on the earth's surface passing from east to west through the poles. These two circles will intersect in two places at right spherical angles and divide the whole earth into four parts, one of which is our habitable region, namely, that which is intercepted between the semicircle drawn from east to west along the equator and the semicircle carried from east to west through the Arctic pole. Nor is that quarter entirely habitable, since parts of it near the equator are uninhabitable because of too great heat, and parts near the pole because of too great cold. Suppose, then, a line parallel to the equator dividing the parts uninhabitable on account of heat from those habitable parts toward the north. And suppose another line equidistant at all points from the Arctic pole dividing the parts which are uninhabitable for cold from the habitable parts toward the equator. Between these two extreme lines suppose six lines parallel to the equator, which, with the two former, divide the whole habitable quarter into seven parts which are called the "seven climes."

First clime.—The middle of the first clime is where the length of the longest day is 13 hours and the pole is elevated above the horizon 16 degrees, and is called the "clime of Meroe." It begins where the length of the longest day is 12¾ hours and the pole is elevated above the horizon 12¾ degrees. And its breadth extends to the place where the length of the longest day is 13¼ hours and the pole is elevated above the horizon 20½ degrees, which distance is 440 miles.

Second clime.—The middle of the second clime is where the longest day is 13½ hours and the elevation of the pole above the horizon is 24¼ degrees, and it is called the "clime of Syene." Its breadth from the end of the first clime to a place where the longest day is 13¾ hours and pole is elevated 27½ degrees, is a distance of 400 miles.

Third clime.—The middle of the third clime is where the length of the longest day is 14 hours, and the elevation of the pole above the horizon is 30¾ degrees, and it is called the "clime of Alexandria." Its breadth is from the end of the second clime to where the longest day is 14¼ hours, and the altitude of the pole 33-2/3 degrees, which is a distance of 350 miles.

Fourth clime.—The middle of the fourth clime is where the longest day is 14½ hours and the altitude of the axis is 36-2/5 degrees, and it is called the "clime of Rhodes." Its breadth is from the end of the third clime to where the longest day is 14¾ hours and the elevation of the pole is 39 degrees, which distance is 300 miles.

Fifth clime.—The middle of the fifth clime is where the major day is 15 hours and the elevation of the pole is 41-1/3 degrees, and it is called the "clime of Rome." Its breadth is from the end of the fourth clime to where the longest day is 15¼ hours and the elevation of the axis is 43½ degrees, which distance is 255 miles.

Sixth clime.—The middle of the sixth clime is where the longest day is 15½ hours and the pole is elevated above the horizon 45-2/5 degrees, and it is called the "clime of Boristhenes." Its breadth is from the end of the fifth clime to where the length of the longest day is 15¾ hours and the elevation of the axis is 47¼ degrees, which distance is 212 miles.

Seventh clime.—The middle of the seventh clime is where the longest day is 16 hours and the elevation of the pole above the horizon is 48-2/3 degrees, and it is called the "clime of Ripheon." Its breadth is from the end of the sixth clime to where the maximum day is 16¼ hours and the pole is elevated above the horizon 50½ degrees, which space of earth is 185 miles.

Beyond it.—Beyond the end of this seventh clime there may be a number of islands and human habitations, yet whatever there is, since living conditions are bad, is not reckoned as a clime. Therefore, the whole difference between the initial limit of the clime and their end is 3½ hours, and of elevation of the pole above the horizon 38 degrees. So then we have made clear the breadth of each clime from its beginning toward the equator to its end toward the Arctic pole, and that the breadth of the first clime is greater than the latitude of the second, and so on. The length of a clime may be said to be the line drawn from east to west parallel to the equator; wherefore the length of the first clime is greater than the length of the second and so on, which happens because the sphere narrows down.

Enlarging Horizons by Travel

Geographical writings of the early and central medieval period reflect the relatively restricted horizons of the European world, with the exception, of course, of the reports on Norse travels in the Atlantic. In the two centuries following the Mongol invasions of the 1240s, however, European laymen and clerics traversed the width of Asia and, most probably, went as far as equatorial East Africa. Reports on travel—undertaken for missionary, diplomatic, or commercial ends —began to accumulate in government offices, in ecclesiastical centers, and in the counting houses of merchants.

The following selections include, first, reports from the period of intensive diplomatic and missionary activity in Central and East Asia during the second half of the thirteenth century; second, commercial travelers' reports of the fourteenth and fifteenth centuries; and third, a set of excerpts from one of the most popular travel books of the later Middle Ages, the *Travels* of John de Mandeville.

67. Directions to cross the sea

The author of the following excerpt, one William Adam, has been identified by C. R. Beazley as being a Dominican who at one time was a missionary in Persia. The passage is considered to be part of a letter dated July 26, 1330, and sent by the author to Philip VI of France. It is but one of several pieces of evidence pointing to European voyages to and beyond the latitude of Madagascar before 1350.

From "Directorium ad faciendum passagium transmarinum," ed. C. R. Beazley, *American Historical Review* 12 (1907): 821–822. Editor's translation.

I traveled far southward, to a place where I could no longer see our "North Pole" [Pole Star], and I saw the "South Pole" [Southern Cross?] about twenty-four degrees above the horizon. I did not travel beyond that place. However, merchants and men worthy of confidence traveled further south to a place where, they maintained, the "South Pole" [Southern Cross?] was fifty-four degrees above the horizon.

Conclusions

First, that there is more inhabited land than the total land area between the southernmost and northernmost "climata"; second, that Asia is greater than commonly assumed; third, that it is neither frivolous nor false to accept the existence of the Antipodes; and fourth, that we, Christians, are not even the tenth part but only the twentieth part of the population of our world.

68. Marco Polo on Asia and its marvels

Marco Polo (1254–1324), a native of Venice, may well be the most famous traveler in the history of Western exploration. His father and his uncle, having visited the court of Kublai, the Great Khan of the Mongols, returned to Europe to ask the Pope to send missionaries to the Mongol capital. On their second journey the two elder Polos were accompanied by Marco, who found favor in Kublai Khan's eyes and traveled widely through his domain. After an absence of more than twenty years, the three Polos returned to Venice.

During his travels, Marco Polo had traversed the whole of Inner Asia and China, and had gained firsthand knowledge of Southeast Asia and India as well. After his return he participated in a Venetian conflict with Genoa and was taken prisoner; during his captivity he dictated the story of his experiences. For more than two centuries afterwards, his views on Asia influenced mapmakers and geographers alike.

Of the city of Lop and the great Desert

Lop is a large town at the edge of the Desert, which is called the Desert of Lop, and situated between east and north-east. It belongs to the Great

---

From *The Book of Ser Marco Polo,* introduction and notes by G. B. Parks (New York: Macmillan, 1929), pp. 65–67, 135–137, 194, 215–216, 220–223, 303. Introduction Copyright 1927 by Macmillan Publishing Co., Inc. By permission of the publisher.

Khan, and the people worship Mahommet. Now, such persons as propose to cross the Desert take a week's rest in this town to refresh themselves and their cattle; and then they make ready for the journey, taking with them a month's supply for man and beast. On quitting this city they enter the Desert.

The length of this Desert is so great that 'tis said it would take a year and more to ride from one end of it to the other. And here, where its breadth is least, it takes a month to cross it. 'Tis all composed of hills and valleys of sand, and not a thing to eat is to be found on it. But after riding for a day and a night you find fresh water, enough mayhap for some 50 or 100 persons with their beasts, but not for more. And all across the Desert you will find water in like manner, that is to say, in some 28 places altogether you will find good water, but in no great quantity; and in four places also you find brackish water.

Beasts there are none; for there is nought for them to eat. But there is a marvellous thing related of this Desert, which is that when travellers are on the move by night, and one of them chances to lag behind or to fall asleep or the like, when he tries to gain his company again he will hear spirits talking, and will suppose them to be his comrades. Sometimes the spirits will call him by name; and thus shall a traveller ofttimes be led astray so that he never finds his party. And in this way many have perished. [Sometimes the stray travellers will hear as it were the tramp and hum of a great cavalcade of people away from the real line of road, and taking this to be their own company they will follow the sound; and when day breaks they find that a cheat has been put on them and that they are in an ill plight.] Even in the day-time one hears those spirits talking. And sometimes you shall hear the sound of a variety of musical instruments, and still more commonly the sound of drums. [Hence in making this journey 'tis customary for travellers to keep close together. All the animals too have bells at their necks, so that they cannot easily get astray. And at sleeping-time a signal is put up to show the direction of the next march.]

So thus it is that the Desert is crossed.

Concerning the city of Cambaluc, and its great traffic and population

You must know that the city of Cambaluc hath such a multitude of houses, and such a vast population inside the walls and outside, that it seems quite past all possibility. There is a suburb outside each of the gates, which are twelve in number; and these suburbs are so great that they contain more people than the city itself [for the suburb of one gate spreads in width till it meets the suburb of the next, whilst they extend in length some three or four miles]. In those suburbs lodge the foreign merchants and travellers, of whom there are always great numbers who have come to bring presents to the Emperor, or to sell articles at Court, or because the city affords so good a mart to attract traders. [There are in each of the suburbs, to a distance of a mile from the city, numerous fine hostelries for the lodgment of merchants from different parts of the world, and a special hostelry is assigned to each description of people, as if we should say there is one for the Lombards, another for the Germans, and a third for the Frenchmen.] And thus there are as many good houses outside of the city as inside, without counting those that belong to the great lords and barons, which are very numerous.

You must know that it is forbidden to bury any dead body inside the city. If the body be that of an Idolater it is carried out beyond the city and suburbs to a remote place assigned for the purpose, to be burnt. And if it be of one belonging to a religion the custom of which is to bury, such as the Christian, the Saracen, or what not, it is also carried out beyond the suburbs to a distant place assigned for the purpose. And thus the city is preserved in a better and more healthy state.

Moreover, no public woman resides inside the city, but all such abide outside in the suburbs. And 'tis wonderful what a vast number of these there are for the foreigners; it is a certain fact that there are more than 20,000 of them living by prostitution. And that so many can live in this way will show you how vast is the population.

[Guards patrol the city every night in parties of 30 or 40, looking out for any persons who may be abroad at unseasonable hours, *i.e.* after the great bell hath stricken thrice. If they find any such person he is immediately taken to prison, and examined next morning by the proper officers. If these find him guilty of any misdemeanour they order him a proportionate beating with the stick. Under this punishment people sometimes die; but they adopt it in order to eschew bloodshed; for their *Bakshis* say that it is an evil thing to shed man's blood.]

To this city also are brought articles of greater cost and rarity, and in greater abundance of all kinds, than to any other city in the world. For people of every description, and from every region, bring things (including all the costly wares of India, as well as the fine and precious goods of Cathay itself with its provinces), some for the sovereign, some for the court, some for the city which is so great, some for the crowds of Barons and Knights, some for the great hosts of the Emperor which are quartered round about; and thus between court and city the quantity brought in is endless.

As a sample, I tell you, no day in the year passes that there do not enter the city 1000 cart-loads of silk alone, from which are made quantities of cloth of silk and gold, and of other goods. And this is not to be wondered at; for in all the countries round about there is no flax, so that everything has to be made of silk. It is true, indeed, that in some parts of the country there is cotton and hemp, but not sufficient for their wants. This, however, is not of much consequence, because silk is so abundant and cheap, and is a more valuable substance than either flax or cotton.

Round about this great city of Cambaluc there are some 200 other cities at various distances, from which traders come to sell their goods and buy others for their lords; and all find means to make their sales and purchases, so that the traffic of the city is passing great.

Concerning the province of Bengal

Bengal is a province towards the south, which up to the year 1290, when the aforesaid Messer Marco Polo was still at the Court of the Great Khan, had not yet been conquered; but his armies had gone thither to make the conquest. You must know that this province has a peculiar language, and that the people are wretched Idolaters. They are tolerably close to India. There are numbers of eunuchs there, insomuch that all the Barons who keep them get them from that province.

The people have oxen as tall as elephants, but not so big. They live on flesh and milk and rice. They grow cotton, in which they drive a great trade, and also spices such as spikenard, galingale, ginger, sugar, and many other sorts. And the people of India also come thither in search of the eunuchs that I mentioned, and of slaves, male and female, of which there are great numbers, taken from other provinces with which those of the country are at war; and these eunuchs and slaves are sold to the Indian and other merchants who carry them thence for sale about the world.

Concerning the city of Chen-chau and the great river Kiang

You must know that when you leave the city of Yang-chau, after going 15 miles south-east, you come to a city called Chen-chau, of no great size, but possessing a very great amount of shipping and trade. The people are Idolaters and subject to the Great Khan, and use papermoney.

And you must know that this city stands on the greatest river in the world, the name of which is Kiang. It is in some places ten miles wide, in others eight, in others six, and it is more than 100 days' journey in length from one end to the other. This it is that brings so much trade to the city we are speaking of; for on the waters of that river merchandise is perpetually coming and going, from and to the various parts of the world, enriching the city, and bringing a great revenue to the Great Khan.

And I assure you this river flows so far and traverses so many countries and cities that in good sooth there pass and repass on its waters a great number of vessels, and more wealth and merchandise than on all the rivers and all the seas of Christendom put together! It seems indeed more like a Sea than a River. Messer Marco Polo said that he once beheld at that city 15,000 vessels at one time. And you may judge, if this city, of no great size, has such a number, how many must there be altogether, considering that on the banks of this river there are more than sixteen provinces and more than 200 great cities, besides towns and villages, all possessing vessels?

Messer Marco Polo aforesaid tells us that he heard from the officer employed to collect the Great Khan's duties on this river that there passed up-stream 200,000 vessels in the year, without counting those that passed down! [Indeed as it has a course of such great length, and receives so many other navigable rivers, it is no wonder that the merchandise which is borne on it is of vast amount and value. And the article in largest quantity of all is salt, which is carried by this river and its branches to all the cities on their banks, and thence to the other cities in the interior.]

The vessels which ply on this river are decked. They have but one mast, but they are of great burthen, for I can assure you they carry (reckoning by our weight) from 4000 up to 12,000 cantars each.

Now we will quit this matter and I will tell you of another city called Kwa-chau. But first I must mention a point I had forgotten. You must know that the vessels on this river, in going up-stream, have to be tracked, for the current is so strong that they could not make head in any other manner. Now the tow-line, which is some 300 paces in length, is made of nothing but cane. 'Tis in this way: they have those great canes of which I told you before that they are some fifteen paces in length; these they take and split from end to end [into many slender strips], and then they twist these strips together so as to make a rope of any length they please. And the ropes so made are stronger than if they were made of hemp.

Description of the great city of Kinsay, which is the capital of the whole country of Manzi

When you have left the city of Kia-hsing and have travelled for three days through a splendid country, passing a number of towns and villages, you arrive at the most noble city of Kinsay, a name which is as much as to say in our tongue "The City of Heaven," as I told you before.

And since we have got thither I will enter into particulars about its magnificence; and these are well worth the telling, for the city is beyond dispute the finest and the noblest in the world. In this we shall speak according to the written statement which the Queen of this Realm sent to Bayan the conqueror of the country for transmission to the Great Khan, in order that he might be aware of the surpassing grandeur of the city and might be moved to save it from destruction or injury. I will tell you all the truth as is was set down in the document. For truth it was, as the said Messer Marco Polo at a later date was able to witness with his own eyes. And now we shall rehearse those particulars.

First and foremost, then, the document stated the city of Kinsay to be so great that it hath an hundred miles of compass. And there are in it twelve thousand bridges of stone, for the most part so lofty that a great fleet could pass beneath them. And let no man marvel that there are so many bridges, for you see the whole city stands as it were in the water and surrounded by water, so that a great many bridges are required to give free passage about it. [And though the bridges be so high the approaches are so well contrived that carts and horses do cross them.]

The document aforesaid also went on to state that there were in this city twelve guilds of the different crafts, and that each guild had 12,000 houses in the occupation of its workmen. Each of these houses contains at least 12 men, whilst some contain 20 and some 40,—not that these are all masters, but inclusive of the journeymen who work under the masters. And yet all these craftsmen had full occupation, for many other cities of the kingdom are supplied from this city with what they require.

Inside the city there is a Lake which has a compass of some 30 miles: and all round it are erected beautiful palaces and mansions, of the richest and most exquisite structure that you can imagine, belonging to the nobles of the city. There are also on its shores many abbeys and churches of the Idolaters. In the middle of the Lake are two Islands, on each of which stands a rich, beautiful and spacious edifice, furnished in such style as to seem fit for the palace of an Emperor. And when any one of the citizens desired to hold a marriage feast, or to give any other entertainment, it used to be done at one of these palaces. And everything would be found there ready to order, such as silver plate, trenchers, and dishes [napkins and table-cloths], and whatever else was needful. The King made this provision for the gratification of his people, and the place was open to every one who desired to give an entertainment. [Sometimes there would be at these palaces an hundred different parties; some holding a banquet, others celebrating a wedding; and yet all would find good accommodation in the different apartments and pavilions, and that in so well ordered a manner that one party was never in the way of another.]

Since the Great Khan occupied the city he has ordained that each of the 12,000 bridges should be provided with a guard of ten men, in case of any disturbance, or of any being so rash as to plot treason or insurrection against him. [Each guard is provided with a hollow instrument of wood and with a metal basin, and with a time-keeper to enable him to know the hour of the day or night. And so when one hour of the night is past the sentry strikes one on the wooden instrument and on the basin, so that the whole quarter of the city is made aware that one hour of the night is gone. At the second hour he gives two strokes, and so on, keeping always wide awake and on the look-out. In the morning again, from the sunrise, they begin to count anew, and strike one hour as they did in the night, and so on hour after hour.

Part of the watch patrols the quarter, to see if any light or fire is burning after the lawful hours; if they find any they mark the door, and in the morning the owner is summoned before the magistrates, and unless he can plead a good excuse he is punished. Also if they find any one going about the streets at unlawful hours they arrest him, and in the morning they bring him before the magistrates. Likewise if in the daytime they find any poor cripple unable to work for his livelihood, they take him to one of the hospitals, of which there are many, founded by the ancient kings, and endowed with great revenues. Or if he be capable of work they oblige him to take up some trade. If they see that any house has caught fire they immediately beat upon that wooden instrument to give the alarm, and this brings together the watchmen from the other bridges to help to extinguish it, and to save the goods of the merchants or others, either by removing them to the towers above mentioned, or by putting them in boats and transporting them to the islands in the lake. For no citizen dares leave his house at night, or to come near the fire; only those who own the property, and those watchmen who flock to help, of whom there shall come one or two thousand at the least . . .

They burn the bodies of the dead. And when any one dies the friends and relations make a great mourning for the deceased, and clothe themselves in hempen garments, and follow the corpse playing on a variety of instruments and singing hymns to their idols. And when they come to the burning place, they take representations of things cut out of parchment, such as caparisoned horses, male and female slaves, camels, armour, suits of cloth of gold (and money), in great quantities, and these things they put on the fire along with the corpse, so that they are all burnt with it. And they tell you that the dead man shall have all these slaves and animals of which the effigies are burnt, alive in flesh and blood, and the money in gold, at his disposal in the next world; and that the instruments which they have caused to be played at his funeral, and the idol hymns that have been chanted, shall also be produced again to welcome him in the next world; and that the idols themselves will come to do him honour.

Concerning the island of Zanzibar

Zanzibar is a great and noble Island, with a compass of some 2000 miles. The people are all Idolaters, and have a king and a language of their own, and pay tribute to nobody. They are both tall and stout, but not tall in proportion to their stoutness, for if they were, being so stout and brawny, they would be absolutely like giants; and they are so strong that they will carry for four men and eat for five.

They are all black, and go stark naked, with only a little covering for decency. Their hair is as black as pepper, and so frizzly that even with water you can scarcely straighten it. And their mouths are so large, their noses so turned up, their lips so thick, their eyes so big and bloodshot, that they look like very devils; they are in fact so hideously ugly that the world has nothing to show more horrible.

Elephants are produced in this country in wonderful profusion. There are also lions that are black and quite different from ours. And their sheep and wethers are all exactly alike in colour; the body all white and the head black; no other kind of sheep is found there, you may rest assured. They have also many giraffes. This is a beautiful creature, and I must give you a description of it. Its body is short and somewhat sloped to the rear, for its hind legs are short whilst the fore-legs and the neck are both very long, and thus its head stands about three paces from the ground. The head is small, and the animal is not at all mischievous. Its colour is all red and white in round spots, and it is really a beautiful object.

69. John of Plano Carpini: a Franciscan papal ambassador journeys to the Mongol court

John of Plano Carpini (1185?–1252) was one of the first companions of St. Francis of Assisi. He served the Franciscan Order in numerous high administrative posts and in 1245, at the age of sixty, was chosen by Pope Innocent IV to be his ambassador to the "King and Nation of the Tartars." Central Europe was just recovering from the Tartar invasion which had taken place five years earlier; the Pope sent Carpini to the Tartar/Mongol ruler in the hope of converting him to Christianity. Carpini and three fellow friars rode across the whole of Europe and most of Asia, covering vast distances in what appears to be a rather short time. They were received by the Great Khan and returned to the Pope in 1248, carrying the Khan's reply. Carpini's journal of his voyage—that of the first Catholic missionary in Central Asia—contains descriptions of the land and the people of the Mongol Empire at the pinnacle of its power.

After leaving this country (the land of the Bashkir) they came to the *Parossits,* who are tall in stature but thin and frail, with a tiny round belly like a little cup. These people eat nothing at all but live on steam, for instead of a mouth they have a minute orifice, and obtain nourishment by inhaling the steam of meat stewed in a pot through a small opening; and as they have no regard for the flesh they throw it to their dogs. The Tartars took no heed of these people, as they thoroughly despise all monstrous things. Next they came to the people called *Samoyeds,* but took no notice of them either, because they are poverty-stricken men who dwell in forests and sustain life only from hunting. Lastly they came to the people called *Ucor-*

From *The Vinland Map and the Tartar Relation* by R. A. Skelton, Thomas E. Marston, and George D. Painter, pp. 74 and 86. Permission granted by Yale University Press. Copyright © 1965 by Yale University.

colon, that is, Ox-feet, because *ucor* is Tartar for ox and *colon* for foot, or otherwise *Nochoyterim,* that is, Dog-heads, *nochoy* being Tartar for dog and *terim* for head. They have feet like oxen from the ankles down, and a human head from the back of the head to the ears, but with a face in every respect like a dog's; and for that reason they take their name from the part of them which is monstrous in form. They speak two words and bark the third, and so can be called dogs for this reason also. They, too, live in forests and are nimble enough when they run . . .

Having described the wars of the Tartars and their origin, I must now explain the position of their country. Part of it is very mountainous, and part very flat. It is infertile because the soil is sandy, and the climate is very intemperate, perhaps owing to the alternation of mountains and plainland. Violent winds are frequent, and they also have lightning, thunder, and storms out of season. They told our friars that in the last few years their climate had undergone a remarkable change, for frequently clouds seem to contend with clouds near the surface of the ground, and they added that a little before the arrival of the friars fire descended from heaven and consumed many thousand horses and cattle with all but a few of their herdsmen. While the friars were present at the election of the Khan or Emperor hail fell in such quantities that more than a hundred and sixty men were drowned by its sudden melting and their property and huts were swept far away, but did no damage whatever to the friars' dwelling which was nearby. While the friars were watching and suffering the same calamity with the others, a violent wind sprang up and raised such a cloud of dust that no one could ride or even stand.

These Tartars are generally of low stature and rather thin, owing to their diet of mare's milk, which makes a man slim, and their strenuous life. They are broad of face with prominent cheekbones, and have a tonsure on their head like our clerics from which they shave a strip three fingers wide from ear to ear. On the forehead, however, they wear their hair in a crescent-shaped fringe reaching to the eyebrows, but gather up the remaining hair, and arrange and braid it like the Saracens.

As to their clothing, one needs to know that men and women wear the same kind of garments and are therefore not easy to tell apart; and as these matter seem more curious than useful I have not troubled to write further about their clothing and adornment.

Their houses are called stations and are of round shape, made of withies and stakes. At the top they have a round window to let out the smoke and let in the daylight. The roof and the door are of felt. They differ in size and are movable insofar as the size permits them to be carried. The "stations" of the Khan and princes are called hordes. They have no towns but are organized in stations in various places. They have one city called Karakorum, from which our friars were a half a day's journey when they were at the Emperor's Sira Ordu or superior court. Owing to shortage of wood neither nobles nor commoners have any fuel but cattle dung and horse dung.

70. William of Rubruck, ambassador of the King of France, on Mongolia

William of Rubruck, or Rubruquis (c. 1220–1293), a Flemish Franciscan, went to the court of Mangu, the Great Khan of the Mongols, as envoy of Louis IX of France. He spent from 1252 to 1255 traveling from Europe to the Khan's residence in Mongolia; his report complements that of his fellow-Franciscan and contemporary, Carpini.

The Tartars and their dwellings

The Tartars have no abiding city nor do they know of the one that is to come. They have divided among themselves Scythia, which stretches from the Danube as far as the rising of the sun. Each captain, according to whether he has more or fewer men under him, knows the limits of his pasturage and where to feed his flocks in winter, summer, spring and autumn, for in winter they come down to the warmer districts in the south, in summer they go up to the cooler ones in the north. They drive their cattle to graze on the pasture lands without water in winter when there is snow there, for the snow provides them with water.

The dwelling in which they sleep has as its base a circle of interlaced sticks, and it is made of the same material; these sticks converge into a little circle at the top and from this a neck juts up like a chimney; they cover it with white felt and quite often they also coat the felt with lime or white clay and powdered bone to make it a more gleaming white, and sometimes they make it black. The felt round the neck at the top they decorate with lovely and varied paintings. Before the doorway they also hang felt worked in multicoloured designs; they sew coloured felt on to the other, making vines and trees, birds and animals. They make these houses so large that sometimes they are thirty feet across; for I myself once measured the width between the wheel tracks of a cart, and it was twenty feet, and when the house was on the cart it stuck out at least five feet beyond the wheels on

From *The Journey of William of Rubruck,* ed. Christopher Dawson, The Makers of Christendom Series (New York: Sheed & Ward, 1955), pp. 93–94, 139–140, 143–145. Copyright Christina Scott. By permission.

each side. I have counted to one cart twenty-two oxen drawing one house, eleven in a row across the width of the cart, and the other eleven in front of them. The axle of the cart was as big as the mast of a ship, and a man stood at the door of the house on the cart, driving the oxen.

Of their temples and idols and how they comfort themselves in the worship of their gods

All their priests shave their heads all over and their beards, and they wear saffron garments and observe chastity from the time they shave their heads; they live together, one or two hundred in one community. The days on which they go into their temple, they bring two benches and they sit on the floor facing each other, choir to choir, holding their books in their hands and these from time to time they put down on the benches. As long as they are in the temple they keep their heads uncovered, and they read to themselves and keep silence. When I went into one of their temples in Caracorum and found them sitting like this, I tried in many ways to provoke them to speech and was in no way able to do so.

They also have in their hands wherever they go a string of one or two hundred beads, just as we carry our rosaries, and they always say these words, "On man baccam," that is "O God, Thou knowest" so one of them translated it for me and they expect to be rewarded by God as many times as they make mention of Him by saying this.

Around their temple they always make a fair courtyard which is well shut in by a wall and to the south they make a large gate where they sit and talk together. Above this gate they erect a long pole which, if possible, rises above the whole town, and by means of this pole one can tell that a building is a pagan temple. These things are to be found among all idolaters.

When, therefore, I went into the aforementioned temple I found priests sitting in the outer doorway. When I caught sight of them they looked to me like Frenchmen, with their shaven beards and wearing Tartar mitres on their heads. The Uigur priests wear the following dress: wherever they go they are always in saffron coats, quite close-fitting and with a belt on top, just like Frenchmen, and they have a cloak on their left shoulder hanging down in folds over the breast and back to their right side as the deacon wears a chasuble in Lent.

The Tartars have adopted their script. They begin to write at the top and continue the line downwards, and they read it in the same way and they increase their lines from left to right. They make great use of paper and characters for their sorcery, consequently their temples are full of short sentences hanging up. Mangu Chan is sending you a letter in the Mongol language and their script.

They burn their dead after the fashion of bygone days and enclose the ashes in the top of a pyramid.

On Cathay

Next is Grand Cathay, the inhabitants of which in ancient times, so I believe, used to be called Seres. From them came the best silken materials, which are named Seric after the people, and the latter get their name from one of their towns. I learned on good authority that in that country there is a town with walls of silver and ramparts of gold. In that land there are many provinces a number of which are still not subject to the Mongols. Between them and India lies the sea.

The inhabitants of Cathay are little men, and when they speak they breathe heavily through their noses; it is a general characteristic of all Orientals that they have a small opening for the eyes. They are very fine craftsmen in every art, and their physicians know a great deal about the power of herbs and diagnose very cleverly from the pulse; on the other hand they do not use urinals nor know anything about urine; I saw this myself, for there were many of them in Caracorum. It has always been their custom that whatever craft the father follows all the sons have to follow the same craft and that is why they pay such a large tribute, for every day they give the Mongols fifteen hundred *iascot;* a *iascot* is a piece of silver weighing ten marks, so that is fifteen thousand marks every day, not counting the silk materials and food-stuffs which they receive from them, and the other services they render them.

All these peoples dwell between the mountains of the Caucasus on the north side and the eastern ocean and the south of that part of Scythia inhabited by the Mongol nomads. They are all subject to the Tartars and are all given to idolatry; they invent numbers of gods, men who have turned into gods, and the genealogy of the gods, as do our poets.

There are Nestorians and Saracens living among them like foreigners as far as Cathay. There are Nestorians in fifteen cities of Cathay and they have a bishopric there in the city called Segin, but beyond that they are pure pagans. The pagan priests of these said people all wear wide saffron cowls. There are also among them certain hermits, so I learned, in the woods and mountains, and they are of wondrous life and austerity.

The Nestorians there know nothing. They say their offices and have their sacred books in Syriac, a language of which they are ignorant, and so they sing like our monks who know no grammar, and this accounts for the fact that they are completely corrupt. In the first place they are usurers and drunkards, and some of them who are with the Tartars even have several wives like them. When they enter a church they wash their lower members like the Saracens; they eat meat on Fridays and have feasting on that day after the Saracen custom.

The bishop puts off coming into these regions; he comes perhaps scarcely once in fifty years. When he does come, they have all the little male children, even those still in their cradles, ordained priests, consequently almost all their men are priests, and after this they marry, which is clearly contrary to the decrees of the Fathers, and they are bigamists, for when their first wife dies these priests take another. They are also all of them simoniacal, administering no sacrament without payment.

They look after their wives and children well, consequently they pay more attention to gaining money than spreading the faith, whence it comes about that when any of them bring up sons of Mongol noblemen, although they teach them the Gospel and the faith, yet by their evil life and greed they rather alienate them from the Christian religion, for the lives of the Mongols themselves and even of the *tuins,* that is the pagans, are more innocent than theirs.

71. John of Monte Corvino, first archbishop of Peking, on the Nestorian Christians and the Tartar Empire

John of Monte Corvino (1247–1328), a Minorite, was the first Latin Christian missionary in India. He also founded Franciscan missions in China and spent the last twenty years of his life as the first Archbishop of Peking. Among his observations on the lands and peoples of Asia, those on the Nestorian Christians are particularly interesting since they bear witness to the survival of this ancient Christian church in China during the reign of the Mongol dynasty. Little trace was left of the Chinese Nestorians by the time the first Portuguese arrived, two hundred years later.

———

I, Brother John of the Order of Friars Minor, departed from Tauris [Tabriz] a city of Persia, in the year of Our Lord 1291 and entered India and I was in the country of India and in the Church of St. Thomas the Apostle for 13 months. And there I baptized about one hundred persons in different places. And my fellow traveller was Brother Nicholas of Pistoia of the Order of the Friars Preachers, who died there and was buried in the same church. And going on further, I reached Cathay, the kingdom of the Emperor of the Tartars, who is called the Great Chan. Indeed I summoned the Emperor himself to receive the Catholic faith of Our Lord Jesus Christ with the letters of the Lord Pope, but he was too far gone in idolatry. Nevertheless he behaves very generously to the Christians and it is now the twelfth year that I have been with him. However the Nestorians, who call themselves Christians, but behave in a very unchristian manner, have grown

———

From *The Letters of John of Monte Corvino,* trans. Christopher Dawson, The Makers of Christendom Series (New York: Sheed & Ward, 1955), pp. 224–227, 231. Copyright Christina Scott. By permission.

so strong in these parts that they did not allow any Christian of another rite to have any place of worship, however small, nor to preach any doctrine but their own. For these lands have never been reached by any apostle or disciple of the apostles and so the aforesaid Nestorians both directly and by the bribery of others have brought most grievous persecutions upon me, declaring that I was not sent by the Lord Pope, but that I was a spy, a magician and a deceiver of men. And after some time they produced more false witnesses, saying that another messenger had been sent with a great treasure to the Emperor and that I had murdered him in India and made away with his gifts. And this intrigue lasted above five years, so that I was often brought to judgment, and in danger of a shameful death. But at last, by God's ordering, the Emperor came to know my innocence and the nature of my accusers, by the confession of some of them, and he sent them into exile with their wives and children.

Now I have been alone in this journeying without a confessor for eleven years, until there came to me Brother Arnold, a German of the Cologne province, more than a year ago. I have built a church in the city of Cambaliech [Khanbalik or Peking] where the chief residence of the king is, and this I completed six years ago and I also made a tower and put three bells in it. Moreover I have baptized about 6,000 persons there up to the present, according to my reckoning. And if it had not been for the aforesaid slanders I might have baptized 30,000 more, for I am constantly baptizing.

Also I have purchased by degrees forty boys of the sons of the pagans, between seven and eleven years old, who as yet knew no religion. Here I baptized them and taught them Latin and our rite, and I wrote for them about thirty psalters and hymnaries and two breviaries by which eleven boys now know the office. And they keep choir and say office as in a convent whether I am there or not. And several of them write psalters and other suitable things. And the Lord Emperor takes much delight in their singing. And I ring the bells for all the Hours and sing the divine office with a choir of "sucklings and infants." But we sing by rote because we have no books with the notes.

Of good King George

A certain king of these parts, of the sect of the Nestorian Christians, who was of the family of that great king who was called Prester John of India, attached himself to me in the first year that I came here. And was converted by me to the truth of the true Catholic faith. And he took minor orders and served my Mass wearing the sacred vestments, so that the other Nestorians accused him of apostasy. Nevertheless he brought a great part of his people to the true Catholic faith, and he built a fine church with royal generosity in honour of God, the Holy Trinity and the Lord Pope, and called it according to my name "the Roman church". This King

George departed to the Lord a true Christian, leaving a son and heir in the cradle, who is now nine years old. But his brothers who were perverse in the errors of Nestorius perverted all those whom King George had converted and brought them back to their former state of schism. And because I was alone and unable to leave the Emperor the Chaan, I could not visit that church, which is distant twenty days' journey. Nevertheless if a few helpers and fellow workers were to come, I hope in God that all could be restored for I still hold the grant of the late King George.

Again I say that had it not been for the aforesaid slanders great results might have followed. If I had even two or three fellow coadjutors, perhaps the Emperor the Chan might have been baptized. I beg for some brethren to come, if any are willing to do so, so being as they are such as are anxious to offer themselves as an example and not to gain notoriety.

As to the road: I report that the way by the land of Cothay, the emperor of the Northern Tartars, is safer and more secure, so that, travelling with envoys, they might be able to arrive within five or six months. But the other route is the most long and perilous since it involves two sea voyages, the first of which is about the distance of Acre from the province of Provence, but the second is like the distance between Acre and England, and it may happen that the journey is scarcely completed in two years. But the first was not safe for a long time on account of the wars, and for twelve years I have not received news of the Roman Curia, and of our Order and of the state of the West. Two years ago there came a certain physician, a surgeon from Lombardy, who infected these parts with incredible blasphemies about the Roman Curia and our Order and the state of the West, whereupon I greatly desire to know the truth.

I ask the brethren who shall receive this letter that they should do their best to bring its contents to the notice of the Lord Pope and Cardinals and the Procurator of our Order in the Curia. I beg the Minister General of our Order for an antiphonary, and the legends of the saints, a gradual and a noted psalter as an example, for I have nothing but a small breviary with shortened lessons and a small missal.

If only I had an example, the aforesaid boys could make copies from it.

Now I am in the act of building another church, so that the boys can be divided among several places.

I have already grown old, and my hair is white from labours and tribulations rather than years, for I am fifty-eight years old. I have an adequate knowledge of the Tartar language and script, which is the usual language of the Tartars, and now I have translated into that language and script the whole of the New Testament and the Psalter and have had it written in beautiful characters. And I bear witness to the Law of Christ and read and preach openly and in public. And I planned with the aforesaid King George, if he had lived, to translate the whole of the Latin Office so that it might be sung through the whole land under his dominion. And while he lived Mass was celebrated in his church according to the Latin rite in their own script and language, both the words of the Canon and the Preface. And the son of the said king is called John after my own name and I hope in God that he will follow in the footsteps of his father.

Now from what I have seen and heard, I believe that there is no king or prince in the world who can equal the Lord Chan in the extent of his land, and the greatness of the population and wealth.

Given in the city of Cambaliech of the kingdom of Cathay, in the Year of Our Lord 1305, the 8th day of January.

Of the Great Tartar Empire

Concerning the lands of the Orientals and especially the empire of the Lord Chaan, I declare that there is none greater in the world. And I have a place in his court and the right of access to it as legate of the Lord Pope, and he honours me above the other prelates, whatever their titles. And although the Lord Chaan has heard much of the Roman Curia and the state of the Latins, yet he greatly desires to see envoys come from those parts.

In this country there are many sects of idolators who have different beliefs, and there are many kinds of monks with different habits, and they are much more austere and strict in their observance than are the Latin religious.

Concerning India, I have seen a great part myself and have made enquiries about the rest, and it would be of great profit to preach the Christian faith to them if Friars would come. But none save thoroughly reliable men should be sent, for these countries are most beautiful, abounding in spices and precious stones, though they have few of our fruits. And owing to the high temperature and the warmth of the climate of this country they go naked with nothing but a small loin cloth and therefore they have no needs of the arts and products of our tailors and shoemakers. There it is always summer and never winter. I baptized there about a hundred persons.

In the same letter Brother John himself says that a stately embassy came to him from Ethiopia, requesting that he should go there to preach or should send good preachers, for from the time of St. Matthew and his disciples they have had no preachers to instruct them in the Christian faith and they greatly desired to come to the true faith of Christ. And if Friars were sent there, they would convert them all to Christ and they would become true Christians. For there are very many in the East who are Christians only in name and believe in Christ, but know nothing of the Scriptures and the doctrines of the Saints, living in simplicity because they have no preachers and teachers.

Brother John also says that since the Feast of All Saints he has baptized more than four hundred people. And because he has heard that a number of Friars of the two Orders have reached Gazzaria and Persia, he exhorts them to preach the faith of our Lord Jesus Christ fervently and to gain fruit of souls.

This letter was written in the city of Cambaliech [Peking] of the Kingdom of Cathay in the year of Our Lord 1306 in the month of February, on Quinquagesima Sunday.

72. Pegolotti's advice to merchants traveling to Asia

Francesco Balducci Pegolotti journeyed in Asia in the service of the great Florentine banking house of the Bardi, between 1324 and 1335. His Book on the Divisions of Countries and the Practice of Trade and other Things Needed to be Known by Merchants was written shortly after his return to Italy. His reports were written as a vade mecum for traders, describing trade routes, currency, business methods, and prices current at the time.

Information regarding the journey to Cathay, for such as will go by Tana and come back with goods

In the first place, from Tana to Gintarchan may be twenty-five days with an ox-waggon, and from ten to twelve days with a horse-waggon. On the road you will find plenty of *Moccols,* that is to say, of *gens d'armes.* And from Gittarchan to Sara may be a day by river, and from Sara to Saracanco, also by river, eight days. You can do this either by land or by water; but by water you will be at less charge for your merchandize.

From Saracanco to Organci may be twenty days' journey in camel-waggon. It will be well for anyone travelling with merchandize to go to Organci, for in that city there is a ready sale for goods. From Organci to *Oltrarre* is thirty-five to forty days in camel-waggons. But if when you leave Saracanco you go direct to Oltrarre, it is a journey of fifty days only, and if you have no merchandize it will be better to go this way than to go by Organci.

From Oltrarre to *Armalec* is forty-five days' journey with pack-asses, and every day you find Moccols. And from Armalec to *Camexu* is seventy days with asses, and from Camexu until you come to a river called . . . is forty-five days on horseback; and then you can go down the river to Cassai, and there you can dispose of the *sommi* of silver that you have with you, for that is a most active place of business. After getting to Cassai you carry

From "Pegolotti's Notices of the Land Route to Cathay," trans. Henry Yule, in *Cathay and the Way Thither* (London: Hakluyt Society, 1866, no. 37) pp. 146–155 passim. © The Hakluyt Society, London.

on with the money which you get for the *sommi* of silver which you sell there; and this money is made of paper, and is called *balishi*. And four pieces of this money are worth one *sommo* of silver in the province of Cathay. And from Cassai to Gamalec [Cambalec], which is the capital city of the country of Cathay, is thirty days' journey.

Things needful for merchants who desire to make the journey to Cathay above described

In the first place, you must let your beard grow long and not shave. And at Tana you should furnish yourself with a dragoman. And you must not try to save money in the matter of dragomen by taking a bad one instead of a good one. For the additional wages of the good one will not cost you so much as you will save by having him. And besides the dragoman it will be well to take at least two good men servants, who are acquainted with the Cumanian tongue. And if the merchant likes to take a woman with him from Tana, he can do so; if he does not like to take one there is no obligation, only if he does take one he will be kept much more comfortably than if he does not take one. Howbeit, if he do take one, it will be well that she be acquainted with the Cumanian tongue as well as the men.

And from Tana travelling to Gittarchan you should take with you twenty-five days' provisions, that is to say, flour and salt fish, for as to meat you will find enough of it at all the places along the road. And so also at all the chief stations noted in going from one country to another in the route, according to the number of days set down above, you should furnish yourself with flour and salt fish; other things you will find in sufficiency, and especially meat.

The road you travel from Tana to Cathay is perfectly safe, whether by day or by night, according to what the merchants say who have used it. Only if the merchant, in going or coming, should die upon the road, everything belonging to him will become the perquisite of the lord of the country in which he dies, and the officers of the lord will take possession of all. And in like manner if he die in Cathay. But if his brother be with him, or an intimate friend and comrade calling himself his brother, then to such an one they will surrender the property of the deceased, and so it will be rescued . . .

Cathay is a province which contains a multitude of cities and towns. Among others there is one in particular, that is to say the capital city, to which is great resort of merchants, and in which there is a vast amount of trade; and this city is called Cambalec. And the said city hath a circuit of one hundred miles, and is all full of people and houses and of dwellers in the said city.

You may calculate that a merchant with a dragoman, and with two men servants, and with goods to the value of twenty-five thousand golden florins, should spend on his way to Cathay from sixty to eighty *sommi* of silver, and not more if he manage well; and for all the road back again from Cathay to Tana, including the expenses of living and the pay of servants, and all other charges, the cost will be about five *sommi* per head of pack animals, or something less. And you may reckon the *sommo* to be worth five golden florins. You may reckon also that each ox-waggon will require one ox, and will carry ten cantars Genoese weight; and the camel-waggon will require three camels, and will carry thirty cantars Genoese weight; and the horse-waggon will require one horse, and will commonly carry six and half cantars of silk, at 250 Genoese pounds to the cantar. And a bale of silk may be reckoned at between 110 and 115 Genoese pounds.

You may reckon also that from Tana to Sara the road is less safe than on any other part of the journey; and yet even when this part of the road is at its worst, if you are some sixty men in the company you will go as safely as if you were in your own house.

Anyone from Genoa or from Venice, wishing to go to the places above-named, and to make the journey to Cathay, should carry linens with him, and if he visit Organci he will dispose of these well. In Organci he should purchase *sommi* of silver, and with these he should proceed without making any further investment, unless it be some bales of the very finest stuffs which go in small bulk, and cost no more for carriage than coarser stuffs would do.

Merchants who travel this road can ride on horseback or on asses, or mounted in any way that they list to be mounted.

Whatever silver the merchants may carry with them as far as Cathay the lord of Cathay will take from them and put into his treasury. And to merchants who thus bring silver they give that paper money of theirs in exchange. This is of yellow paper, stamped with the seal of the lord afore-said. And this money is called *balishi;* and with this money you can readily buy silk and all other merchandize that you have a desire to buy. And all the people of the country are bound to receive it. And yet you shall not pay a higher price for your goods because your money is of paper. And of the said paper money there are three kinds, one being worth more than another, according to the value which has been established for each by that lord.

And you may reckon that you can buy for one *sommo* of silver nineteen or twenty pounds of Cathay silk, when reduced to Genoese weight, and that the *sommo* should weigh eight and a half ounces of Genoa, and should be of the alloy of eleven ounces and seventeen deniers to the pound.

You may reckon also that in Cathay you should get three or three and a half pieces of damasked silk for a *sommo;* and from three and a half to five pieces of *nacchetti* of silk and gold, likewise for a *sommo* of silver.

73. Nicolò Conti on India in the early 1400s

Nicolò Conti, a Venetian, traveled across the Near East and reached India during the early 1400s. His report describes lands and peoples who were to become much better known half a century later when the first Portuguese ships would land on the shores of the Indian Ocean. Conti's description of India did complement earlier ones, however, such as that by Marco Polo.

Because he was compelled to renounce his Christian faith during his journeys, Conti was ordered to do penance by telling his story to Poggio Bracciolini, the secretary of Pope Eugenius IV. It is from Bracciolini's writings that the following selection is taken.

A certain Venetian named Nicolò, who had penetrated to the interior of India, came to pope Eugenius (he being then for the second time at Florence) for the purpose of craving absolution, inasmuch as, when, on his return from India, he had arrived at the confines of Egypt, on the Red Sea, he was compelled to renounce his faith, not so much from the fear of death to himself, as from the danger which threatened his wife and children who accompanied him. I being very desirous of his conversation (for I had heard of many things related by him which were well worth knowing), questioned him diligently, both in the meetings of learned men and at my own house, upon many matters which seemed very deserving to be committed to memory and also to writing. He discoursed learnedly and gravely concerning his journey to such remote nations, of the situation and different manners and customs of the Indians, also of their animals and trees and spices, and in what place each thing is produced. His accounts bore all the appearance of being true, and not fabrications. He went farther than any former traveller ever penetrated, so far as our records inform us. For he crossed the Ganges and travelled far beyond the island of Taprobana, a point which there is no evidence that any European had previously reached, with the exception of a commander of the fleet of Alexander the Great, and a Roman citizen in the time of Tiberius Claudius Cæsar, both of whom were driven there by tempests.

Nicolò, being a young man, resided as a merchant in the city of Damascus in Syria. Having learnt the Arabic language, he departed thence with his merchandise in company with six hundred other merchants (who formed what is commonly called a caravan), with whom he passed over the

---

From *The Travels of Nicolò Conti in the East in the Early Part of the Fifteenth Century*, trans. from the original of Poggio Bracciolini by J. Winter Jones (London, Hakluyt Society, no. 22, 1858), pp. 3–6, 7–8. © The Hakluyt Society, London.

deserts of Arabia Petræa, and thence through Chaldæa until he arrived at the Euphrates. He says that on reaching the border of these deserts, which are situated in the midst of the province, there happened to them a very marvellous adventure; that about midnight, while they were resting, they heard a great noise, and thinking that it might be Arabs who were coming to rob them, they all got up, through fear of what might be about to happen. And while they stood thus they saw a great multitude of people on horseback, like travellers, pass in silence near their tents without offering them any molestation. Several merchants who had seen the same thing before, asserted that they were demons, who were in the habit of passing in this manner through these deserts.

On the river Euphrates there is a noble city, a part of the ancient city of Babylon, the circumference of which is fourteen miles, and which is called by the inhabitants thereof by the new name of Baldochia. The river Euphrates flows through the centre of the city, the two parts of which are connected together by a single bridge of fourteen arches, with strong towers at both ends. Many monuments and foundations of buildings of the ancient city are still to be seen. In the upper part of the city there is a very strong fortress, and also the royal palace.

Sailing hence for the space of twenty days down the river, in which he saw many noble and cultivated islands, and then travelling for eight days through the country, he arrived at a city called Balsera, and in four days' journey beyond at the Persian Gulf, where the sea rises and falls in the manner of the Atlantic Ocean. Sailing through this gulf for the space of five days he came to the port of Colcus, and afterwards to Ormuz (which is a small island in the said gulf), distant from the mainland twelve miles. Leaving this island and turning towards India for the space of one hundred miles, he arrived at the city of Calacatia, a very noble emporium of the Persians. Here, having remained for some time, he learned the Persian language, of which he afterwards made great use, and also adopted the dress of the country, which he continued to wear during the whole period of his travels. Subsequently he and some Persian merchants freighted a ship, having first taken a solemn oath to be faithful and loyal companions one to another.

Sailing in this wise together, he arrived in the course of a month at the very noble city of Cambay, situated in the second gulf after having passed the mouth of the river Indus. In this country are found those precious stones called sardonixes. It is the custom when husbands die, for one or more of their wives to burn themselves with them, in order to add to the pomp of the funeral. She who was the most dear to the deceased, places herself by his side and her arm round his neck, and burns herself with him; the other wives, when the funeral pile is lighted, cast themselves into the flames. These ceremonies will be described more at length hereafter.

Proceeding onwards he sailed for the space of twenty days, and arrived at two cities situated on the sea shore, one named Pacamuria, and the other Helly. In these districts grows ginger, called in the language of the country *beledi, gebeli,* and *neli.* It is the root of a shrub, which grows to the height of two cubits, with great leaves, similar to those of the blue lilies called *Iris,* with a hard bark. They grow like the roots of reeds, which cover the fruit. From these the ginger is obtained, on which they cast ashes and place it in the sun for three days, in which time it is dried ...

In the middle of the gulf there is a very noble island called Zeilam, which is three thousand miles in circumference, and in which they find, by digging, rubies, saffires, garnets, and those stones which are called cats' eyes. Here also cinnamon grows in great abundance. It is a tree which very much resembles our thick willows, excepting that the branches do not grow upwards, but are spread out horizontally: the leaves are very like those of the laurel, but are somewhat larger. The bark of the branches is the thinnest and best, that of the trunk of the tree is thicker and inferior in flavour. The fruit resembles the berries of the laurel: an odoriferous oil is extracted from it adapted for ointments, which are much used by the Indians. When the bark is stripped off the wood is used for fuel.

74. Mandeville's Travels: notes of an armchair geographer

The *Travels* of John de Mandeville were among the most widely read books on travel during the latter Middle Ages and the Renaissance. The introduction to the Hakluyt edition of the *Travels* notes that there seems to be general agreement that "Mandeville was a man who had lived in Liège, in Belgium; a physician known as 'the man with the beard'; an Englishman who fled from England with a possible charge of homicide hanging over his head; and that in order to conceal his identity he assumed the name of de Bourgogne. These statements are attributed to Jean d'Outremeuse, notary and chronicler of Liège." The date of the work's composition is probably shortly after 1360. Mandeville, if that was his real name, must have been a well-read, well-informed man; his reference to the astrolabe is the earliest recorded use of the word.

Men go to the land of Lombe by sea. In that land grows pepper in a forest, which is called Combar; and it grows in no place of the world but only in that forest. That forest is twenty-four journeys in length. And there

---

From *Mandeville's Travels, Texts, and Translations* by Malcolm Letts (London: Hakluyt Society, 2d ser., no. 101, 1953), pp. 128–129, 120–121, 131–132. © The Hakluyt Society, London.

are two good cities, of the which the tane hight Flabryne and the tother Zinglauns, and in both these cities dwell Christian men and Jews in great number, for the country is right plentifous and good. But it is right hot and therefore there is great abundance of divers nedders and worms. And ye shall understand that pepper grows in manner of wild vines besides the trees of the forest, for to be suppoweld [supported] by them. The fruit thereof hangs in great clusters in manner of bobs of grapes, and they hang so thick that, but if they were suppoweld by other trees they might not bear their fruit. When the fruit is ripe it is all green like the berries of woodbine, and then they gather the fruit and dry it at the sun and syne lay it upon a floor till it become black and wrinkled. And so they have their three manners of pepper growing on a tree, that is to say long pepper ripe of the one kind, white pepper not burnt ne birstled with fire ne with heat of the sun, and black pepper dried with heat of the fire or of the sun. Long pepper call they Spotyn, black Fulphul, and white Bonoile. First comes out the long pepper when the leaves begin to spring; and it is like unto the flower of the hazel, that springs out before the leaves. Then comes out white pepper with the leaves in great clusters, as it were green grapes, the which, when it is gathered, is white, and it is somewhat less than black pepper. Syne springs out black pepper in great abundance. Of the white pepper sell they but little til other countries or else not, but keep it til their own use; for it is better and more profitable and of more attempre [temperate] working than the other, and longer will be kept in his virtue. And ye shall understand that, aye the heavier pepper is, the better it is and the newer. Nevertheless it falls oft-time that merchants sophisticate pepper when it is old, as Isidorus tells. For they take old pepper and steep it, and strew upon it spume of silver or of lead and dry it again, and of other worms because of the great heat of the country and also of none of these three manners of pepper so great abundance as of the black. In that country, as I said before, are many divers manners of nedders and of other worms because of the great heat of the country and also of the pepper. And some men say that a certain time of the year, when they go for to gather this pepper, they make fires here and there, for to burn the nedders or else make them to flee thence. But save their grace, it is not so. For if they thus made fires about the pepper, they should burn the pepper and the trees that it grows on, or else dry them so that they should no more bear fruit; and that is not true. But they anoint their hands and their feet and other places of their bodies with an ointment made of the juice of a fruit that they call lemons mingled with other certain things, and then they go boldly for to gather the pepper. And the nedders and venomous worms, when they feel the reflaire of the ointment, flee away; and on this wise in soothfastness get they the pepper . . .

And ye shall understand that in this land, and in many other thereabout, men may not see the star that is called *Polus Articus*, which stands even north and stirs never, by which shipmen are led, for it is not seen in the south. But there is another star which is called Antarctic, and that is even against the tother star; and by that star are shipmen led there, as shipmen are led here by *Polus Articus*. And right as that star may not be seen here, on the same wise this star may not be seen there. And thereby may men see well that the world is all round; for parts of the firmament which may be seen in some country may not be seen in another. And that may men prove thus. For if a man might find ready shipping and good company and thereto had his health and would go to see the world, he might go all about the world, both above and beneath. And that prove I thus after that I have seen. For I have been in Brabant and seen by the astrolabe that the Pole Arctic is there fifty-three degrees high and in Almayne towards Bohemia it has fifty-eight degrees, and furthermore towards the north it has sixty-two degrees of height and some minutes. All this I perceived by the astrolabe. And ye shall understand that in the south, even anent this star, is the star that is called Pole Antarctic. These two stars stir nevermore; and about them moves the firmament, as a wheel does about an axletree. And so the line that is between these two stars departs all the firmament in two parts, either alike mickle. Afterwards I went toward the south, and I found that in Libya see men first the star Antarctic; and as I went further, I found that in high Libya it has in height eighteen degrees and some minutes, of which minutes sixty make a degree. And so, passing by land and by sea toward the country that I spake of, and other lands and isles that are beyond, I found that this star Antarctic had in height thirty-two degrees. And if I had had company and shipping that would have gone further, I trow forsooth that we should have seen all the roundness of the firmament, that is to say both the hemispheres, the uppermore and the nethermore. For, as I said you before, half the firmament is between these two stars; the which I have seen. And of the tother I saw a part toward the north, that is to say sixty-two degrees and ten minutes under the Pole Arctic; and another part I saw toward the south, that is to say thirty-three degrees and sixteen minutes under the Pole Antarctic. And half the firmament contains but nine score degrees, of which I have seen sixty-two degrees of Arctic and ten minutes, and of Antarctic towards the south I have seen thirty-three degrees and sixteen minutes. These are four score and fifteen degrees and near half a degree. And so there lack but four score and four degrees and more than half a degree, that I ne have seen all the firmament. For the fourth part contains four score and ten degrees. And so the three parts have I seen and five degrees more and near a half. And therefore I say sickerly that a man might go all the world about, both above and beneath, and come again to his own country, so that he had his health, good shipping and good company, as I said before. And alway he should find men, lands, isles and cities and towns, as are in their countries . . .

For the earth is right great and large, and it contains in roundness about, above and beneath, twenty thousand four hundred and twenty-five mile, after the opinion of old wise men that say it, which I will not reprove, but after my feeble wit methinks, save their grace, that it is mickle more about. And for to make you to understand how, I imagine a figure of a great compass; and about the point of that compass, which is called the centre, be another little compass departed by lines in many parts and that all those lines meet sammen [together] on the centre, so that as many parts or lines as the great compass has be on the little compass, if all the space be less. Now be the great compass set for the firmament, the which by astronomers is divided in twelve signs, and ilk a sign is divided in thirty degrees; this is three hundred and sixty degrees, that it is about. Now, be the less compass set for the earth and departed in als many parts as the firmament; these are in all seven hundred and twenty. Now, be these all multiplied three hundred times and sixty, and it shall amount in all til thirty-one thousand mile and five, ilk a mile of eight furlongs, as miles are in our country. And so mickle has the earth in roundness all about, after mine opinion and mine understanding. And ye shall understand, after the opinion of old wise philosophers and astronomers, that England, Scotland, Wales ne Ireland are not reckoned in the height of the earth, as it seems well by all the books of astronomy. For the height of the earth is departed in seven parts, the which are called seven climates after the seven planets that are called climates; and til ilk one of these planets is appropriated one of the climates. And these countries that I spake of are not in those climates, for they are downward toward the west. And also isles of India, which are even against us, are not reckoned in the climates, for they are toward the east. These climates environ all the world. Nevertheless some astronomers appropriate these foresaid countries to the moon, which is the lowest planet, and swiftliest makes his course.

Physical Geography in the Later Middle Ages

The thirteenth century saw the expansion of the geographical horizons of Europe by travel; it was also a time of focusing on studies of the environment. Certain writers of the time, Giraldus Cambrensis among them, concentrated on accurate descriptions of nearby lands, rather than distant ones. Others, such as Albert the Great and Roger Bacon, attempted to articulate the laws that govern the physical universe and to consider the earth as a whole.

75. Giraldus Cambrensis on Ireland and Wales

Giraldus Cambrensis, or Gerald de Barri (c. 1146–1220), a native of Wales, was a cleric as well as a historian. He had a long and checkered ecclesiastical career and provided us with excellent descriptions of Ireland and Wales in his historical writings.

On the prevalence of winds and rain; and their causes

The crops which the spring brings forth, and the summer nourishes and advances, are harvested with difficulty, on account of the autumnal rains. For this country is exposed more than others to storms of wind and deluges of rain. A wind blowing transversely from the northwest, and more frequent and violent than any other winds, prevails here; the blast either bending or uprooting all the trees standing on high ground in the western districts, which are exposed to its sweep. This arises from the land, surrounded on

---

From *The Historical Works of Giraldus Cambrensis*, rev. and ed. Thomas Wright (London, 1863), pp. 20–21, 28, 124, 453–455.

all sides by a vast sea and open to the winds, not having in those parts any solid shelter and protection, either distant or near. Add to this, that the waters attracted in clouds, and collected together by the high temperature of that region, and yet neither exhaled by fiery atmospheric heat, nor congealed by the coldness of the air and converted into snow or hail, at last burst in copious showers of rain. In short, this country, like other mountainous regions, generates and nourishes most abundant rains. For the heat evaporating from the high lands by excessive wet, the moisture which they attract is easily converted into its native element. And it is usually distinguished by various names, according to its various elevations. While yet hanging about the hills, it is called mist; when it rises higher, and, floating in the atmosphere, is quite disengaged from the earth, it becomes clouds; again descending in drops or particles, it is called snow or rain, according as it is solid or liquid. Thus, Ireland, Wales, and Scotland are subject to much rain.

Of the surface of Ireland, and its inequalities; and of the fertility of the soil

Ireland is a country of uneven surface, and mountainous; the soil is friable and moist, well wooded, and marshy; it is truly a desert land, without roads, but well watered. Here you may see standing waters on the tops of the mountains, for pools and lakes are found on the summits of lofty and steep hills. There are, however, in some places very beautiful plains, though of limited extent in comparison with the woods. On almost all sides, and towards the sea-coast, the land is very low, but in the interior it rises into hills of various elevations and mountains of vast height; not only the surrounding country, but also the central districts, being rather sandy than rocky.

The tillage land is exuberantly rich, the fields yielding large crops of corn; and herds of cattle are fed on the mountains. The woods abound with wild animals; but this island is more productive in pasture than in corn, in grass than in grain. The crops give great promise when in the blade, still more in the straw, but less in the ear; for the grains of wheat are shrivelled and small, and can hardly be separated from the chaff by dint of winnowing. The fields are luxuriantly covered, and the barns loaded with the produce. The granaries only show scanty returns.

On the lack of agriculture in Ireland

The Irish are a rude people, subsisting on the produce of their cattle only, and living themselves like beasts—a people that has not yet departed from the primitive habits of pastoral life. In the common course of things, mankind progresses from the forest to the field, from the field to the town, and to the social conditions of citizens; but this nation, holding agricultural labour in contempt, and little coveting the wealth of towns, as well as being exceedingly averse to civil institutions—lead the same life their fathers did in the woods and open pastures, neither willing to abandon their old habits or learn anything new. They, therefore, only make patches of tillage; their pastures are short of herbage; cultivation is very rare, and there is scarcely any land sown. This want of tilled fields arises from the neglect of those who should cultivate them; for there are large tracts which are naturally fertile and productive. The whole habits of the people are contrary to agricultural pursuits, so that the rich glebe is barren for want of husbandmen, the fields demanding labour which is not forthcoming.

Of the mountains of Eryri

I must not pass over in silence the mountains called by the Welsh Eryri, but by the English Snowdon, or Mountains of Snow, which gradually increasing from the land of the sons of Conan, and extending themselves northwards near Deganwy, seem to rear their lofty summits even to the clouds, when viewed from the opposite coast of Anglesey. They are said to be of so great an extent, that according to an ancient proverb, "As Mona could supply corn for all the inhabitants of Wales, so could the Eryri mountains afford sufficient pasture for all the herds, if collected together." Hence these lines of Virgil may be applied to them:—

> Et quantum longis carpent armenta diebus,
> Exigua tantum gelidus ros nocte reponet.
> And what is cropt by day the night renews,
> Shedding refreshful stores of cooling dews.

On the highest parts of these mountains are two lakes worthy of admiration. The one has a floating island in it, which is often driven from one side to the other by the force of the winds; and the shepherds behold with astonishment their cattle, whilst feeding, carried to the distant parts of the lake. A part of the bank naturally bound together by the roots of willows and other shrubs may have been broken off, and increased by the alluvion of the earth from the shore; and being continually agitated by the winds, which in so elevated a situation blow with great violence, it cannot reunite itself firmly with the banks. The other lake is noted for a wonderful and singular miracle. It contains three sorts of fish—eels, trout, and perch, all of which have only one eye, the left being wanting; but if the curious reader should demand of me the explanation of so extraordinary a circumstance, I cannot presume to satisfy him. It is remarkable also, that in two places in Scotland, one near the eastern, the other near the western sea, the fish called mullets possess the same defect, having no left eye. According to vulgar tradition, these mountains are frequented by an eagle who, perching on a fatal stone every fifth holiday, in order to satiate her hunger with the carcases of the slain, is said to expect war on that same day, and to have almost perforated the stone by cleaning and sharpening her beak.

76. Roger Bacon on the shape of the universe and the size of the earth; on the Nile and on China

Roger Bacon (c. 1219–c. 1292), called "Doctor mirabilis," is one of the leading figures in the history of medieval science. He was a controversial figure during much of his life, coming into conflict with his own Franciscan Order. He wrote extensively on astronomy, astrology, and alchemy, and in his *Opus Majus* he concerned himself with the universe and with the shape and dimensions of the earth. One of his suggestions—that the distance between Spain and India is small (a statement he attributes to Aristotle)—was quoted by Cardinal Pierre d'Ailly in his *Imago Mundi;* it may well have been among the several such statements that convinced Columbus of the feasibility of a westward voyage from Spain to India.

The shape of the universe

Since of necessity the bodies of the universe are many and divisible and are quantities, they must have a form required for the existence of the universe. Now form is a property of matter, and is found in things by reason of matter, just like quantity also. For figure in one aspect is quantity inclosed by lines; in another it is called the limitation of quantity. It is necessary, in fact, that the universe on the outside should have a spherical shape. For should any other form be given to it, there would be a vacuum or the possibility for it. But nature does not endure a vacuum nor a possibility in respect to it. For if it were of some angular form, then of necessity there would be a vacuum in its motion; for where at one moment there was an angle, there would be nothing until another angle came to the same place. Other figures especially suitable would be either of oval shape or those like it or of lenticular shape and those formed like it, according to Aristotle in his book on the Heavens and the World. But he states that the heavens do not have a shape of this kind, but he does not give a reason. A lenticular shape is that of the vegetable called lentil. For it has gibbous lateral surfaces, and is not a true sphere, owing to a shorter diameter passing through those sides. If the universe had the shape of an oval, cone, cylinder, or other figure of this kind, and revolved about the shorter diameter, there would actually be

---

From *The Opus Majus of Roger Bacon,* trans. Robert Belle Burke, (Philadelphia: University of Pennsylvania Press, 1928), pp. 173–174, 176–177, 311–312, 325–327, 338–340, 387.

a vacuum; but if it revolved about the longer one, there would be no vacuum actually, but there would be the possibility of a vacuum, since there would be just as great a possibility, as far as the figure is concerned, that it would revolve about the shorter diameter as about the longer. If, however, it were of lenticular shape or of the form of a cheese or of similar shape, and moved about the longer diameter, there would actually be a vacuum, and if about the shorter, there would not actually be a vacuum, but the possibility of it would remain. For it is just as possible as a matter of fact that it should revolve about one diameter as the other. Nor can we say that if it were oval in shape it would revolve always about the longer diameter, and if it were lenticular it would revolve about the shorter, in order that there might be no vacuum. We must refute that statement just as we did above, namely, that it is impossible that a pure negation should be the cause of an affirmation; but the statement, that there may be no vacuum is a pure negation. Therefore the universe must be of spherical form, in which solid alone are all diameters equal, so that it can revolve freely on every diameter, and thus no inconvenience results . . .

Moreover, among all isoperimetric figures the sphere has the greatest content, as the eighth proposition of the book on Isoperimeters states. Plane figures are called isoperimetric, as the triangle, quadrangle, and circle, when the sides of the triangle extended continuously and in a straight line have the same length as the four sides of the quadrangle when so extended, and the same length as the circumference of the circle if it were extended, and the same thing is true of all other plane figures. Whence the world isoperimeter is formed from ἴσον, equal, περί, around, and μετρόν, measure, meaning, as it were, of equal circummensuration. And among all these isoperimetric surfaces the circle is the maximum, as the seventh proposition on isoperimeters states. Solids, moreover, are called isoperimetric; as the sphere, cube, cylinder, and any other, when the surface of the sphere extended continuously and in the same plane contains as much in length and width as the six surfaces of the cube and as much as the surface of the cylinder, and so of the others. But among all those figures the sphere has the maximum content, as is demonstrated in the book mentioned above. Since, therefore, the heavens must contain all things, it was necessary that it should be of spherical form. Likewise the nobleness of the universe and the excellence of this figure correspond. For this figure is the first of the solid figures, because it is contained in a single surface, while all the others have more. Therefore it is adapted to the first body, the heavens. Moreover, this figure is the simplest, since it is without angles, vertex, sides, and free from every difference. Therefore it belongs to the simplest body, namely, the heavens. It is likewise most fitted for motion. Therefore it belongs to the *primum mobile*. It is likewise removed from chances and hindrances, because it has no angle against which something might strike. Therefore it is especially fitted for a body that cannot have hindrance or chance of accident. It is, moreover, the most perfect, because nothing can be added to it; but to all others something can be added. Therefore it belongs to the most perfect body.

The size of the earth

In the Almagest in the second book he [Ptolemy] states that habitation is not known except in a quarter of the earth, namely, in that in which we dwell; whose length is from east to west and is one half of the equinoctial circle; and its width is from the equinoctial circle to the pole, and is one fourth of the colure. But Aristotle maintains at the end of the second book of the Heavens and the World that more than a fourth is inhabited. And Averroës confirms this. Aristotle says that the sea is small between the end of Spain on the west and the beginning of India on the east. Seneca in the fifth book on Natural History says that this sea is navigable in a very few days if the wind is favorable. And Pliny teaches in his Natural History that it was navigated from the Arabic Gulf to Cadiz; whence he states a man fled from his king in fear and entered a gulf of the Red Sea which is called Arabic, which is distant about the space of a year's voyage from the Indian Sea according to Jerome in his letter, as will be explained below. Therefore the width of the earth through which the Red Sea extends is very great; from which fact it is clear that the beginning of India in the east is far distant from us and from Spain, since the distance is so great from the beginning of Arabia toward India. From the end of Spain beneath the earth the sea is so small that it cannot cover three quarters of the earth . . .

I say that although the habitable portion known to Ptolemy and his followers is reduced to less than a quarter, yet the habitable portion is more. And Aristotle was able to know more, because on the authority of Alexander he sent two thousand men to investigate the things of this world, as Pliny states in the eighth book of the Natural History. Alexander himself journeyed as far as the boundary of the east, and as is clear from the History of Alexander and from the letters which he wrote to Aristotle, he always sent him word about all the wonderful and unusual things he found in the East. Therefore Aristotle could attest more than Ptolemy could. And Seneca likewise; because the Emperor Nero, his pupil, in similar fashion sent him to explore the doubtful things in this world, as Seneca states in the Questions on Nature. Therefore according to these facts the extent of the habitable portion is great and what is covered by water must be small. For toward the poles of the world the water must abound, because those parts are cold, owing to their distance, but cold multiplies moisture, and therefore from pole to pole the water runs down into the body of the sea and extends for no great width between the end of Spain and the beginning of India, and is called the ocean; so that the beginning of India can be far beyond half of the equinoctial circle beneath the earth quite close to the boundary of Spain . . .

For certain places in our summer cast a shadow to the south when they have the sun to the north. In our winter they make a shadow to the north since the sun is in the south, according to Pliny in the second book; and alternation of shadow is through six months, as Pliny states in the sixth book, chapter XIX. This cannot be except on the equinoctial circle, since although those who dwell between the tropic of Cancer and the equinoctial circle have a manifold variety in casting shadows, now to the north, now to the south, yet they cannot have this variety through six months, but they will cast the shadow more to the north than to the south, because they have the sun more to the south than to the north. But those who are on the equinoctial circle have the sun equally to the north and to the south, namely, both for six months . . .

We discover through him [Pliny] that there is habitation on the tropic of Capricorn or beyond it. For a region in India is called Pathalis with a very noted port, as he says, where the shadows fall only in the south; therefore dwellers here have the sun always to the north. He states the same thing in the sixth book respecting the island of Taprobane in India, from which men coming to Rome in the reign of Claudius were surprised that their shadows fell to the north and that the sun rose from the south; and therefore among these people shadows fall always to the south and the sun always rises to the north.

But not only the philosophers, but also the sacred writers, like Ambrose in the Hexaemeron and Basil, agree in the matter of this diversity of shadows. For in the fourth book Ambrose says, "There are those who through two whole days of the year are without shadow in southern parts, because having the sun above their heads they are illuminated on all sides; whence they are called *ascii,* that is, without shadow; and *amphiscii,* that is, shaded round about: These people dwell on the equinoctial circle and about it on either side, who when the sun is not above their heads cast sometimes the shadow to the north, sometimes to the south, according to the sun's position now to the north of them and now to the south." He adds, moreover, "There are people located to the south in this world we inhabit who are seen to cast their shadow to the south." Because he speaks expressly of shadows to the south, he can be understood to have reference to those who have shadows falling to the south only, namely, dwellers on the tropic of Capricorn and beyond it; because the sun is always north of them except at one time when it is over their heads on the tropic of Capricorn.

The Nile

The emperor Nero sent two centurions to explore the source of the Nile, and when they came to the first king of the Aethiopians they were instructed and aided by him, in order that the other kings of the Aethiopians might grant them safe conduct. At length they came to shallow marshes filled with vegetation, the extent of which was unknown to the inhabitants; and they despaired of ascertaining it. For they could explore neither in a boat owing to the shallowness of the water, nor could the muddy ground support the weight of a man. The inhabitants therefore believed that the head of the Nile was in this place. Therefore Pliny's statement that the Nile rises at the confines of the west near Mount Atlas not far from the sea is not to be believed.

Now according to Pliny and others a peculiarity of the Nile is the fact that it inundates at certain times and waters the plains of Egypt. According to the overflow of the river is fertility granted or denied to Egypt. For if it passes beyond its natural limits by only twelve cubits, Egypt feels hunger; if for thirteen cubits, Egypt is no longer hungry; fourteen cubits bring joyousness, fifteen security, sixteen delight. If its overflow is still greater yet in moderation, it rouses the inhabitants to an abuse of pleasure. If its overflow is excessive, harm is done, as Seneca states. The Nile begins to increase, authorities states, at any new moon after the solstice sensibly, that is, by degrees and moderately, while the sun is passing through Cancer, and most abundantly while he is in Leo, and the river rises no further while the sun is in Virgo. In the same way in which it rose is it called back within its banks while the sun is in Libra and on the hundredth day from the beginning of its flow. It is difficult to assign the reasons of this inundation and increase, since it is very strange occurring in the heat of summer as it does when waters are consumed more than at other times. But no other river has an inundation of this kind, according to Aristotle in his treatise on the Nile; according to Pliny no river except the Euphrates.

China

Beyond these is Great Cathay, which is called Seres by the philosophers. It is in the extreme east, to the north with respect to India, divided from it by an arm of the sea and by mountains. Here excellent Seric garments are made in great quantities and are exported to other lands. This people breathes much through the nostrils. They are excellent workmen in every art, and there are good physicians among them in all matters except as regards the urine, about which they form no judgment, but make an excellent diagnosis by means of the pulse and by other symptoms and are well acquainted with the virtues of herbs and the potency of the whole medical art. Many of them are among the Tartars. The common money of these people of Cathay is a card of mulberry bark on which are stamped certain lines. Nor is this strange, since the Ruscani who are near us have as money the face of squirrels. This Cathay is not twenty days' distant from the land in which the emperor lives. In that land are lofty cliffs on which certain creatures dwell with a human form in every respect; they do not, however, bend their knees, but walk by leaping. They are not over a cubit in height, have the whole body covered with hair, and do not speak. Hunters bring beer and make holes in the rocks like cups, and those animals come and drink the beer, fall asleep, and so are captured. The hunters bind their hands and feet, and opening a vein in their necks draw three or four drops of blood, and then untie them and let them go. This blood is most valuable as a purple.

77. Albert the Great on the nature of places

Saint Albert the Great (Albert von Bollstädt, c. 1200–1280), called "Doctor universalis," was a member of the Dominican Order and shares with Bacon the position of leadership in thirteenth-century European scientific thought. Albert insisted on the introduction of Greek and Arabic science into the medieval universities and was one of the first empirical investigators. The following selections are from his major work in geography, De Natura Locorum.

Concerning uninhabitable regions and land

Under the pole there is a region which is perpetually cold and continually dark. The stars and the sun never appear there, although it may be day there for half the year and night for half the year. Therefore it happens that the excessive cold of that region continually thickens the air and it turns it into fog (mist).

Explorers sailing on the inhabitable parts of the northern part of the ocean tell this. When they see the darkness, they flee calculating that they cannot sail further. On account of the mistiness of the heavy fog which is there, they are not able to direct themselves. Their sight cannot wander outside the ship to the nearest places; and therefore, if they enter the fog, they don't know how to return to the area of the light.

On account of the cold of those regions there are cold animals in those lands with quite small bodies, as the ass, and animals of such kind. Moreover birds of prey, on account of the heat of their compressed bodies and spirits (souls) are strongest in those habitable regions near the pole. The animals which are there tend to be white, as bears, and lions, and such like. Vegetation there grows a little at a time, for example, oats and barley; but wheat can never grow strong there. If wheat is sometimes planted it breaks down into a grain of another nature.

From Albert the Great, De Natura Locorum, as translated by Sister Jean Paul Tilmann, O. P., in An Appraisal of the Geographical Works of Albertus Magnus and His Contributions to Geographical Thought (Ann Arbor: Michigan Geographical Publication No. 4, 1971), pp. 65–67, 77, 79–80, 81, 86, 87–89, 101–105. By permission of the author.

Concerning the difference of the temperature of climates

The place of exceeding cold is not the seventh clime, but rather an uninhabitable place beyond the northern region. Similarly, the place exceeding in heat is not beneath the equator and a little to this side in the latitude of the first clime, but rather beneath the tropic, which latitude is semicircular. Therefore, it seems that the midpoint between the coldest place which is 90° in latitude, or a little less, and the hottest which is 24°, has a latitude of more than 40°. These are the sixth and seventh climes; therefore those climes seem to have temperatures greater than the fourth clime. Of this place there seems to be evidence that men of those climes are very handsome of body, are of noble and fair stature, and they are beautiful in color, while the men of the fourth clime are small and dark. The place where men grow strongest seems to be most suitable for habitation. For where men are more generally handsome and brave and noble of stature, there man thrives more readily. This is, as it appears, in the sixth and seventh climes. The place which is more agreeable to the habitation of man seems simply to be more temperate, since man is more moderate in his disposition among others which spring from contrary ones, for his habitat approaches more toward the median, and recedes from the highest degrees of contrary ones . . .

There is, however, even now another reason for the difference in nature of the climates. Since the earth is a round figure, as I pointed out in the book *Coeli et mundi,* the stars which project their rays over the earth, do not project over another (place). Northern stars send their rays over the last clime and not over the first. The southern stars send their rays over the first, and not over the last. Since the lights of those stars are of diverse natures, as noted in the book, *Coeli et mundi,* the result is that the climates affected by those rays, are of different natures.

Concerning the nature and arrangement of the lower hemisphere

Now we have to investigate the other half of the earth, that which is found in the lower hemisphere. The philosophers have on this subject very diversified opinions and even contradictory ones. Ordinarily they teach that nobody has passed from our regions into the other hemisphere. That is the reason why all those who have made observations on the different sections of the earth and on the stars have operated or worked on the upper hemisphere. That is especially manifest in their writings on lunar eclipses. A lunar eclipse takes place in fact for the whole world in the same manner and at the same time. Although philosophers who were on different longitudes had noted the moment of the eclipse, we do not ever see any difference between the observations of some and others—we do not see that it exceeds twelve hours. Just as each hour corresponds to a movement of 15° of the sky, it follows that in twelve hours the sky will have accomplished a revolution of 180°. That is exactly half of the celestial sphere to which corresponds a half of the earth. The consequence is therefore that the different observers of lunar eclipses cannot be separated some from the others by more than half of the earth. For example, if for an inhabitant in the east a lunar eclipse takes place at the first hour of night, for the inhabitant of the west, it will be twelve hours difference. From that the philosophers concluded that there are inhabitants only on a half of the earth.

Concerning the difference in nature of places due to the influence of mountains, the seas, and forests

The natures of places differ according to the influence of the seas and ponds and other waters, and because of forests and mountains. For the sea is warm in its own way due to the combustible earth mixed with it, and on account of the great spreading of rays over it. And therefore, it is necessary that the places near it be either hot and dry, or hot and damp with an abundance of moisture, or cold and humid. For southern places located above the seas are sandy, hot, and dry as if they are scorched from the reflections of the rays of the sun from the multiplication of the light near the shores. And the sea distributed over those places is even turned into salt because of the burning rays of the sun, and it makes those places terrestrial (dries up) from the scorching heat and the parched dryness. And if wines are produced in such places, they are very fiery and dry, producing intolerable heat in those drinking, just as the wines produced on the island of Cyprus and in other similar places . . .

The nearness of the mountains gives different properties to the places according to the position and altitude of the mountains. For if the mountains are low, they change but little the nature of places, and if they are very high, then they are frequently snow covered, in which case they are exposed either toward the north closing off the place, or toward the south or the west. And if, indeed, they face the south, then the place will be hot by nature, because the reflection of the ray (which reflection is toward the mountains) makes that place very hot. And, therefore, strong wines are produced in such mountains as these on the southern side, unless the mountains are very high; in which case they will have perpetual snow, from which those places are made either temperate or cold.

If however, the mountains are near the south and facing the north, and the mountains are both high and numerous, the place will be very cold, as much from the snow on the mountains, as from the fact that the north wind blows here, and the place is both closed off and protected from the south wind. And because the north wind disperses clouds, that place will be cold and dry and a healthful place in which to live, unless some other environmental influence prevents habitation.

On the other hand, an elevated place, not mountainous, however, is the best especially for human beings to live, since it is not too cold and not too dry, but in reality it has more cold than heat. Therefore, its air is frequently pure, since the cold represses the gases and the vapors which make the air impure, and does not have impurities and mist except after a snowfall and then not much because of its great purity. This air is heated quickly in the summer, and cools quickly in the winter. For this air is best, which quickly takes on a change of seasons, and in such a place a good number of elements are changed according to the seasons. And therefore there is long life and health in these places, granted perhaps that the wines of such a region are not many nor very strong.

Concerning the diversity of the accidents of those things which are generated according to predetermined diversities

Those things which are born in the hottest places are the hottest, and exceedingly wrinkled from dryness, as a pepper seed (peppercorn), and very black on account of their heat as are the Ethiopians whose first seed of generation is hot, and so is the womb of women hot and dry, and the semen which is conceived is burned by the very strong heat, and their bodies grow dark on account of the scorching of the body; for it gives off a fine moisture and it burns the earthly mass which remains, and generates blackness. The earthly members which are inside them, as bones, become very white as is apparent in their teeth. Their flesh is suffused with blood as if they are glowing coals, as is apparent in their tongues and throats when their mouths are open. And they have prominent mouths, thick lips, reddened eyes, veins and eye lids on account of the heat. And because their bodies are surrounded by very hot air, it is necessary that they be porous and dry, since the moisture evaporates continually from them. For this reason their bodies are light and agile. Because of evaporation their hearts are made timid and cold having few humors. For this reason they do not fear fevers since nature takes care of itself because there is not much matter in them of corrupted moisture. And such as these age very quickly on account of the defect of natural strength which evaporates with the spirit, so their life span is thirty years; and then they are old and feeble. Because they have thin bodies, the moisture evaporates on account of the heat of the place, as we have said, and the earthy, dry, burned material remains, which if it is light and can come forth makes a red bile; and if it is unnaturally scorching, then it makes a black bile (choleric). The bodies of the Ethiopians are characterized more by these two humors, than by the phlegmatic and the melancholic. Indeed warm bodies due to the heat of the place are always relaxed and exceedingly pliable, for which reason, their women easily give birth, although due to dryness and weakness, they do not easily conceive. On account of the refinement of their spirits, they thrive more than the living things who are under the equator, and they excel more in ingenuity on account of the moving heat, and the keenness of their spirits. Of which this is the sign (proof): there have been distinguished philosophers in India especially in mathematics and the magic arts, on account of the power of the stars over those climates over which the planets project perpendicular rays. This, however, is in the first clime under the equator, and not under the second clime which is under Cancer, on account of the intemperance of the heat as we have shown above. The very black Ethiopians are indeed slight in body, frivolous in mind on account of the weakness and evaporation of the spirit. Moreover, the dryness and the heat also are the reasons why their hairs are thin and curly in the manner of grains of mustard. Granted, however, black people of this kind are sometimes born in other climes as in the fourth and fifth, nevertheless, they take their blackness from their first ancestors who are complexioned in the first and second climes, and a little at a time, they are altered to whiteness when they are transferred to other climes.

Conversely, the Goths and the Dacians from the west, and the Slavs from the east, having been born on the boundary of some clime and beyond are white on account of the cold, and because their bodies are not porous, and because the place in which they live is cold, and the cold constricts their bodies, much moisture remains in them. And this extends their bodies and makes them fleshy and phlegmatic. Since the vapor generated in the place of digestion, cannot evaporate because of the constriction of the body and pores, it is reflected back to the stomach. Thus it makes a watery fluid in it, as in boiling pots steam is reflected to the cover and is converted to water, and is distilled on the pot from which steam has been raised. Their bodies have been rendered fat. Therefore their stomachs are hot and their digestion is good, and the bodies solid. For this reason the members of the child bearing women are constricted and they are of firm flesh, and therefore, they give birth with difficulty, and many of them are endangered in child bearing on account of the firmness of their bodies. The cold constricts the flow and their veins, and then presses forth the blood that is in them. Therefore the northern women are rarely cleansed by menstrual flow, and frequently they bleed from the nose. For this reason philosophers have said that these women rarely conceive.

But we see an exception in the German women, who conceive almost more readily than any other yet they give birth with difficulty, and a good number of them rarely menstruate. Undoubtedly it happens because of the cold of the place, and the constriction of their bodies impedes the evaporation of their life spirits and bodily fluids. Because of this their power remains strong, and allows them to conceive when they are not perfect nor cleansed by menstruation. And this is also the reason why their bodies are so very hot. Therefore, they are bold because heat always abounds in the blood and bodily spirit. On account of which (as if nature were conscious of itself) they fear fevers very much, because of the small evaporation of bodily fluid. They do not fear wounds since there is an abundance of blood in them. And they have a thick head of hair and it is straight and not curly. On account of the heaviness of their bodies, they do not engage in spirited activities. Their humor is thick and bodily spirit does not respond to the motion and receptivity of mental activity. Therefore, they are dull-witted and stupid. They have not been exposed to the exercise of study. But when they are moved to study they persevere longer and they are much better by far after mental exercise . . .

The fourth clime and the fifth one which is neighboring it are laudable, and are midway between those excellences, having the laudable middle properties of both regions; this can be easily understood by anyone who knows that the mean is determined by the extremes. Moreover, the life of those there is long and their functions as natural as very laudable activities (living beings), and their customs good, and their pursuits praiseworthy, unless they have been misled from the virtuous life to a wicked life. But the customs of the northern people are wolfish on account of the heat in their hearts. The people of the south are light-hearted. The middle people, however, between these easily cultivate justice, keep their word, embrace peace, and love the society of men. For this reason Vitruvius, the architect, says that the rule of the Romans had endured longer than other reigns; because as the exposed middle has lain between the southerners who contrive with keenness of ingenuity and reason, and the northerners who lay hold of things, with eagerness and without discretion, the northerners on the one hand having audacity, and on the other, the southerners have political acumen; and so they equalled the former and the latter in those things by which they sought to shake the Roman kingdom.

Geographical Writings
of the Age of Discovery

The Age of Discovery is usually seen as the time when Europeans "discovered" the rest of the planet Earth. More accurately, this was a time—during the late fifteenth and early sixteenth centuries—when European ships managed to circumnavigate Africa, thus establishing firmly the open nature of the Indian Ocean and opening a direct sea route to the spice lands of Southern Asia. No less significant, European ships found a New World to the west, later called America, and by sailing around its southernmost extremity, they managed to sail around the whole world as well.

The first selection in this chapter is taken from the *Cosmography* of Martin Waldseemüller, a humanist-scholar of the turn of the sixteenth century. It typifies the traditional conceptual framework of geography, inherited from the Middle Ages and only slightly affected by the discoveries. Following this are descriptions of the Portuguese voyages along the African coast and into the Indian Ocean; a letter from an Italian humanist containing the theoretical basis of Columbus' westward journey; Columbus' own reports; and the passage in the writings of Waldseemüller in which he names the New World.

While the early period of discoveries was dominated by the two Iberian nations, Portugal and Spain, England gradually fought its way into the inner circle of colonial-seafaring powers. This process is illustrated by selections from the Elizabethan Age and by documents concerning the voyages of the greatest of the later English explorers, Captain Cook.

78. Waldsemüller's Cosmography: the state of the art in 1507

Martin Waldseemüller (c. 1470–1518) was an astronomer and map-maker, one of a small circle of scholars residing at the court of the René, Duke of Lorraine, in St. Dié, at the turn of the sixteenth century. He is known for his *Cosmography,* a compendium representative of the state of geography during the early part of the Age of Discovery; for his maps which include a world map and a globe; and for the first updated edition of the *Geography* and maps of Ptolemy.

The reader who consulted this short treatise was secure in knowing that its precepts were firmly founded on tradition and approved by the Church. In fact, were it not for the brief mention of "recently discovered" lands in Africa and of the "fourth part of the earth . . . Amerige," the *cosmography* could well have been written several centuries earlier.

Before any one can obtain a knowledge of cosmography, it is necessary that he should have an understanding of the material sphere. After that he will more easily comprehend the description of the entire world which was first handed down by Ptolemy and others and afterwards enlarged by later scholars, and on which further light has recently been thrown by Amerigo Vespucci.

A sphere, as Theodosius defines it in his book on spheres, is a solid and material figure bounded by a convex surface, in the center of which there is a point, all straight lines drawn from which to the circumference are equal to one another. And while, according to modern writers, there are ten celestial spheres, there is a material sphere like the eighth (which is called the fixed sphere because it carries the fixed stars), composed of circles joined together ideally by a line and axis crossing the center, that is, the earth.

The axis of a sphere is a line passing through the center and touching with its extremities the circumference of the sphere on both sides. About this axis the sphere whirls and turns like the wheel of a wagon about its axle, which is a smoothly rounded pole, the axis being the diameter of the circle itself. Of this Manilius speaks as follows:

> Through the cold air a slender line is drawn,
> Round which the starry world revolves.

From Martin Waldseemüller, *Cosmographiae Introductio,* trans. Joseph Fischer and Franz Von Wieser, ed. Charles George Hebermann (New York: The United States Catholic Historical Society, 1907), pp. 39–41, 60–63. By permission of the publisher.

The poles, which are also called *cardines* (hinges) and *vertices* (tops), are the points of the heavens terminating the axis, so fixed that they never move, but always remain in the same place. What is said here about the axis and the poles is to be referred to the eighth sphere, since for the present we have undertaken the limitation of the material sphere, which, as we have said, resembles the eighth sphere. There are accordingly two principal poles, one the northern, also called *Arcticus* (arctic) and *Borealis* (of Boreas), the other the southern, also called *Antarcticus* (antarctic). Of these Vergil says:

> The one pole is always above us, but the other
> The black Styx and the deep shades see 'neath our feet.

We who live in Europe and Asia see the arctic pole always. It is so called from *Arctus,* or *Arcturus,* the Great Bear, which is also named *Calisto, Helice,* and *Septentrionalis,* from the seven stars of the Wain, which are called *Triones;* there are seven stars also in the Lesser Bear, sometimes called *Cynosura.* Wherefore Baptista Mantuanus says:

> Under thy guidance, Helice, under time, Cynosura,
> We set sail over the deep, etc.

Likewise, the wind coming from that part of the world is called *Borealis* and *Aquilonicus* (northern). Sailors are accustomed to call *Cynosura* the star of the sea.

Opposite to the arctic pole is the antarctic, whence it derives its name, for ἀντί in Greek is the equivalent of *contra* in Latin. This pole is also called *Noticus* and *Austronoticus* (southern). It can not be seen by us on account of the curvature of the earth, which slopes downward, but is visible from the antipodes (the existence of which has been established). It should be remarked in passing that the downward slope of a spherical object means its swelling or belly; that convexity is the contrary of it and denotes concavity.

There are, besides, two other poles of the zodiac itself, describing two circles in the heavens, the arctic and the antarctic. Since we have made mention of the zodiac, the arctic, and the antarctic (which are circles in the heavens), we shall treat of circles in the following chapter.

Although the word *climate* properly means a region, it is here used to mean a part of the earth between two equidistant parallels, in which from the beginning to the end of the climate there is a difference of a half-hour in the longest day. The number of any climate, reckoned from the equator, indicates the number of half-hours by which the longest day in that climate exceeds the day that is equal to the night. There are seven of these climates, although to the south the seventh has not yet been explored. But toward the north Ptolemy discovered a country that was hospitable and habitable, at a distance represented by seven half-hours. These seven climates have obtained their names from some prominent city, river, or mountain.

1. The first climate is called Dia Meroes (of Meroe, modern Shendi), from διά, which in Greek means *through* and governs the genitive case, and Meroe, which is a city of Africa situated in the torrid zone 16° on this side of the equator, in the same parallel in which the Nile is found. Our world map, for the better understanding of which this is written, will clearly show you the beginning, the middle, and the end of this first climate and also of the rest, as well as the hours of the longest day in every one of them.

2. Dia Sienes (of Syene, modern Assuan), from Syene, a city of Egypt, the beginning of the province of Thebais.

3. Dia Alexandrias (of Alexandria), from Alexandria, a famous city of Africa, the chief city of Egypt, founded by Alexander the Great, of whom is has been said by the poet:

> One world is not enough for the youth of Pella.
>
> (Juvenal, x, 168)

4. Dia Rhodon (of Rhodes), from Rhodes, an island on the coast of Asia Minor, on which in our time there is situated a famous city of the same name, which bravely resisted the fierce and warlike attacks of the Turks and gloriously defeated them.

5. Dia Rhomes (of Rome), from a well-known city of Europe, the most illustrious among the cities of Italy and at one time the famous conqueror of all nations and the capital of the world. It is now the abode of the great Father of Fathers.

6. Dia Borysthenes (of Borysthenes, modern Dnieper), from a large river of the Scythians, the fourth from the Danube.

7. Dia Rhipheon (of the Rhiphæan Mountains), from the Rhiphæan mountains, a prominent range in Sarmatian Europe, white with perpetual snow.

From these prominent places, through which approximately the median lines of the climates pass, the seven climates established by Ptolemy derive their names.

The eighth climate Ptolemy did not locate, because that part of the earth, whatever it is, was unknown to him, but was explored by later scholars. It is called Dia Tyles (of Thule, modern Iceland or Shetland), because the beginning of the climate, which is the twenty-first parallel from the equator, passes directly through Thule. Thule is an island in the north, of which our poet Vergil says:

> The farthest Thule will serve.
>
> (Georgics, i, 30)

So much for the climates north of the equator. In like manner we must speak of those which are south of the equator, six of which having corresponding names have been explored and may be called Antidia Meroes

(Anti-climate of Meroe), Antidia Alexandrias, Antidia Rhodon, Antidia Rhomes, Antidia Borysthenes, from the Greek particle αντί, which means *opposite* or *against*. In the sixth climate toward the antarctic there are situated the farthest part of Africa, recently discovered, the islands Zanzibar, the lesser Java, and Seula (Sumatra?), and the fourth part of the earth, which, because Amerigo discovered it, we may call Amerige, the land of Amerigo, so to speak, or America. It is of these southern climates that these words of Pomponius Mela, the geographer, must be understood, when he says:

The habitable zones have the same seasons, but at different times of the year. The Antichthones inhabit the one, and we the other. The situation of the former zone being unknown to us on account of the heat of the intervening zone, I can speak only of the situation of the latter. (Perieg. i, I, 9)

Here it should be remarked that each one of the climates generally bears products different from any other, inasmuch as the climates are different in character and are controlled by different influences of the stars. Wherefore Vergil says:

Nor can all climes all fruits of earth produce.
. . . . .
Here blithelier springs the corn, and here the grape,
Their earth is green with tender growth of trees
And grass unbidden. See how from Tmolus comes
The saffron's fragrance, ivory from Ind,
From Saba's weakling sons their frankincense,
Iron from the naked Chalybs, castor rank
From Pontus, from Epirus the prize-palms
O' the mares of Elis.

(Georgics, i, 54–59)

79. Zurara on the early Portuguese voyages to western Africa

Gomes Eannes de Zurara, or Azurara, completed his *Chronicle of the Discovery of Guinea* shortly before the death of the patron of that great adventure, the Infante Henrique of Portugal, now known as Henry the Navigator (d. 1460). In the half-century preceding that date, ships built in Portugal were gradually extending European knowledge of the northwest coast of Africa. By 1460 these vessels

From Gomes Eannes de Azurara, *The Chronicle of the Discovery of Guinea*, trans. C. R. Beazley and Edgar Prestage (London: Hakluyt Society, no. 95, 1896), pp. 173–178, 229–232.

were well acquainted with the Gulf of Guinea and by 1488 a Portuguese captain, Bartolomeu Diaz, had sailed around the southernmost tip of Africa, found the coast trending northeastward, and reported his discovery to the King of Portugal.

And so those six caravels having set out, pursued their way along the coast, and pressed on so far that they passed the land of Sahara, belonging to those Moors which are called Azanegues, the which land is very easy to distinguish from the other (which they had now come to) by reason of the extensive sands that are there, and after it by the verdure which is not to be seen in it (the Sahara) on account of the great dearth of water there, which causeth an exceeding dryness of the soil. And to this land resort usually all the swallows, and also all the birds that appear at certain times in this our kingdom, to wit, storks, quails, turtle-doves, wry-necks, nightingales and linnets and other birds of various species. And many are there, by reason of the cold of the winter, that go from this land (Portugal) and journey to that one for the sake of its warmth. But other kinds of birds leave it in the winter, such as falcons, herons, ring-doves, thrushes, and other birds that breed in that land, and afterwards they come and take refuge in this because of the food they find here suitable to their nature. And of these birds the men of the caravels found many upon the sea, and others on land at their breeding-places. And since I have begun to speak of this matter, I will not omit to say a little more about the divers other kinds of birds and fishes that I hear are to be found in that land: among which we may speak first of all of some birds called flamingoes, which are of the same size as herons, with necks as long, but with short feathers; also their heads are small in comparison with their bodies, but their beaks are huge, though short, and so heavy that their necks are not well able to support the weight of them, in such wise that for the aid of these same necks they always have their beaks against their legs and rested upon them, or else upon their feathers for the residue of the time. And there also are other birds larger than swans, called hornbills, of which I have already spoken. And as for the fishes of these parts, there are some that have mouths three or four palms long, some smaller and others larger, in which mouths there are teeth both on the one side and on the other, so close together that a finger could not be put between one and another, and all are of fine bone, a little larger than those of a saw and farther apart; and these fish are some as large as and others greater than sharks, and the jaw-bones of these are in size not greater than those of other fish. And there is another kind of fish there, as small as mullet, that have, as it were, crowns on their heads, like gills, through which they breathe; and if they are turned over and put with these crowns below in a basin, they lay hold so firmly that on attempting to withdraw them they lift the basin with them, even as the lampreys do with their mouths while they are quite alive. And there are also many other birds and animals and fish in that land whose appearance we do not care to describe at length, as it would be an occasion of wandering too far from our history.

How those caravels arrived at the river of Nile, and of the Guineas that they took

Now these caravels having passed by the land of Sahara, as hath been said, came in sight of the two palm trees that Dinis Diaz had met with before, by which they understood that they were at the beginning of the land of the Negroes. And at this sight they were glad indeed, and would have landed at once, but they found the sea so rough upon that coast that by no manner of means could they accomplish their purpose. And some of those who were present said afterwards that it was clear from the smell that came off the land how good must be the fruits of that country, for it was so delicious that from the point they reached, though they were on the sea, it seemed to them that they stood in some gracious fruit garden ordained for the sole end of their delight. And if our men showed on their side a great desire of gaining the land, no less did the natives of it show their eagerness to receive them into it; but of the reception they offered I do not care to speak, for according to the signs they made to our men from the first, they did not intend to abandon the beach without very great loss to one side or the other. Now the people of this green land are wholly black, and hence this is called Land of the Negroes, or Land of Guinea. Wherefore also the men and women thereof are called "Guineas," as if one were to say "Black Men." And when the men in the caravels saw the first palms and lofty trees as we have related, they understood right well that they were close to the river of Nile, at the point where it floweth into the western sea, the which river is there called the Senegal. For the Infant had told them that in little more than 20 leagues after the sighting of those trees they should look out for the same river, for so he had learnt from several of his Azanegue prisoners. And so, as they were going along scanning the coast to see if they could discern the river, they perceived before them, as it might be about two leagues of land measure, a certain colour in the water of the sea which was different from the rest, for this was of the colour of mud. And they thought that this might arise from shoals, so they took their soundings for the safety of their ships, but they found no difference in this place from the others in which there was no such movement, and at this they were all amazed, especially by the difference in colour. And it happened that one of those who were throwing in the sounding lead, by chance and without any certain knowledge, put his hand to his mouth and found the water sweet. "Here we have another marvel," cried he to the others, "for this water is sweet;" and at this they threw a bucket forthwith into the sea and put the water to the test, all drinking of it as a thing in which nothing was wanting to make it as good as possible. "Of a surety," said they, "we are near the river of Nile, for it seemeth that this water belongeth to the same, and by its great might the stream doth cut through the sea and so entereth into it." Thereat they made signs to the other caravels, and all of them began to coast in and look for the river, and they were not very long in arriving at the estuary.

How the author beginneth to speak of the manner of that land

It is well that we should here leave these matters at rest for a space and treat of the limits of those lands through the which our people journeyed in the labours of which we have spoken, in order that you may have an understanding of the delusion in which our forefathers ever lived who were affrighted to pass that Cape for fear of those things of which we have told in the beginning of this book; and also that you may see how great praise our Prince deserveth, by bringing their doubts before the presence not only of us who are now living, but also of all others who will be born in the time to come. And because one of the things which they alleged to be a hindrance to the passage into these lands consisted of the very strong currents that were there, on account of which it was impossible for any ship to navigate those seas, you now have a clear knowledge of their former error in that you have seen vessels come and go as free from danger as in any part of the other seas. They further alleged that the lands were all sandy and without any inhabitants, and true it is that in the matter of the sands they were not altogether deceived, but these were not so great as they thought; while as to the inhabitants, you have clearly seen the contrary to be the fact, since you witness the dwellers in those parts each day before your eyes, although their inhabited places are chiefly villages and very few towns. For from the Cape of Bojador to the kingdom of Tunis there will not be in the whole, what with towns and places fortified for defence, as many as fifty. They were no less at fault as regards the depth of the sea, for they had it marked on their charts that the shores were so shallow that at the distance of a league from the land there was only a fathom of water; but this was found not to be so, for the ships have had and have sufficient depth for their management, except for certain shoals; and thus dwellings were made that exist on certain sandbanks, as you will find now in the navigating charts which the Infant caused to be prepared.

In the land of the Negroes there is no walled place save that which they call Oadem, nor are there any settlements except some by the water's edge, of straw houses, the which were emptied of their dwellers by those that went there in the ships of this land. True it is that the whole land is gen- erally peopled, but their mode of living is only in tents and carts, such as we use here when our princes do happen to go upon a warlike march; and those who were captured there gave testimony of this, and also John Fernandez, of whom we have already spoken, related much concerning the same. All their principal study and toil is in guarding their flocks, to wit, cows and sheep and goats and camels, and they change their camp almost every day, for the longest they can rest in one spot will be eight days. And some of their chief men possess tame mares, of which they breed horses, though very few.

Their food consisteth for the great part of milk, and sometimes a little meat and the seeds of wild herbs that they gather in those mountains, and some who have been there have said that these herbs (but of them there are few) seem to be the millet of that land. Also they eat wheat when they can obtain it, in the same way that we in this land eat confetti. And for many months of the year they and their horses and dogs maintain themselves by no other thing except the drinking of milk. And those that live by the sea shore eat nothing save fish, and all for the most part without either bread or anything else, except the water that they drink, and they generally eat their fish raw and dried. Their clothing consisteth of a skin vest and breeches of the same, but some of the more honourable wear bournouses; and some pre-eminent men, who are almost above all the others, have good garments, like the other Moors, and good horses and good saddles, and good stirrups, but these are very few.

The women wear bournouses which are like mantles, with the which they only cover their faces, and by that they think they have covered all their shame, for they leave their bodies quite naked. "For sure," saith he who compiled this history, "this is one of the things by the which one may discern their great bestiality, for if they had some particle of reason they would follow nature, and cover those parts only which by its shewing ought to be covered, for we see how naturally in each one of these shameful parts it placeth a circle of hair in proof that it wished to hide them; and also some naturalists hold that if those hairs be let alone, they will grow so much as to hide all the parts of your shame." And the wives of the most honourable men wear rings of gold in their nostrils and ears, as well as other jewels.

80. Camoens' poetic description of da Gama's voyage to India

The discovery of the Cape of Good Hope by Bartolomeu Diaz in 1488 was followed by other Portuguese explorations of the east coast of Africa until, in 1498, a Portuguese squadron commanded by Vasco

From Luis Vaz De Camoens, *The Lusiads*, trans. William C. Atkinson (Penguin Classics, 1953), pp. 46–47, 124–125, 164–166, 237–238, 242–245. Copyright © William C. Atkinson, 1953. Reprinted by permission of Penguin Books, Ltd.

da Gama sailed around Africa to present-day Somalia, crossed the Indian Ocean, and landed on the Malabar (west) coast of India. That great voyage, which laid the foundations for Portugal's African and Asian empires, was the subject of *Os Lusiadas,* the epic poem by Luis Vaz de Camoens published in 1572. The narrative of *The Lusiads* follows the history of da Gama's expedition closely. Based on contemporary reports, it narrates the principal episodes of the journey, including the Portuguese landing in India, and describes the lands surrounding the Indian Ocean.

The first and third parts of this selection are in simple narrative form. The second and fourth parts are spoken by Calliope, the Muse of epic poetry.

As they left Cape Corrientes behind more islands came into view. To Vasco da Gama, their stalwart leader—and a right proper man he was for such an enterprise, proud and indomitable of heart, always smiled on by fortune—there was no cause here to break their journey. The islands appeared to be deserted and he decided to continue on his course. But the matter turned out otherwise.

For straightway, from the one that seemingly lay nearest to the mainland, there put out several small craft with long sails, which made swiftly towards them. The Portuguese were greatly excited, and with keen expectation began asking what this could portend. "Of what race can these people be?" they said to one another. "What will their customs be like? And their religion? And whose subjects are they?"

The canoes were long, narrow, and very fast, with sails made of palm-leaves woven like mats. Those who manned them were of burnt complexion and wore cotton loin- and shouldercloths, some white, some striped in various colours, and light caps. The shouldercloths they now tucked jauntily under their arms. For the rest they were naked. As weapons they had daggers and short swords; and as they approached they blew on long trumpets.

Waving their arms and cloths, they made signs to the vessels to await them. The Portuguese were already turning their prows to run in under the islands, every man lending a hand as if the end of their labours were in sight. Soon the sails were furled, the upper yards struck, and the anchors were thrown overboard with a loud splash.

They were scarcely at anchor before the islanders were clambering up the ropes. Their expression was cheerful, and the Captain gave them a friendly reception, ordering tables to be spread and glasses filled. They drank the wine without hesitation, and, eating merrily, enquired in Arabic who the strangers were, from what land and over what seas they had come, and what they sought.

The others replied: "We are Portuguese from the West: we seek the lands of the East. We have sailed the seas that reach from the North to the South Pole, skirting the length of Africa and nodding acquaintance with many a land and sky. We are subjects of a powerful king, so esteemed and beloved by everyone that in his service we would gladly brave not merely the broad ocean but the very waters of the under-world. It is at his command we have sailed into the unknown, with only the ungainly seals for company, and seek those eastern lands where the Indus flows. And now," they ended, "it is only fair that you tell us, truthfully, of what race you are and what land this is you dwell in. Tell us too whether you know anything of India."

One of the islanders spoke for the rest. "We are foreigners here, too, and have nothing in common with the natives. They are pagans, and uncivilized. We are Moslems of the true faith, like the rest of the world. This small island where we live is called Mozambique, and is a port of call for all shipping up and down the coast, whether from Kilwa, Mombasa or Sofala. That is why we have settled here. And seeing that you have come so far, and are bound for India, you will have need of a pilot. That we can supply, and refreshment. But our Governor will wish to see you. He will look to any further needs you may have."

With this the Moslem and his fellows took polite leave of da Gama and his men, and returned to their canoes. The sun was setting.

Calliope speaks

On past all these we sailed, heading ever to the south: past the forbidding Sierra Leone, past the promontory we call Cape Palmas, on into the mighty Gulf of Guinea. The River Niger too, at whose mouth the sea pounds on shores well known to fame, and where there were already Portuguese settlements, we soon left behind, and with it the famous island of São Thomé, recalling the apostle who put his hand in Jesus's side.

Then came the mighty kingdom of the Congo, that my countrymen had even then converted to Christianity, with the great Congo River, unknown to antiquity, flowing through it.

And here at length we crossed the equator, and said good-bye to the familiar constellations of the northern world. In this new southern hemisphere we had already discovered one constellation, the Southern Cross, that before us had never been seen by any. The heavens here sparkle less brilliantly and, having fewer stars, impress less with their beauty. And still we could not tell whether the ocean stretched on for ever or would give way eventually to some other continent. Twice in the year the sun crosses over these regions, giving two summers and two winters in its passage from one to the other extreme. And as we sailed on, now becalmed, now tempest-tossed and sore oppressed when the angry winds belaboured the waters, we saw the Great and the Little Bear sink into the waves in spite of Juno.

But even if I had a voice of iron, it would be both wearisome and irrelevant to recount to you at length all the perils that befall and baffle men at sea, the sudden terrifying peals of thunder that threaten to bring the world about one's ears, the lightning-flashes setting the heavens ablaze, the black rainstorms, the dark nights. Men of learning, able through native perception or scientific training to penetrate the hidden secrets of the universe, often reject as mere illusion or misapprehension things that rude mariners, who have known no other school than experience and can only judge by appearances, accept as true. Well, I have seen such things.

One—I had a clear view of it—was our blessed St Elmo's Fire, that is sometimes to be seen in time of storm and raging winds, when the heavens lower and even strong men give way to tears. And it was no less demonstrably a miracle to all of us, a thing to strike terror to our hearts, to see the clouds drinking up, as through a long spout, the waters of the ocean.

First a thin smoky vapour formed in the air and began to swirl in the breeze; then out of it there took shape a kind of tube stretching right up to the sky, but so slender that one had to strain one's eyes to see it—made, as it were, of the very stuff of clouds. Gradually it grew and swelled until it was thicker than a masthead, bulging here, narrowing there as it sucked up the water in mighty gulps, and swaying with the ocean swell. At its summit a thick cloud formed, that bellied heavier and heavier with the mass of water it absorbed.

Sometimes a beast, drinking rashly from an inviting spring, will pick up a leech that fastens on its lip and there sates its thirst with the animal's blood: it sucks and sucks, and swells and swells. In the same way this mighty column waxed ever mightier, and with it the black cloud that crowned it; until at length, sated too, it drew up its lower extremity from the sea and drifted off across the sky, spattering the ocean with its own water returned as rain, restoring to the waves what it had stolen, minus its salty savour.

The landing in India

As they drew in to the strange shore they came upon small fishing-smacks hailing from Calicut. These told them how to get there, and to Calicut accordingly they set their course; for among all the fine cities of Malabar this was the finest, as well as being the seat of the king who ruled over the whole land.

Beyond the Indus, lying between it and the Ganges, there extends a vast territory, not unknown to fame, whose southern boundaries reach to the sea and whose northern to the cavernous Himalayas. Divers kings rule over it, and there are in consequence divers religions: some worship the infidel Mahomet, some bow down to idols, others to native animals. And in that mighty Himalayan range that divides a whole continent—for it stretches right across Asia, taking different names in different regions—are to be found the sources of these rivers which empty their huge volume into the Indian Ocean, embracing the entire land, peninsula-wise, in their course.

To the south the country is shaped in the form approximately of a pyramid, ending in a point that reaches the sea over against the island of Ceylon.

Not far from the source of the Ganges, according to ancient report, there dwelt a people who lived by inhaling the perfume of flowers. To-day the names and customs of the inhabitants have changed and multiplied. There are the people of Delhi and the Pathans—these two are the most numerous and hold the greatest territories—those of the Deccan and of Orissa, who pin their hope of salvation on the murmuring waters of the Ganges, and the Bengalis, with the fertility of whose soil no other can compare. There is the warlike kingdom of Cambay—it was here, they say, that the great king Porus ruled—and the kingdom of Narsinga, more notable for its gold and precious stones than for the valour of its people.

Here, over towards the sea, there is a lofty mountain range stretching for a considerable distance that serves as bastion to Malabar against attack from Canara. The natives speak of it as the Ghats. At its foot there runs for some little way a narrow coastal strip buffeted by the ocean breakers, and it is on this that Calicut stands, a rich and beautiful city and unchallenged capital of the empire. Its lord is known as the Samorin.

As soon as the fleet had reached this seat of wealth and power, one of the Portuguese was sent ashore to inform the pagan ruler of their arrival from such distant parts. The messenger made his way up the river-mouth, and the strangeness of his aspect—for everything about him, colour, features, clothes, was new—straightway drew the whole population to see the sight.

Among the crowd was a Moslem born in Barbary, where the giant Antaeus once held sway, and there, before fate bore him to such distant exile, he had either had occasion to visit the neighbouring kingdom of Portugal or had made the acquaintance of its people in battle. At sight of the messenger his face brightened, and, speaking in Spanish, he hailed him with a "And who brought you to this other world, so far from your native Portugal?"

"We have come across the mighty deep," the other replied, "where none has ever sailed before us, in search of the Indus. Our purpose is to spread the Christian faith." The Moor, Monsaide by name, listened as the other told him of the long voyage and the many perils that had attended it, and was filled with amazement.

Calliope speaks

But enough of the Strait of Ormuz and the well-known Ras Jaskah, once called Carpela, whose entire territory—the Carmania of old—is so ill-favoured that Nature has bestowed on it none of her customary gifts.

See now the graceful Indus, with its source in yonder height, and the Ganges flowing from another height near by. Look at this extremely fertile land of Sind, and the deep indentation made by the Gulf of Cutch with the sudden inrush of its flood-tide and the impetuous haste with which it ebbs. Cambay, where the sea has again eaten deeply into the coast, is likewise a land fruitful in the extreme. And there are a thousand cities besides that I pass over now, but which are reserved for you Portuguese.

This celebrated coast of India, as you see, continues to run southward till it ends in Cape Comorin, once Cape Cori, facing Taprobana or Ceylon. Everywhere along these shores Portuguese soldiers still to come will win victories, lands, and cities, and here for long ages they will make their abode. The territories that lie between the two great rivers are beyond compute, and are apportioned among various nations, some Moslem, others pagan, their beliefs deriving from the devil.

And here are the three Arabias, their wide expanse entirely peopled by swarthy nomads. This is the home of the thoroughbred so highly esteemed by the warrior for its fire and pace. Note the line of the coast, how it determines here the Strait of Ormuz and projects into Cape Fartak, so called from the well-known city of that name. This is Dofar, renowned as the source of the most aromatic of all altar-incense.

But look: it is here, just on the other side of Ras el-Hadd with its sterile strands, that the kingdom of Ormuz begins. Its fame will date from the day when the fleet of dread Turkish galleys encounters off its shores the ire of Pedro de Castelo Branco. This next cape is Asaboro, that mariners now call Musandum, at the very entrance to the Persian Gulf, with Arabia on one side and the fertile lands of Persia on the other. And this is Bahrein Island, where the ocean-bed is adorned with rich pearls, their hue the colour of dawn. The Tigris and the Euphrates, joining, here reach the sea together.

This Persia is a great and noble empire, its people always waging war, for ever in the saddle. They hold it beneath their dignity to fight with cannon, preferring that their hands should be calloused with the wielding of nobler weapons. Observe in this island of Gerum what the passage of time can do. On it once stood the city of Armuza; that vanished, and now a second city, Ormuz, has inherited its name and fame.

Felipe de Meneses will give proof here of his outstanding bravery in the field when, with a mere handful of Portuguese, he overthrows a large Persian force from Lar. And there are other blows and reverses to come, at the hand of Pedro de Sousa, whose prowess will have been seen earlier in the razing of Ampaza by the sword alone . . .

Let us return to our globe with its picture of the coast-line. Here, by this same illustrious city of Meliapor, it begins to curve into the Bay of Bengal. We pass out of the great and wealthy kingdom of Narsinga now, and into Orissa with its busy looms. Where the curve goes deepest the famous Ganges enters the sea. The natives here all bathe in its waters before they die, being persuaded that so their sins, however great, will be washed away and they will be purified. This is Chittagong, one of the finest cities in all Bengal, a land that boasts of its abundance.

Notice now how the coast swings round and heads once more to the south. This is the kingdom of Arakan, and this Pegu, once peopled with the monstrous breed of its solitary first inhabitants, a woman and a dog. The men here wear small tinkling bells on their genitals, a subtly effective measure introduced by one of their queens against the unspeakable sin.

The city of Tavoy marks the beginning of the great empire of Siam. Tenasserim comes next, and then Kedah, the chief centre for the production of pepper in these parts, though far from the only one. Farther on lies Malacca, that your countrymen will make known as a great emporium for the wealth and merchandise of all the territories bordering on this vast ocean.

This noble island of Sumatra was said in ancient times to be joined to the mainland, until ocean breakers eating into the coast drove a wedge between. The Chersonese, the region was called, to which was added the adjective "Golden" from its valuable seams of the metal. Some have imagined it to be the site of Solomon's Ophir. And here, on the very tip of this peninsula, lies Singapore, where vessels find their passage narrowed to a strait.

The coast now curves north again, and then eastwards, and you can see the kingdoms of Pahang and Patani and the long extent of Siam, to which they and others are subject. This is the River Menam, that rises in the great Lake Chiamai.

And here now is a vast territory dotted, as you can see, with the names of a thousand nations you have never even heard of: the Laos, powerful in extent and in numbers, the Avas and Bramas, who have their homes in great mountain-ranges, the Karens, savage tribes inhabiting the remoter hills who eat the flesh of their enemies and cruelly tattoo their own with red-hot irons.

Through Cambodia, you will observe, flows the Mekong River, its name meaning "prince of waters." Like the chilly Nile, it floods even in summer from the volume of its tributaries, turning great stretches of country into lakes and causing much anxiety. The inhabitants of these parts believe in their ignorance that for the entire animal creation there exist a heaven and a hell after death . . .

This part of the coast is known as Tsian-Pa; in its forests grows the sweet-smelling aloe. Here you may see Cochin-China, still but little known, and Hainan, in the as yet undiscovered Gulf of Tongking.

And now you are looking at the mighty empire of China, that boasts greater territories and riches than it knows of, its sway reaching from the tropics to the frigid zone. Observe this Great Wall, an incredible thing to build as frontier between one empire and another, proof incontestable and accepted by all of its sovereign power, its pride and wealth. The emperor of the Chinese is not born a prince, nor does the office descend from father to son: he is elected by the people for his outstanding wisdom, virtue, and nobility.

There is much more of the earth's surface that must still remain hidden from you: in its proper time it shall all be revealed. But do not forget these islands where Nature has chosen to show her greatest wonders. This, that you can only half distinguish where it lies far out to sea, facing China—it is from China that it is to be sought—is Japan, famous for its fine silver, and soon to be famous too through the spreading of Christianity among its people.

And there are other countless islands, as you see, scattered over the seas of the East. Here is Tidor, and here Ternate, with its volcano spurting columns of fire. You can recognize the clove-trees, that Portugal must still pay for with her blood. This is the abode of the bird of paradise, which never alights on the ground and is only to be found dead. These others are the Banda Islands, gay with the many-hued flower of the nutmeg, from which the numerous species of birds exact their tribute. And here is Borneo, where the trees shed tear-drops of a thick, resinous substance known as camphor, that has made the name of the island famous.

Yonder lies Timor with its aromatic and health-giving sandalwood. And yonder Java, so large that the southern, more mountainous part has not yet been explored. The natives of the interior say there is a river there with miraculous qualities: provided its waters be not mingled with those of another, it will petrify any piece of wood that falls into it.

And now we are back to Sumatra, that has only become an island with time. Here too volcanoes erupt vapour and quivering flames, and you can see a spring that runs oil, and the strange spectacle of a tree weeping tears of fragrant benzoin, more fragrant than all the myrrh of Arabia. And note that this island, that lacks nothing to be found in others, produces also delicate silk and fine gold. In Ceylon observe this mountain, Adam's Peak, so high it overtops the clouds, or so it seems. The natives hold it sacred by reason of a rock on it that bears the imprint of a human foot.

It is in the Maldive Islands that the stately coco-palm is found growing under the water, its fruit being esteemed an excellent antidote to poison. And here, opposite the strait leading to the Red Sea, is Socotra, famous for its bitter aloes. There are still other islands in this sea, likewise subject to your country, that washes the sandy shores of Africa, where the valuable and little-known ambergris is found, its perfume the rarest of all. And see here the famous island of São Lourenço, that some call Madagascar.

Such are the new regions of the East that you Portuguese are now adding to the known world by throwing open the gates of the mighty ocean over which you sail with such fortitude. But it is fitting that you should glance too at one achievement in the West of another of your race who, offended with his monarch, will blaze [under an alien flag] such a trail as none had ever thought of.

81. Toscanelli on sailing westward to the Indies

Paolo del Pozzo Toscanelli (1397–1482), a Florentine physician, astronomer, and humanist, was considered to be one of the leading astronomers of his day. An acquaintance of his, a cleric of Lisbon, asked him in 1474 to set forth his views on the west-east extent of the known world; the inquiry was initiated by Alfonso V, the King of Portugal, who was then engaged in exploring a seaway to the Indies by way of the coast of Africa.

Toscanelli's response took the form of a letter and a map; the map has been lost, but the letter is reproduced here. Toscanelli bases his ideas on the Ptolemaic concept of the west-east dimension of the known world, corrects it according to the views of Marinus of Tyre, and then predicts that sailing westward from Portugal the shores of the Far East can be reached in a short time. Shortly before his death in 1481, Toscanelli sent a copy of this letter to Columbus, who used it to bolster his proposal to open a westward seaway to the Indies.

The letter is a curious mixture of views based on classical geography (especially on the writings of Ptolemy, then considered the greatest of the geographers of antiquity), of descriptions taken from the travel reports of Marco Polo, and of references to a conversation with an Asian that may have taken place during the Council of Ferrara-Florence (1438–1445).

From Henry Vignaud, *The Letter and Chart of Toscanelli* (New York: E. P. Dutton & Co., 1902), pp. 275–293.

Copy sent to Christopher Columbus by Paul the Physician, together with a chart of navigation.

To Ferdinand Martins, Canon of Lisbon, Paul the physician gives greeting.

It was pleasing to me to have intelligence concerning your health, and concerning your favour and familiar friendship with that most generous and magnificent prince, your King. Whereas I have spoken with you elsewhere concerning a shorter way of going by sea to the lands of spices, than that which you are making by Guinea; the most serene King now wishes that I should give some explanation thereof, or rather that I should so set it before the eyes of all, that even those who are but moderately learned might perceive that way and understand it.

But though I know that this could be shown by the spherical form, which is that of the world; nevertheless I have determined to show it in the way in which charts of navigation show it, and this both that it may be more readily understood, and that the work may be easier.

Wherefore I send to His Majesty a chart, made by my hands, wherein your shores are shown, and the islands from which you may begin to make a voyage continually westwards, and the places whereunto you ought to come, and how much you ought to decline from the pole or from the equinoctial line, and through how much space, *i.e.*, through how many miles you ought to arrive at the places most fertile in all spices and gems. And do not wonder if I call those places where the spices are western, whereas they are commonly called eastern: because to those that sail by subterranean navigation those places are ever found in the west. For if (we go) by land and by the upper ways they will always be found in the east.

The straight lines, therefore, marked lengthwise in the chart, show the distances from east to west, but those which are transverse show the space from south to north.

But I have marked in the chart divers places where you might arrive, and this indeed for the better information of navigators if they should come by the winds or by some chance where they did not think to come: but this is partly in order that they may show the inhabitants that they have some knowledge of that country—which will surely be no little pleasure to them.

It is said only merchants stay in these islands; for here there is so great an abundance of men sailing with merchandise, that in all the rest of the world they are not as they are in a most noble port called Zaiton, for they say that every year a hundred large ships of pepper are brought into that port, without (counting) other ships bearing other spices. That country is very populous, and very rich, with a multitude of provinces and kingdoms and cities without number, under one prince who is called the Great Kan (*sic*), which name in Latin means *rex regum* (king of kings), whose seat and residence are chiefly in the province of Katay. His ancestors desired to have fellowship with the Christians. For it is now two hundred years since they sent to the Pope and asked for several men learned in the faith, in order that they might be enlightened. But those who were sent went back, being hindered on their journey. In the time of (Pope) Eugenius, also, one came to Eugenius and spoke of (their) great goodwill towards Christians. And I held speech with him for a long time on many things, on the greatness of the royal buildings, and on the greatness of the rivers of wondrous breadth and length, and on the multitude of cities on the banks of the rivers; and how on one river there are established about two hundred cities, and marble bridges of great breadth and length adorned with columns on every side. This country is worthy of being sought by the Latins, not only because from thence may be obtained vast gains of gold and silver and gems of every kind, and of spices that are never brought to us; but also because of the wise men, learned philosophers and astrologers, by whose genius and arts that mighty and magnificent province is governed, and wars are also waged. These things (I write) to give some little satisfaction to your demand, in so far as the shortness of the time allowed, and my occupations suffered; being ready to satisfy your Royal Majesty in the future as much further as may be desired. Given at Florence, 25th June 1474.

From the city of Lisbon in a direct line to the westward, unto the most noble and very great city of Quinsay, there are 26 spaces marked in the chart, each one of them containing 250 miles. For it (the aforementioned city) is a hundred miles round and has ten bridges, and its name means *cita del cielo* (*sic*), City of Heaven, and many wondrous things are told of it, of the multitude of its works and its resources. This space is almost a third part of the whole sphere. This city is in the provice of Mangi, that is to say, nigh unto the province of Katay, wherein is the royal residence of the country. But from the island of Antilia, which is known to you, unto the most noble island of Cippangu there are 10 spaces. For that island is most fertile in gold and in pearls and gems, and they cover the temples and the royal houses with solid gold. Thus by the unknown ways there are not great spaces of the sea to be passed. Many things perchance ought to be explained more openly. But for these things he that considereth diligently will be able to see the rest for himself.

Farewell, beloved.

82. Columbus describes the first glimpse of the West Indies

The following passages from *The Journal of Christopher Columbus* (1451?–1506) are taken from the version preserved by the Spanish

From *The Journal of Christopher Columbus,* trans. Cecil Jane, rev. L. A. Vigneras (London: Hakluyt Society, 1960), pp. 22–24, 70, 96–97, 100–101. © The Hakluyt Society, London.

historian Las Casas. They include the description of Columbus' first landfall on October 12, 1492, and his remarks on the appearance and the inhabitants of the West Indies.

Thursday, October 11th / He navigated to the west-south-west; they had a rougher sea than they had experienced during the whole voyage. They saw petrels and a green reed near the ship. Those in the caravel *Pinta* saw a cane and a stick, and they secured another small stick, carved, as it appeared, with iron, and a piece of cane, and other vegetation which grows on land, and a small board. Those in the caravel *Niña* also saw other indications of land and a stick loaded with barnacles. At these signs, all breathed again and rejoiced. On this day, to sunset, they went twenty-seven leagues. After sunset, he steered his former course to the west; they made twelve miles an hour, and up to two hours before midnight they had made ninety miles, which are twenty-two leagues and a half. And since the caravel *Pinta* was swifter and went ahead of the admiral, she found land and made the signals which the admiral had commanded. This land was first sighted by a sailor called Rodrigo de Triana, although the admiral, at ten o'clock in the night, being on the sterncastle, saw a light. It was, however, so obscured that he would not affirm that it was land, but called Pero Gutierrez, butler of the King's dais, and told him that there seemed to be a light, and that he should watch for it. He did so, and saw it. He said the same also to Rodrigo Sanchez de Segovia, whom the King and Queen had sent in the fleet as *veedor,* and he saw nothing since he was not in a position from which it could be seen. After the admiral had so spoken, it was seen once or twice, and it was like a small wax candle, which was raised and lowered. Few thought that this was an indication of land, but the admiral was certain that they were near land. Accordingly, when they had said the *Salve,* which all sailors are accustomed to say and chant in their manner, and when they had all been gathered together, the admiral asked and urged them to keep a good look out from the forecastle and to watch carefully for land, and to him who should say first that he saw land, he would give at once a silk doublet apart from the other rewards which the Sovereigns had promised, which were ten thousand maravedis annually to him who first sighted it. Two hours after midnight land appeared, at a distance of about two leagues from them. They took in all sail, remaining with the mainsail, which is the great sail without bonnets, and kept jogging, waiting for day, a Friday, on which they reached a small island of the Lucayos, which is called in the language of the Indians "Guanahaní." Immediately they saw naked people, and the admiral went ashore in the armed boat, and Martin Alonso Pinzón and Vicente Yañez, his brother, who was captain of the *Niña.* The admiral brought out the royal standard, and the captains went with two banners of the Green Cross, which the admiral flew on all the ships as a flag, with an F and a Y, and over each letter their crown, one being on one side of the ✠ and the other on the other. When they had landed, they saw very green trees and much water and fruit of various kinds. The admiral called the two captains and the others who had landed, and Rodrigo de Escobedo, secretary of the whole fleet, and Rodrigo Sanchez de Segovia, and said that they should bear witness and testimony how he, before them all, took possession of the island, as in fact he did, for the King and Queen, his Sovereigns, making the declarations which are required, as is contained more at length in the testimonies which were there made in writing. Soon many people of the island gathered there. What follows are the actual words of the admiral, in his book of his first voyage and discovery of these Indies.

"I," he says, "in order that they might feel great amity towards us, because I knew that they were a people to be delivered and converted to our holy faith rather by love than by force, gave to some among them some red caps and some glass beads, which they hung round their necks, and many other things of little value. At this they were greatly pleased and became so entirely our friends that it was a wonder to see. Afterwards they came swimming to the ships' boats, where we were, and brought us parrots and cotton thread in balls, and spears and many other things, and we exchanged for them other things, such as small glass beads and hawks' bells, which we gave to them. In fact, they took all and gave all, such as they had, with good will, but it seemed to me that they were a people very deficient in everything. They all go naked as their mothers bore them, and the women also, although I saw only one very young girl. And all those whom I did see were youths, so that I did not see one who was over thirty years of age; they were very well built, with very handsome bodies and very good faces. Their hair is coarse almost like the hairs of a horse's tail and short; they wear their hair down over their eyebrows, except for a few strands behind, which they wear long and never cut. Some of them are painted black, and they are the colour of the people of the Canaries, neither black or white, and some of them are painted white and some red and some in any colour that they find. Some of them paint their faces, some their whole bodies, some only the eyes, and some only the nose. They do not bear arms or know them, for I showed to them swords and they took them by the blade and cut themselves through ignorance. They have no iron. Their spears are certain reeds, without iron, and some of these have a fish tooth at the end, while others are pointed in various ways. They are all generally fairly tall, good looking and well proportioned. I saw some who bore marks of wounds on their bodies, and I made signs to them to ask how this came about, and they indicated to me that people came from other islands, which are near, and wished to capture them, and they defended themselves. And I believed and still believe that they come here from the mainland to take them for slaves. They should be good servants and of quick intelligence, since I see that they very soon say all that is said to them, and I believe that they would easily be made Christians, for it appeared to me that they had no creed. Our Lord willing, at the time of my departure I will bring back six of them to Your Highnesses, that they may learn to talk. I saw no beast of any kind in this island, except parrots." All these are the words of the admiral . . .

Sunday, November 25th / Before sunrise, he entered the boat and went to examine a cape or point of land to the south-east of the flat island, a matter of a league and a half, because it seemed to him that there should be a good river there. Directly at the beginning of the cape, on the south-eastern side, having gone two crossbow shots, he saw a large stream of very fine water flowing; it came down from the mountain and made a great noise. He went to the river and saw some stones shining in it, on which were some veins of the colour of gold, and he remembered that in the river Tagus, at the mouth of it, near the sea, gold is found, and it seemed to him to be certain that there must be gold here, and he ordered some of those stones to be collected to take them to the Sovereigns. While they were there, the ships' boys shouted that they saw pines. He looked towards the mountain and saw them, so tall and wonderful that he could not overstate their height and straightness, like spindles, thick and slender. From these he realised that ships could be built, and a vast quantity of planks secured and masts for the largest ships in Spain. He saw oaks and strawberry trees, and a good river, and means for constructing saw-mills. The land and the breezes were more temperate than any so far, owing to the height and beauty of the mountain ranges. He saw on the beach many other stones, the colour of iron, and others which some said came from silver mines. All these were brought down by the river . . .

They said also, concerning the beauty of the lands which they saw, that the best lands in Castile for beauty and fertility could not be compared with these. And the admiral saw that this was so from those lands which he had visited and from those which he had before him, and they told him that these could not be compared with those of that valley, which the plain of Cordoba did not equal, the two being as different as day and night. They said that all those lands were cultivated, and that through the middle of that valley there flowed a river, very wide and great, which could water all the lands. All the trees were green and full of fruit, and the plants all flowering and very tall, the roads very wide and good, the breezes like those of Castile in April. The nightingale and other small birds were singing as they do in that month in Spain, and he says that it was the greatest delight in the world. At night some small birds sang sweetly, and many crickets and frogs were heard. The fish were as those of Spain; they saw much mastic and aloe and cotton trees. They found no gold, nor is this surprising in the very short while that they were there. Here the admiral tested the length of the day and night, and from sunset to sunrise was twenty half-hour glasses, although he says that there may have been some error, either because they were not turned quickly enough or because the sand did not all pass through. He also says that he found by the quadrant that he was thirty-four degrees from the equinoctial line . . .

This land is very cool and the best that tongue can describe. It is very lofty and on the highest mountain oxen could plough and all could be made like the plains and valleys. In all Castile there is no land which could be compared to this for beauty and fertility; all this island and that of Tortuga is as cultivated as the plain of Cordoba. They have them sown with *ajes,* which are certain slips which they plant, and at the foot of them grow some roots like carrots, which serve as bread, and they grate them, knead them and make bread of them. Afterwards they again plant the same slip in another place and it again produces four or five of these roots, which are very savoury and have the exact taste of chestnuts. Those here are the largest and best that he had seen in any land, for he also says that they are found in Guinea; these here were as thick as a leg. He says of these people that they were all stout and valiant and not feeble like the others whom he had previously found, and with very pleasant voices; they have no creed. He says that the trees there grew so luxuriantly that their leaves were no longer green but dark coloured. It was a thing of wonder to behold those valleys and those rivers and fair springs of water, and the lands suited for growing bread, for raising stock of all kinds, of which they have none, for gardens, and for everything in the world that a man could desire.

83. Columbus describes his first voyage to America: the formal report to Ferdinand and Isabella

On his return from his first voyage, Columbus wrote a report of the expedition to the Catholic Kings, Ferdinand and Isabella. (The actual addressee of the letter was the Treasurer of the Kings, Gabriel Sánchez.) The letter is the oldest authentic document of the European presence in the New World; only one copy of the first edition is known to exist.

The letter of Christopher Columbus, to whom our age owes a great debt, on the recent discovery of the islands of India beyond the

------

The Columbus Letter of 1493, a new translation in English by Frank E. Robbins (Ann Arbor: The Clements Library Associates, 1952). By permission of the Clements Library.

Ganges, to look for which he had been sent eight months before under the auspices and at the expense of the most invincible Ferdinand and Isabella, sovereigns of the Spains; sent to the eminent lord Gabriel Sánchez, treasurer of the said most serene sovereigns; which the noble and learned gentleman, Leander di Cosco, translated from Spanish into Latin the third day before the Kalends of May,[1] 1493, in the first year of the pontificate of Alexander VI.

As I know that it will please you that I have carried to completion the duty which I assumed, I decided to write you this letter to advise you of every single event and discovery of this voyage of ours. On the thirty-third day after I left Cadiz;[1] I reached the Indian Sea, there I found very many islands, inhabited by numberless people, of all of which I took possession without opposition in the name of our most fortunate king by making formal proclamation and raising standards; and to the first of them I gave the name of San Salvador,[2] the blessed Savior, through dependence on whose aid we reached both this and the others. The Indians however call it Guanahani. I gave each one of the others too a new name; to wit, one Santa Maria de la Concepción,[3] another Fernandina,[4] another Isabella,[5] another Juana,[6] and I ordered similar names to be used for the rest.

When we first put in at the island which I have just said was named Juana, I proceeded along its shore westward a little way, and found it so large (for no end to it appeared) that I believed it to be no island but the continental province of Cathay; without seeing, however, any towns or cities situated in its coastal parts except a few villages and rustic farms, with whose inhabitants I could not talk because they took to flight as soon as they saw us.

I went on further, thinking that I would find a city or some farmhouses. Finally, seeing that nothing new turned up, though we had gone far enough, and that this course was carrying us off to the north (a thing which I myself wanted to avoid, for winter prevailed on those lands, and it was my hope to hasten to the south) and since the winds too were favorable to our desires, I concluded that no other means of accomplishment offered, and thus reversing my course I returned to a certain harbor which I had marked and from that point sent ashore two men of our number to find out whether there was a king in that province, or any

---

1. April 29.
1. Translator's error for "the Canaries," where Columbus put in on his westward voyage. He did not touch at Cadiz at all.
2. Watling's Island in the Bahamas.
3. Rum Cay.
4. Long Island.
5. Crooked Island.
6. Cuba. Columbus called it Juana in honor of the Infant Don Juan.

cities. These men proceeded afoot for three days and found countless people and inhabited places, but all small and without any government; and therefore they returned.

In the meantime I had already learned from some Indians whom I had taken aboard at this same place that this province was in fact an island; and so I went on toward the east, always skirting close to its shores, for 322 miles, where is the extremity of the island. From this point I observed another island to eastward, 54 miles from this island Juana, which I immediately called Hispana.[7] I withdrew to it, and set my course along its northern coast, as I had at Juana, to the east for 564 miles.

The before-mentioned island Juana and the other islands of the region, too, are as fertile as they can be. This one is surrounded by harbors, numerous, very safe and broad, and not to be compared with any others that I have seen anywhere; many large, wholesome rivers flow through this land; and there are also many very lofty mountains in it. All these islands are most beautiful and distinguished by various forms; one can travel through them, and they are full of trees of the greatest variety, which brush at the stars; and I believe they never lose their foliage. At any rate, I found them as green and beautiful as they usually are in the month of May in Spain; some of them were in bloom, some loaded with fruit, some flourished in one state, others in the other, each according to its kind; the nightingale was singing and there were countless other birds of many kinds in the month of November when I myself was making my way through them. There are furthermore, in the before-mentioned island Juana, seven or eight kinds of palm trees, which easily surpass ours in height and beauty, as do all the other trees, grasses, and fruits. There are also remarkable pines, vast fields and meadows, many kinds of birds, many kinds of honey, and many kinds of metals, except iron.

There are moreover in that island which I said above was called Hispaniola fine, high mountains, broad stretches of country, forests, and extremely fruitful fields excellently adapted for sowing, grazing, and building dwelling houses. The convenience and superiority of the harbors in this island and its wealth in rivers, joined with wholesomeness for man, is such as to surpass belief unless one has seen them. The trees, coverage, and fruits of this island are very different from those of Juana. Besides, this Hispaniola is rich in various kinds of spice and in gold and in mines, and its inhabitants (and those of all the others which I saw, and of which I have knowledge) of either sex always go as naked as when they were born, except some women who cover their private parts with a leaf or a branch of some sort, or with a skirt of cotton which they themselves prepare for the purpose.

---

7. Hispaniola (San Domingo and Haiti). Columbus greatly overestimated the dimensions of these islands.

They all of them lack, as I said above, iron of whatever kind, as well as arms, for these are unknown to them; nor are they fitted for weapons, not because of any bodily deformity, for they are well built, but in that they are timid and fearful. However, instead of arms they carry reeds baked in the sun, in the roots of which they fasten a sort of spearhead made of dry wood and sharpened to a point. And they do not dare to use these at close quarters; for it often happened that when I had sent two or three of my men to certain farmhouses to talk with their inhabitants a closely packed body of Indians would come out and when they saw our men approach they would quickly take flight, children deserted by father and vice versa; and that too not that any hurt or injury had been brought upon a single one of them; on the contrary, whenever I approached any of them and whenever I could talk with any of them I was generous in giving them whatever I had, cloth and very many other things, without any return being made to me; but they are naturally fearful and timid.

However when they see that they are safe and all fear has been dispelled they are exceedingly straightforward and trustworthy and most liberal with all that they have; none of them denies to the asker anything that he possesses; on the contrary they themselves invite us to ask for it. They exhibit great affection to all and always give much for little, content with very little or nothing in return. However I forbade such insignificant and valueless things to be given to them, as pieces of platters, dishes, and glass, or again nails and lace points;[8] though if they could acquire such it seemed to them that they possessed the most beautiful trinkets in the world. For it happened that one sailor got in return for one lace point a weight of gold equivalent to three golden solidi,[9] and similarly others in exchange for other things of slighter value; especially in exchange for brand-new blancas, certain gold coins, to secure which they would give whatever the seller asks, for example, an ounce and a half or two ounces of gold, or thirty or forty [pounds] of cotton by weight, which they themselves had spun;[10] likewise they bought pieces of hoops,[11] pots, pitchers, and jars for cotton and gold, like dumb beasts. I forbade this, because it was clearly unjust, and gave them free many pretty and acceptable objects that I had brought with me, in order more easily to win them over to me, and that they might become Christians, and be inclined to love our King and Queen and Prince and all the peoples of Spain, and to make them diligent to seek out and accumulate and exchange with us the articles in which they abound and which we greatly need.

---

8. The tips of the laces used to secure the hose to the upper garment.

9. *Castellanos* in the Spanish version.

10. *Nouerant,* "had known," which stands in the text, is a mistake for *neuerant,* "had spun."

11. The Spanish text has here "hoops of winecasks," which Leander probably intends by the word *arcuum,* "bows."

They know nothing of idolatry; on the contrary they confidently believe that all might, all power, all good things, in fact, are in the heavens; they thought that I too had descended thence with these ships and sailors, and in that opinion I was received everywhere after they had rid themselves of fear. Nor are they slow or ignorant; on the contrary, they are of the highest and keenest wit; and the men who navigate that sea give an admirable account of each detail; but they have never seen men wearing clothes, or ships of this sort. As soon as I came to that sea I forcibly seized some Indians from the first island, so that they might learn from us and similarly teach us the things of which they had knowledge in those parts; and it came out just as I had hoped; for we quickly taught them, and then they us, by gestures and signs; finally they understood by means of words, and it paid us well to have them. The ones who now go with me persist in the belief that I leaped down out of the skies, although they have associated with us for a long time and are still doing so today; and they were the first to announce that fact wherever we landed, some of them calling out loudly to the others, "Come, come, and you will see the men from heaven." And so women as well as men, children and grown people, youths and old men, laying aside the fear they had conceived shortly before, vied with each other in coming to look at us, the great crowd of them clogging the road, some bringing food, others drink, with the greatest manifestation of affection and unbelievable good will.

Each island possesses many canoes of solid wood, and though they are narrow, nevertheless in length and shape they are like our double-banked galleys, but faster. They are steered with oars alone. Some of these are large, some small, some of medium size; a considerable number however are larger than the galley which is rowed by eighteen benches. With these they cross to all the islands, which are innumerable, and with them they ply their trade, and commerce is carried out between them. I saw some of these galleys or canoes which carried seventy or eighty oarsmen.

In all the islands there is no difference in the appearance of the people, nor in their habits or language; on the contrary, they all understand each other, which circumstance is most useful to that end which I think our most serene sovereigns especially desire, namely, their conversion to the holy faith of Christ, to which indeed as far as I could see they are readily submissive and inclined.

I have told how I sailed along the island of Juana on a straight course from west to east 322 miles; from this voyage and the length of the course I can say that this Juana is larger than England and Scotland together; for beyond the aforesaid 322 miles, in the western part, there are two more provinces which I did not visit, one of which the Indians call Anan, whose inhabitants are born with tails. These provinces extend to a length of 180 miles, as I have found out from these Indians whom I am bringing with me, who are well acquainted with all these islands.

The circumference of Hispaniola, indeed, is more than all Spain from

Catalonia to Fuenterrabia. And this is easily proved by this fact, that the one of its four sides which I myself traversed on a straight course from west to east measures 540 miles. We should seek possession of this island and once gained it is not to be thrown away; for although, as I said, I formally took possession of all the others in the name of our invincible King and their sovereignty is entirely committed to that said King, nevertheless in this island I took possession in a special way of a certain large village in a more favorable situation, suitable for all sorts of gain and trade, to which I gave the name Navidad del Señor; and I gave orders to erect a fort there at once. This should by now be built, and in it I left behind the men who seemed necessary with all kinds of arms and suitable food for more than a year, furthermore, one caravel,[12] and for the construction of others men skilled in this art as well as in others; and, besides, an unbelievable goodwill and friendship on the part of the king of that island toward the men. For all those peoples are so gentle and kind that the aforesaid king took pride in my being called his brother. Even if they should change their minds and want to injure the men who stayed in the fort they cannot, since they have no arms, go naked, and are extremely timid; and so if our men only hold the said fort they can hold the whole island, with no hazard on the part of the people to threaten them as long as they do not depart from the laws and government which I gave them.[13]

In all those islands, as I understood it, each man is content with only one wife, except the princes or kings, who may have twenty. The women seem to do more work than the men. I could not clearly make out whether they have private property, for I noted that what an individual had he shared with others, especially food, meats, and the like. I did not find any monsters among them, as many expected, but men of great dignity and kindliness. Nor are they black, like the Negroes; they have long, straight hair; they do not live where the heat of the sun's rays shines forth; for the strength of the sun is very great here, since apparently it is only twenty-six degrees from the equator.[14] On the mountain peaks extreme cold reigns, but this the Indians mitigate both by being used to the region and by the aid of very hot foods upon which they dine often and luxuriously.

And so I did not see any monsters, nor do I have knowledge of them anywhere with the exception of a certain island called Charis,[15] which is

---

12. The *Santa Maria,* which was wrecked off this coast.

13. But the Spaniards did not behave themselves and the whole colony was destroyed before Columbus returned to it on his Second Voyage.

14. The text is corrupt and the translation follows the obvious sense. Probably *ubi videntur* is a mistake for *ut videtur;* the punctuation is wrong, too.

15. Probably Puerto Rico. The name Charis is derived from that of the savage tribe here mentioned, the Caribs, who were much feared by the Tainos found by Columbus in the Bahamas, Cuba, and Hispaniola.

the second as you sail from Spain toward India and which a tribe inhabits that is held by its neighbors to be extremely savage. These feed on human flesh. The aforesaid have many kinds of galleys in which they cross to all the Indian islands, rob, and steal all they can. They differ in no respect from the others, except that in feminine fashion they wear their hair long; and they use bows and arrows with shafts of reeds fitted as we said at the thicker end with sharpened arrowheads. On that account they are held to be savage, and the other Indians are afflicted with constant fear of them, but I do not rate them any more highly than the rest. These are the ones who cohabit with certain women who are the only inhabitants of the island of Mateunin,[16] which is the first encountered in the passage from Spain toward India. These women moreover, do not occupy themselves with any of the work that properly belongs to their sex, for they use bows and arrows just as I related of their husbands; they protect themselves with copper plates of which there is an ample supply in their land. They assure me that there is another island larger than the above-mentioned Hispaniola; its inhabitants are hairless, and it abounds in gold more than all the others. I am bringing with me men from this island and the others that I saw who bear testimony to what I have said.

Finally, to compress into a few words the advantage and profit of our journey hence and our speedy return, I make this promise, that supported by only small aid from them I will give our invincible sovereigns as much gold as they need, as much spices, cotton, and the mastic, which is found only in Chios,[17] as much of the wood of the aloe, as many slaves to serve as sailors as their Majesties wish to demand; furthermore, rhubarb and other kinds of spices which I suppose those whom I left in the before-mentioned fort have already discovered and will discover, since indeed I lingered nowhere longer than the winds compelled, except at the village of Navidad while I took care to establish the fort and to make all safe. Though these things are great and unheard of, nevertheless they would have been much greater if the ships had served me as they reasonably should.[18]

---

16. Martinique. Columbus was prepared to find an island inhabited by men and another inhabited by women in the Indian Ocean (where he thought he was), because he had read about them in his copy of Marco Polo's travels.

17. Columbus probably visited Chios in a Genoese trading ship in 1474 and 1475 (Morison, *Admiral of the Ocean Sea,* I, 31).

18. This may refer to the conduct of Martín Alonso Pinzón and to the carelessness that permitted *Santa Maria* to run aground. Pinzón, never too cooperative, took Pinta on an exploring and gold-hunting trip (November 22, 1492–January 6, 1493) without the Admiral's permission. Columbus made *Santa Maria* his flagship, but Juan de la Cosa was her master and part owner, and officer of the watch at the time of the disaster. He seems to have been not only negligent but also cowardly and insubordinate (see Morison, *op. cit.,* I, 388–389).

Indeed this outcome was manifold and marvelous, and fitting not to my own claims to merit, but rather to the holy Christian faith and the piety and religion of our sovereigns, for what the human mind could not comprehend, that the divine mind has granted to men. For God is accustomed to listen to his servants, and to those who love his commands, even in impossible circumstances, as has happened to us in the present instance, for we have succeeded in that to which hitherto mortal powers have in no wise attained. For if others have written or spoken of these islands, they have all done so by indirection and guesses; no one claims to have seen them, whence it seemed to be almost a fable. Therefore let the King and Queen, the Prince, their happy realms, and all other provinces of Christendom give thanks to the Savior, our Lord Jesus Christ, who has granted us so great a victory and reward; let processions be celebrated; let solemn holy rites be performed; and let the churches be decked with festival branches; let Christ rejoice on earth as He does in heaven when He foresees that so many souls of peoples hitherto lost are to be saved. Let us too rejoice, both for the exaltation of our faith and for the increase in temporal goods in which not only Spain but all Christendom together are to share. As these things were done, so have they been briefly narrated. Farewell.

Lisbon, the day before the Ides of March.[19]

CHRISTOPHER COLUMBUS
Admiral of the Ocean Fleet

84. Waldseemüller names the New World "America"

Among the contemporaries of Columbus, the Italian explorer-navigator Amerigo Vespucci (1454–1512) was widely known from the reports of his voyages to the Caribbean and northern South America. It was on the basis of Vespucci's exploits that Martin Waldseemüller suggested naming the New World "America." Waldseemüller, who made this suggestion in his *Cosmography* (1507), also applied the name "America" to the present South America in a large world map and a set of globe gores, both printed in 1507. The map disappeared soon after its publication; the only surviving copy was

---

From Martin Waldseemüller, *Introduction to Cosmography, together with those Principles of Geometry and Astronomy which are Necessary to It. Furthermore, the four voyages of Amerigo Vespucci* . . . (St. Dié, 1507), passim. Editor's translation from Latin text.

19. March 14.

found in 1900. Waldseemüller's little book survived, however, and his suggestion was accepted by everyone.

A large part of the earth, hitherto unknown, was recently discovered by Amerigo Vespucci . . .

The fourth part of the world, recently found by Americus, could be called the Land of Americus, or America . . .

Today, in fact, these parts [of the world, that is, Europe, Africa, and Asia] have been of late explored, and the fourth part was discovered by Amerigo Vespucci, as it shall be heard further on. I see no reason why, and by what right, this land of Amerigo should not be named after that wise and ingenious man who discovered it, America, since both Europe and Asia had been allotted the names of women.

85. Pigafetta on the first circumnavigation of the earth

In 1519 a Spanish squadron of ships sailed from Seville, commanded by Ferdinand Magellan (Fernão Magalhães), a Portuguese sea captain in Spanish service. Magellan's task was to sail around South America and reach the Spice Islands of the East—using the westerly rather than the easterly route—and to establish their exact location. Ever since the world had been divided between Spain and Portugal by papal decree in 1493, and the division confirmed by a Spanish-Portuguese treaty in 1494, the two nations had disagreed as to the exact location of the line of the division on the "far side" of the globe, that is, half-way around the earth from the line drawn between the Old World and the Americas in 1493 and 1494.

Among those aboard Magellan's squadron was an Italian nobleman, Antonio Pigafetta of Vicenza (?1480/1490–1534). Pigafetta was a gentleman-in-waiting to the papal nuncio in Spain when the voyage was organized, and joined it for the adventure of sailing around the world. His report was first published in a French translation in 1524, only two years after the sole surviving ship of Magellan's squadron, the *Vitoria*, returned to Seville.

Of the passages included here, the first describes the crossing of the Pacific by Magellan's squadron; it includes a detailed discussion of the exact location of the line of demarcation set by the Treaty of Tordesillas between Spain and Portugal, on the "other," (Pacific) side of the globe. The second excerpt describes the death of Magellan in the Philippines; the third, the return of the *Vitoria* to Spain.

From *The First Voyage Round the World by Magellan*, trans. Lord Stanley of Alderney (London, Hakluyt Society, no. 52, 1874), pp. 64–67, 101–102, 161–162.

Crossing the Pacific

Wednesday, the twenty-eighth of November, 1520, we came forth out of the said strait, and entered into the Pacific sea, where we remained three months and twenty days without taking in provisions or other refreshments, and we only ate old biscuit reduced to powder, and full of grubs, and stinking from the dirt which the rats had made on it when eating the good biscuit, and we drank water that was yellow and stinking. We also ate the ox hides which were under the main-yard, so that the yard should not break the rigging: they were very hard on account of the sun, rain, and wind, and we left them for four or five days in the sea, and then we put them a little on the embers, and so ate them; also the saw-dust of wood, and rats which cost half-a-crown each, moreover enough of them were not to be got. Besides the above-named evils, this misfortune which I will mention was the worst, it was that the upper and lower gums of most of our men grew so much that they could not eat, and in this way so many suffered, that nineteen died, and the other giant, and an Indian from the county of Verzin. Besides those who died, twenty-five or thirty fell ill of divers sicknesses, both in the arms and legs, and other places, in such manner that very few remained healthy. However, thanks be to the Lord, I had no sickness. During those three months and twenty days we went in an open sea, while we ran fully four thousand leagues in the Pacific sea. This was well named Pacific, for during this same time we met with no storm, and saw no land except two small uninhabited islands, in which we found only birds and trees. We named them the Unfortunate Islands; they are two hundred leagues apart from one another, and there is no place to anchor, as there is no bottom. There we saw many sharks, which are a kind of large fish which they call Tiburoni. The first isle is in fifteen degrees of austral latitude, and the other island is in nine degrees. With the said wind we ran each day fifty or sixty leagues, or more; now with the wind astern, sometimes on a wind or otherwise. And if our Lord and his Mother had not aided us in giving us good weather to refresh ourselves with provisions and other things, we should all have died of hunger in this very vast sea, and I think that never man will undertake to perform such a voyage.

When we had gone out of this strait, if we had always navigated to the west we should have gone without finding any land except the Cape of the Eleven Thousand Virgins, which is the eastern head of the strait in the ocean sea, with the Cape of Desire at the west in the Pacific sea. These two capes are exactly in fifty-two degrees of latitude of the antarctic pole.

The antarctic pole is not so covered with stars as the arctic, for there are to be seen there many small stars congregated together, which are like to two clouds a little separated from one another, and a little dimmed, in the midst of which are two stars, not very large, nor very brilliant, and they move but little: these two stars are the antarctic pole. Our compass needle still pointed a little to its arctic pole; nevertheless it had not as much power as on its own side and region. Yet when we were in the open sea, the captain-general asked of all the pilots, whilst still going under sail, in what direction they were navigating and pointing the charts. They all replied, by the course he had given, punctually [pricked in]; then he answered, that they were pointing falsely (which was so), and that it was fitting to arrange the needle of navigation, because it did not receive so much force as in its own quarter. When we were in the middle of this open sea we saw a cross of five stars, very bright, straight, in the west, and they are straight one with another.

During this time of two months and twelve days we navigated between west and north-west (maestral), and a quarter west of north-west, and also north-west, until we came to the equinoctial line, which was at [a point] one hundred and twenty-two degrees distant from the line of re-partition. This line of delimitation is thirty degrees distant from the meridian, and the meridian is three degrees distant from the Cape Verde towards the east.

The death of Magellan

He then, in order to disperse this multitude and to terrify them, sent some of our men to set fire to their houses, but this rendered them more ferocious. Some of them ran to the fire, which consumed twenty or thirty houses, and there killed two of our men. The rest came down upon us with greater fury; they perceived that our bodies were defended, but that the legs were exposed, and they aimed at them principally. The captain had his right leg pierced by a poisoned arrow, on which account he gave orders to retreat by degrees; but almost all our men took to precipitate flight, so that there remained hardly six or eight of us with him. We were oppressed by the lances and stones which the enemy hurled at us, and we could make no more resistance. The bombards which we had in the boats were of no assistance to us, for the shoal water kept them too far from the beach. We went thither, retreating little by little, and still fighting, and we had already got to the distance of a crossbow shot from the shore, having the water up to our knees, the islanders following and picking up again the spears which they had already cast, and they threw the same spear five or six times; as they knew the captain they aimed specially at him, and twice they knocked the helmet off his head. He, with a few of us, like a good knight, remained at his post without choosing to retreat further. Thus we fought for more than an hour, until an Indian succeeded in thrusting a cane lance into the captain's face. He then, being irritated, pierced the Indian's breast with his lance, and left it in his body, and trying to draw his sword he was unable to draw it more than half way, on account of a javelin wound which he had received in the right arm. The enemies seeing this all rushed against him, and one of them with a great sword, like a great scimitar gave him a great blow on the left leg, which brought the captain down on his face, then the Indians threw themselves upon him, and ran him through with lances and scimitars, and all the other arms which they had, so that they deprived of life our mirror, light, comfort, and true guide. Whilst the Indians were thus overpowering him, several times he turned round towards us to see if we were all in safety, as though his obstinate fight had no other object than to give an opportunity for the retreat of his men. We who fought to extremity, and who were covered with wounds, seeing that he was dead, proceeded to the boats which were on the point of going away. This fatal battle was fought on the 27th of April of 1521, on a Saturday; a day which the captain had chosen himself, because he had a special devotion to it. There perished with him eight of our men, and four of the Indians, who had become Christians; we had also many wounded, amongst whom I must reckon myself. The enemy lost only fifteen men.

He died; but I hope that your illustrious highness will not allow his memory to be lost, so much the more since I see revived in you the virtue of so great a captain, since one of his principal virtues was constance in the most adverse fortune. In the midst of the sea he was able to endure hunger better than we. Most versed in nautical charts, he knew better than any other the true art of navigation, of which it is a certain proof that he knew by his genius, and his intrepidity, without any one having given him the example, how to attempt the circuit of the globe, which he had almost completed.

Return of the *Vitoria*

Constrained by extreme necessity, we decided on touching at the Cape Verde Islands, and on Wednesday the 9th of July, we touched at one of those islands named St. James's. Knowing that we were in an enemy's country, and amongst suspicious persons, on sending the boat ashore to get provision of victuals, we charged the seamen to say to the Portuguese that we had sprung our foremast under the equinoctial line (although this misfortune had happened at the Cape of Good Hope), and that our ship was alone, because whilst we tried to repair it, our captain-general had gone with the other two ships to Spain. With these good words, and giving some of our merchandise in exchange, we obtained two boat-loads of rice.

In order to see whether we had kept an exact account of the days, we charged those who went ashore to ask what day of the week it was, and we were told by the Portuguese inhabitants of the island that it was Thursday, which was a great cause of wondering to us, since with us it was only Wednesday. We could not persuade ourselves that we were mistaken; and I was more surprised than the others, since having always been in good health, I had every day, without intermission, written down the day that was current. But we were afterwards advised that there was no error on our part, since as we had always sailed towards the west, following the course of the sun, and had returned to the same place, we must have gained twenty-four hours, as is clear to any one who reflects upon it.

The boat, having returned for rice a second time to the shore, was detained, with thirteen men who were in it. As we saw that, and, from the movement in certain caravels, suspected that they might wish to capture us and our ship, we at once set sail. We afterwards learned, some time after our return, that our boat and men had been arrested, because one of our men had discovered the deception, and said that the captain-general was dead, and that our ship was the only one remaining of Magellan's fleet.

At last, when it pleased Heaven, on Saturday the 6th of September of the year 1522, we entered the bay of San Lucar; and of sixty men who composed our crew when we left Maluco, we were reduced to only eighteen, and these for the most part sick. Of the others, some died of hunger, some had run away at the island of Timor, and some had been condemned to death for their crimes.

From the day when we left this bay of San Lucar until our return thither, we reckoned that we had run more than fourteen thousand four hundred and sixty leagues, and we had completed going round the earth from East to West.

Monday the 8th of September, we cast anchor near the mole of Seville, and discharged all the artillery.

Tuesday, we all went in shirts and barefoot, with a taper in our hands to visit the shrine of St. Maria of Victory, and of St. Maria de Antigua.

Then, leaving Seville, I went to Valladolid, where I presented to his Sacred Majesty Don Carlos, neither gold nor silver, but things much more precious in the eyes of so great a Sovereign. I presented to him among other things, a book written by my hand of all the things that had occurred day by day in our voyage.

86. Roger Barlow, the first Englishman to sail to South America, reports on the New World

English ships, mostly out of Bristol, had sailed to North America rather frequently in the late fifteenth and early sixteenth centuries, but it was only in 1526 that Roger Barlow embarked on a voyage to

From Roger Barlow, *A Brief Summe of Geographie*, ed. E. G. R. Taylor (London, Hakluyt Society, 2d ser., no. 69, 1932), pp. 180–182. © The Hakluyt Society, London.

"the Spiceries" in a ship commanded by Sebastian Cabot. His report, that of the first Englishman known to have traveled to Brazil and to the River Plate, was written in 1540–1541 and was included in his *Brief Summe of Geographie,* a translation of Martin Fernandez de Enciso's book of the same title, prepared in 1518 for Charles V of Spain.

The Northwest Passage

Now that I have spoken of that parte of the world from the ilond of ferro toward ponyent and austro, we shall speke of the part of the lond that is in the second parte toward septentrion wch is called the new founde lande, which was fyrst discouered by marchantes of brystowe where now the bretons do trat thider everie yere a fyshing, and is called the bacaliaus. It lieth west northwest of galisia and hathe many good portes and ilondes and northest of it it hathe the lond of laboradoris wch stondeth in 57 degrees. What comoditie is within this lande as yet it is not knowen for it hath not ben labored, but it is to be presupposed that ther is no riches of gold, spyces nor preciose stones, for it stondeth farre aparted from the equinoctiall whereas the influens of the sonne doth norishe and bryng fourth gold, spices, stones and perles. But whereas o^r englishe marchantes of brystowe dyd enterpryse to discover and discovered that parte of the land, if at that season thei had folowed toward the equinoctiall, no dowt but thei shuld have founde grete riches of gold and perle as other nations hathe done sence that tyme.

Now by this your grace maie well apperceve what parte of the universal is discouered and what ther resteth for to dycouer. Hit is clerely sene by the cosmographia that of iiij partes of the worlde the iij partes be discovered, for out of spayne thei saile all the indies and sees occidentales, and from portingale thei saylle all the indies and sees orientalles, so that betwene the waie of the orient and the waie of the occydent thei have compassed all the world, the tone departeing from spayne toward occydent and the other out of portingale toward orient thei have mette together. And also by the waie of the meridian there is a grete parte discouered by the spaniards, so ther resteth this waie of the northe onelie for to discover which resteth onto your graces charge, for that the situation of this realme toward that partie is more apte for it then eny other, and also for that your grace hath takyn farre enterprise to discover this part of the world already, and suche an enterprise ought not to be left of, although the followyng thereof hathe not succedyd as your gracis wil and desire was, for in the people, shippes, derotas and provicions suche ordre maye be taken, that without dout and if it please god, it shal folowe unto your gracys purpose. And for such an enterprise no man shuld thinke upon the cost in comparison to the grete profyght that maye therby succede, nor thinke the labour grete where so moche profyt honor and glory maye folow unto this o^r naturall realme and king, and as for iopardies and parells, this waie of navigation well considered and pondred shall seme moche lesse perill then all the other navigations as it maie be proved by verie evident resons. And for to speke somewhat of the comoditie and utilitie of this navigation and discovering, it is very evident that the sees wheras every man will saie that ther is difficultie and daunger and that thei take for impossible to saille, those sees thei maie saile with light and daie alwais, without darknes or eny night, wherefore there is diffrence betwene thes perelles and navigatyon wheras contynuallie thei mai se round about them, and on the contrary when in every 24 hours thei shal saile the most parte in darknes and nyght, and at that tyme thei must saile at aventure, for thei shal see no thinge about them. I thinke ther is none so ignorant but this dothe perceave, and specyallie what avauntage is this for those that dyscovereth new contreis, for to saile alwaies by light and daie. As for the costes and sees alredie dyscovered, wheras the waie is knowen it semeth lesse peryll to saile by nyght, but in those parties for to discover, it is very dificill to saile by nyght. And yet thei have not lefte with darkenes to procure to discover londes and sees unknowen. How moche more shuld thei count us for ferefull and of litil stomak to leve of suche an enterprise which maie be done with contynuall light. Moche more passing this litle space of navigation which is countyd daungerous maie be iij C leges before thei come to the pole and other as moche after thei have passed the pole, it is clere that from thens forward the sees and lond is temperat as it is here in england, and then it shalbe in the will of them that discover to chose the cold countreis, temperat or hote in the degree that thei will. ffor ones passe the pole thei maie chose at ther plesure to declyne to what part thei lyste, for and if thei will take toward orient, thei shall enioye of the regions of the tartarians, wch have ther vertentes toward the southe, and from thens folowing the cost thei shal go to the londes of the china, and from thens forward to the cataio oriental, wch is of the mayne lond the most orientall countyng from o^r habitacion. And if from thens thei wyl contynue ther navigation folowing the cost wch turneth toward occydent, thei shal come to melaca and from thens to all the indies that we calle orientales, and so contynuying that coste thei maye come home by the cape bon espera and so to have gone allmost round aboute the worlde. And if thei will saylle, in passing the pole, toward occident, thei shal go on the backside of all the new found land that is discouered by your graces subiectes, till thei come into the southe see on the backe side of the indies occidentales, and so contynueng ther navigation thei maie turne by the stret of magalianas unto this realme, and so to compasse the worlde about by that part. And if in passing the pole artike thei will saille streite toward the pole antartike, thei shall enclyne to the londes and ilondes that have ther situacion betwene the tropicons and under the equinoctiall, which without dout be the most richest londes and ilondes in the worlde, for all the golde, spices, aromatikes and pretiose stones, with all other thinges that we have in estimation, from thens thei come. And beside all this yet, the comoditie of this navigation by this waie is of so grete avantage over the other in shorting of half the waie, for the other must saile by grete circuites and compasses and thes shal saile by streit wais and lines.

Southern South America

Along the river of parana is a goodlie plaine contreie, and goodlie woodes of divers kyndes of trees that be alwais grene wynter and somer. Ther be many wylde beasts and a straunge facion of shepe, oystriges, and red dere, wch the indies do hunte by divers waies, but not with dogges for ther be none in the contreie but certein mastifes that we brought with us out of spayne. One maner of ther huntyng is this. They wil go togither iij or iiij hundred indies wheras thei se thes beastes feding in the playn, and wil go betwen them and the mountaines and compasse them about saving one waie toward the river, and then every man setteth fire in the drie gras and when thes bestes se the fier and smoke behinde them thei leve the covert and renne toward the ryver, and the indies folowe them til thei come almost to the river side and then thei presse upon them with ther bowes and arowes and force them to take the water, for backe ageyne thei shal have moche adoo to skape, and when thei be in the water ther be indies redy with ther canoos, and so chase them up and downe in the river and kyll them w^t ther bowes and arowes.

This river of parana is a marvelous goodlie rever and a grete, for of iij hundreth leges and above that we went up in it, the narowest place from one shore to an other was above ij or iij leges brede. This river is full of goodlie ilondes and plesant, for thei be full of trees of divers sortes and the levis of them alwais grene, and the bowis hange downe into the water and many straunge birdes brede in them. In one ilond that we came to there were no manner of birdes in it but onlie all white hernes, where we went alande, and in less time then ij houres we killed with staves and bowes above a thousande, for thei wold not voide, but flie crieng about our heades, and some with axes hewed downe the bowes of the trees and threwe downe the nestes with the yong, for the trees were laden with them and thei were fatte and wonderflie swete. And by other ilondes we passed wherin was none other birdes as we coude perceave but popyngayes and turtil doves and an other sort of smal byrdes which be no bigger of bodie then the toppe of a mans thombe but thei have the goodliest colored fethers that ever man might se, the colours wold chaunge in moving of them as it were chaungeable silke. We toke one of them alive and kept it in a cage and was verie tame, it had a verie swete smell like muske, but it lived not long for lacke of knolege to diet it or other keping, and after it was dead we toke of the skynne, the heade, fete and feathers as nie as we coude, and stuffed it with drie mosse and put it in a coffer, and it wold make all the coffer to smell wonders swete.

From Cape sant marie to cape blanco, wch stondeth on the other side of the river solys, is 35 leges and from cape blanco to the stret of magalianos is 300 leges. The stret of magalianas stondeth in 52 degrees toward the pole antartik, and england stondeth in the same degree toward the pole artike, and by this strete out of the ocean see thei passe into the southe see, wch was found and discovered in the yere of our lord 1520 by a portingale called magalianas, wch departed out of sivil with his armie in the yere 1519 for to discover the idondes of the malucus that waie. And of v shipps that he had w^t him one was loste befor thei came to the seid strete and an other returned backe ageine, and he w^t ij more passed the strete and went by many riche ilonds, and or he came to the malucus he was slayne in an ilond called mata by his own folie, and so were many of his people, and those that rested in the iij shippes seing ther captaine and many of ther company slayne and had not companie to furnishe ther iij shippes, thei toke out of one of them wch was most feblest all the ordinance and suche things as thei might save, and after set the ship on fire bycause ther enemies should have no profitte on her, and w^t ther ij shippes that rested thei went into the ilondes of malucus, where thei were lade with clowes and cynamone and other spices. And one of thes shippes came home to civill laden with spices by the waie of cape bona spera in the yere of o^r lord god 1523, and the other ship was taken there by the portugales as afterward it was knowen, so that this ship that came home to civill which was called the victoria had gone est and west almost round about the worlde, for he had departed from cyvill and passed the stret of the magalianos by the west and came home to cyvil ageine by the cape of bon espera which is at easte.

87. From Hakluyt's Voyages

Richard Hakluyt (1552–1616) was the chronicler of the great voyages of discovery undertaken by Englishmen during the reign of Elizabeth I. He was educated at Oxford, served briefly as a diplomat in France, and devoted most of his energy to the encouraging of English exploration and colonization. His book, *The Principal Navigations, Voyages,*

From Richard Hakluyt, *The Principal Voyages . . . of the English Nation* (New York: E. P. Dutton & Co., 1907), I, 1, 270–271; II, 290–291; VIII, 56–57; VII, 225, 228, 247.

Traffiques & Discoveries of the English Nation Made by Sea or Overland to the Remote & Farthest Distant Quarters of the Earth at any Time within the Compasse of these 1600 Yeares was published in one volume in 1589, the year after the Spanish Armada's attempt to conquer England. The final version, in three volumes, was issued in 1599. It is a classic on Elizabethan exploration and discovery.

The first selection from Hakluyt's *Voyages* is part of the dedication of the work to Walsingham, Queen Elizabeth's Secretary of State. The other passages include a description of the departure of the ships of Willoughby and Chancellor on the first English voyage to northernmost Russia, in 1553; part of the report of Elizabeth I's ambassador to Russia on the "Russe Commonwealth" in 1588; a description of the Strait of Magellan, part of the report of Francis Drake's circumnavigation of the earth, begun in 1577; and examples of the style and contents of an English pilot's handbook, or *ruttier* (from the Portuguese *roteiro* and the French *routier*) of the West Indies.

The epistle dedicatorie in the first edition, 1589

To the Right Honorable Sir Francis Walsingham Knight, Principall Secretarie to her Majestie, Chancellor of the Duchie of Lancaster, and one of her Majesties most honourable Privie Councell.

Right Honorable, I do remember that being a youth, and one of her Majesties scholars at Westminster that fruitfull nurserie, it was my happe to visit the chamber of M. Richard Hakluyt my cosin, a Gentleman of the Middle Temple, well knowen unto you, at a time when I found lying open upon his boord certeine bookes of Cosmographie, with an universall Mappe: he seeing me somewhat curious in the view therof, began to instruct my ignorance, by shewing me the division of the earth into three parts after the olde account, and then according to the latter, & better distribution, into more: he pointed with his wand to all the knowen Seas, Gulfs, Bayes, Straights, Capes, Rivers, Empires, Kingdomes, Dukedomes, and Territories of ech part, with declaration also of their speciall commodities, & particular wants, which by the benefit of traffike, & entercourse of merchants, are plentifully supplied. From the Mappe he brought me to the Bible, and turning to the 107 Psalme, directed mee to the 23 & 24 verses, where I read, that they which go downe to the sea in ships, and occupy by the great waters, they see the works of the Lord, and his woonders in the deepe, &c. Which words of the Prophet together with my cousins discourse (things of high and rare delight to my yong nature) tooke in me so deepe an impression, that I constantly resolved, if ever I were preferred to the University, where better time, and more convenient place might be ministred for these studies, I would by Gods assistance prosecute that knowledge and kinde of literature, the doores whereof (after a sort) were so happily opened before me.

The newe Navigation and discoverie of the kingdome of Moscovia, by the Northeast, in the yeere 1553

But after much adoe and many things passed about this matter, they grewe at last to this issue, to set downe and appoynt a time for the departure of the shippes: because divers were of opinion, that a great part of the best time of the yeere was already spent, and if the delay grewe longer, the way would bee stopt and bard by the force of the Ice, and the colde climate: and therefore it was thought best by the opinion of them all, that by the twentieth day of May, the Captaines and Mariners should take shipping, and depart from Radcliffe upon the ebbe, if it pleased God. They having saluted their acquaintance, one his wife, another his children, another his kinsfolkes, and another his friends deerer then his kinsfolkes, were present and ready at the day appoynted: and having wayed ancre, they departed with the turning of the water, and sailing easily, came first to Greenewich. The greater shippes are towed downe with boates, and oares, and the mariners being all apparelled in Watchet or skie coloured cloth, rowed amaine, and made way with diligence. And being come neere to Greenewich (where the Court then lay) presently upon the newes therof, the Courtiers came running out, and the common people flockt together, standing very thicke upon the shoare: the privie Counsel, they lookt out at the windowes of the Court, and the rest ranne up to the toppes of the towers: the shippes hereupon discharge their Ordinance, and shoot off their pieces after the manner of warre, and of the sea, insomuch that the tops of the hilles sounded therewith, the valleys and the waters gave an Eccho, and the Mariners, they shouted in such sort, that the skie rang againe with the noyse thereof. One stoode in the poope of the ship, and by his gesture bids farewell to his friendes in the best manner hee could. Another walkes upon the hatches, another climbes the shrowds, another stands upon the maine yard, and another in the top of the shippe. To be short, it was a very triumph (after a sort) in all respects to the beholders.

The description of the countrey of Russia

The whole Countrey differeth very much from it selfe, by reason of the yeere: so that a man would marveile to see the great alteration and difference betwixt the Winter, and the Summer Russia. The whole Countrey in the Winter lieth under snow, which falleth continually, and is sometime of a yard or two thicke, but greater towards the North. The Rivers and other waters are all frosen up a yard or more thicke, how swift or broade so ever they bee. And this continueth commonly five moneths, viz. from the beginning of November till towardes the ende of March, what time the snow beginneth to melt. So that it would breede a frost in a man to looke abroad at that time, and see the Winter face of that Countrey. The sharpenesse of the aire you may judge of by this: for that water dropped downe or cast up into the aire congealeth into yce before it come to the ground. In the extremitie of Winter, if you holde a pewter dish or pot in your hand, or any other mettall (except in some chamber where their warme stoaves bee) your fingers will friese fast unto it, and drawe off the skinne at the parting. When you passe out of a warme roome into a colde, you shall sensibly feele your breath to waxe starke, and even stifeling with the colde, as you drawe it in and out. Divers not onely that travell abroad, but in the very markets, and streetes of their Townes, are mortally pinched and killed withall: so that you shall see many drop downe in the streetes; many travellers brought into the Townes sitting dead and stiffe in their Sleds. Divers lose their noses, the tips of their eares, and the bals of their cheekes, their toes, feets, &c. Many times (when the Winter is very hard and extreeme) the beares and woolfes issue by troupes out of the woods driven by hunger, and enter the villages, tearing and ravening all they can finde: so that the inhabitants are faine to flie for safegard of their lives. And yet in the Sommer time you shal see such a new hiew and face of a Countrey, the woods (for the most part which are all of firre and birch) so fresh and so sweete, the pastures and medowes so greene and well growen, (and that upon the sudden) such varietie of flowers, such noyse of birdes (specially of Nightingales, that seeme to be more lowde and of a more variable note then in other Countreys) that a man shall not lightly travell in a more pleasant Countrey.

And this fresh and speedy growth of the Spring there seemeth to proceede from the benefits of the snow: which all the Winter time being spread over the whole Countrey as a white robe, and keeping it warme from the rigour of the frost, in the Spring time (when the Sunne waxeth warme, and dissolveth it into water) doeth so throughly drench and soake the ground, that is somewhat of a sleight and sandie mould, and then shineth so hotely upon it againe, that it draweth the hearbes and plants foorth in great plentie and varietie, in a very short time. As the Winter exceedeth in colde, so the Sommer inclineth to over much heat, specially in the moneths of June, July and August, being much warmer then the Sommer aire in England.

The Famous voyage of Sir Francis Drake into the South sea, and therehence about the whole Globe of the earth

The 17. day of August we departed the port of S. Julian, & the 20. day we fell with the streight or freat of Magellan going into the South sea, at the Cape or headland whereof we found the bodie of a dead man, whose flesh was cleane consumed.

The 21. day we entred The streight, which we found to have many turnings, and as it were shuttings up, as if there were no passage at all, by meanes whereof we had the wind often against us, so that some of the fleete recovering a Cape or point of land, others should be forced to turne backe againe, and to come to an anchor where they could.

In this streight there be many faire harbors, with store of fresh water, but yet they lacke their best commoditie: for the water is there of such depth, that no man shal find ground to anchor in, except it bee in some narow river or corner, or betweene some rocks, so that if any extreme blasts or contrary winds do come (whereunto the place is much subject) it carieth with it no small danger.

The land on both sides is very huge & mountainous, the lower mountains whereof, although they be monstrous and wonderfull to looke upon for their height, yet there are others which in height exceede them in a strange maner, reaching themselves above their fellows so high, that betweene them did appeare three regions of cloudes.

These mountaines are covered with snow: at both the Southerly and Easterly partes of the streight there are Islands, among which the sea hath his indraught into the streights, even as it hath in the maine entrance of the freat.

This streight is extreme cold, with frost and snow continually; the trees seem to stoope with the burden of the weather, and yet are greene continually, and many good and sweete herbes doe very plentifully grow and increase under them.

The bredth of the streight is in some place a league, in some other places 2. leagues, and three leagues, and in some other 4. leagues, but the narrowest place hath a league over.

The 24. of August we arrived at an Island in the streights, where we found great store of foule which could not flie, of the bignesse of geese, whereof we killed in lesse then one day 3000. and victualled our selves throughly therewith.

The 6. day of September we entred the South sea at the Cape or head shore.

An excellent ruttier for the Islands of the West Indies ....

The course that a man must keepe departing in winter for the Indies from Sant Lucar

Departing from Sant Lucar in winter thou shalt goe West and by South keeping along the coast, because if thou goe farre from the coast, thou shalt meete with the wind off the sea untill thou be as high shot as Cape Cantin, which is a low flat cape with the sea. And thou shalt see a great wood before thou come at this cape, called Casa del Cavallero. And from thence thou shalt steere thy olde course, that is Southwest and by South for the Isles of Alegranza, and Lancerota; and when thou art North and South with Alegranza, thou shalt steere thence Southwest, and so thou shalt see the Canaria, which is a round high land, and standeth in twentie eight degrees.

The course from Santo Domingo to go for Nueva Espanna
I advise thee that if thou wilt goe from Santo Domingo for Nueva Espanna, thou shalt goe Southwest and by South, and so thou shalt have sight of Punta de Nizao, which is a low point, and is the end of the hilles called Sierras de las minas Viejas, and towards the Northwest of them thou shalt see a lowe land, and to goe into Hocoa thou shalt stirre from this poynt of Nizao Westnorthwest, and thou shalt see the point of Puerto Hermoso, and the Bay that it maketh: and thou must be sure to keepe neere the shore to find a good road, and feare not to go neere the land: for all is deepe water, and cleare ground, and let not fall thine anker til thou be past all the rivers; and beware of the land, for it thou ride much without, thy anker wil come home, because it is rocky and flatte ground. And thou must be ready, that when thine anker commeth home, thou have thy moarings readie in thy boat to carry on shore with foure or five men, and if thou thinke good, thou mayest let them fall on land with a rope. And when thou are come to anker thou mayest send on shore to moare, so shalt thou be best moared.

A principal ruttier . . . to sail to . . . Puerto Rico, Hispaniola and Cuba

The course from the Canaries to the West Indies
If you set saile from any of the Islands of the Canaries for the West Indias, you must stirre away 30. or 40 leagues due South, to the ende you may avoid the calmes of the Island of Fierro: and being so farre distant from the said Island, then must you stirre away West Southwest, untill you finde your selfe in 20. degrees, and then saile West and by South untill you come to 15. degrees and ½. And then thence stirre away West and by North; and so shall you make a West way by reason of the Northwesting of the Compasse: which West way will bring you to the Island of Deseada.

The markes of the Island of Deseada
This Island Deseada lieth East Northeast, and West Southwest, having no trees upon it, and it is proportioned like a Galley, and the Northeast end: thereof maketh a lowe nose like the snowt of a galley; and by comming neere it, and passing by the Norther ende thereof, you shall perceive white broken patches like heapes of sand with red strakes in them: & the South-

west end of this Island maketh like the tilt of a galley. And this Island standeth in 15. degrees and ½.

Markes of the Island of Monserate

Monserate is an high Island, and round, full of trees, and upon the East side thereof you shall perceive certain white spots like sheetes: and being upon the South side at the very point of the Island, somewhat off the land, it maketh like a litle Island, and putting your selfe either East or West from that point, in the midst thereof will appeare a great broken land.

Markes of the Island of Marigalanta

Marigalanta is a smooth Island, and full of wood or trees, and as it were of the fashion of a galley upon her decke: and being on the Southeast side about half a league off you shall make certaine homocks of blacke stones, and certain white patches: but on the West side appeare faire white sandy shores or plaines.

88. William Bourne presents the basic rules of navigation to his fellow seamen

William Bourne (c. 1535–1582) was a gunner from Gravesend, a town on the Thames below London. He published his *Almanacke and Prognostication* and his *Regiment for the Sea* in the 1570s for the use of English seamen. In his own words, he felt that "seamen of the meanest sort— . . . those that sail for single hires—[should] practice one of these three faculties, that is to say: either be a gunner, a carpenter, or else to be a navigator." Bourne's teachings, written to be easily understood, are represented here by passages concerning the principal rules of navigation: how to be a good navigator; how to take the altitude of the sun and know the elevation of the North Pole; how to use the astrolabe and compass and know the variation of the latter instrument; and how to reckon longitude.

The table of the contentes of these Rules folowing

1   The first Rule is of a good Nauigator.
2   The seconde Rule treateth of the .32. wyndes, belonging to a Nauigator, otherwyse called the 32. pointes of the Compas.

From William Bourne, *A Regiment for the Sea*, ed. E. G. R. Taylor (London: Hakluyt Society, 2d ser., no. 121, 1963), pp. 55–59, 83–85, 93–95, 168–169, 170–171. © The Hakluyt Society, London.

3   The thirde Rule treateth of the golden number or Prime, shewing the Epact, & by the Epact to knowe the age of the Moone.

4   The fourth Rule teacheth you to know by the age of the moone whan it doth flowe at any place where you doe knowe what moone maketh a full Sea, with a tabell of tides ioyned therto.

5   The fifth rule treateth of yᵉ Sunnes & Moones course in the zodiake, and how that you shal know at what houre that the moone shall ryse and set, & at what point, and wynde, with other necessarye thinges.

6   The sixt rule is of a table of declination for .3 yeares, exactly calculated for euery day af yᵉ mōth.

7   The seuenth rule sheweth you how to take the altitude of the Sūne, and by the height of yᵉ Sun to knowe the Equinoctiall, and by the altitude of the Equinoctiall, to know the eleuatiō of the pole artike, and howe that you shall behaue yourselfe with your Astralabe, with an ensample of grauesende, and also howe to gette the true meridiā, and also of the Northeasting and Noothweastinge of the compas, necessary for nauigation, otherwise called the variation of the compass.

8   The eigh rule of the north starre, & howe that she should be taken vpon any of the eight principal wyndes or pointes of the compas with an obseruation of the balastela or crosestaffe.

9   The ninth rule is of the sayling of vpon one of the quarters of the compas in howe farre sayling you doe rayse or laye a degree, and what you doe departe from your meridian.

10   The .x. rule treateth of the soūdinges, coming from any place out of the Occidentall Sea.

For to seke vshant or the lezarde, and also all a longest till you doe come to the coast of Flaūders.

11   The .xi. rule treateth of the logitude, although that it be very tedious.

12   The .xii. rule sheweth howe many myles wyll in square to one degree of longitude in euery seuerall latitude betwene the Equinoctiall and any of the twoo poles.

13   The .xiii. rule teacheth of the longitude & the latitude of certaine of the moste notable townes in Englande, and also how lōg that the moone doth chaunge at the one towne before the other, and also the diuersitie of the longest date in Sommer from Southhampton to the North or moste place in Scotlande.

14   The .xiiii. rule is of the longitude and declination of .12. notable fixed starres for nauigation with tables of their shining, and at what point of your compas that they doe both rise and sette, and also tables for euery moneth in yᵉ yeare, declaring at what houre and minute that thei be south running from the first daye of euery moneth to the .15. and from the .15. daye to the last day, and will continue this .100. yeare without much error.

15   The .xv. rule, how to sayle by the Globe.

16   The .xvi. how to know the houre of the day by the Compas.

The first Rule is of a good Nauigator

Of all sciences that is vsed with us in Englād, Nauigation is one of the principall & most necessary for the benefite of our Realme and natiue countrey, and also most defencible against our enemies, because we bee enuironed rounde aboute with the sea, and there be a great nomber of good and wittie Nauigators, but notwithstāding some simple, and therefore I would wishe those that be Sea men of the meanest sorte (I meane those that sayle for single hires) to practise one of these three faculties, that is to saie: either to be a Gonner, a Carpenter or els to be a Nauigator. For I do see a great nomber that do occupie the Sea, that haue no sight almoste at all in their science, although they have occupied the Sea a long time, whiche is a straunge case, that those men that haue had the dealing therewith to be vtterly voide of knoweledge, their maisters to haue both knowledge and conning, and they to bee altogether ignoraunt, which is occasion that I do thinke that the expert maisters in deede, doe seldome instruct their companie: For if they did, there is no doubte but the multitude would haue more vnderstanding. But wherefore should I waste time in wryting hereof, seing it is instructions that must amend this fault: Wherefore that you may the better vnderstande or knowe my purpose, you shall vnderstande that the good and wittie Nauigator, doth consider and knowe by experience, that the moone doth rule the flouddes, or els he knoweth that in one Moone there is two springes, & two nepes, the one spring to be at the full moone or three daies after, the other at the chaunge, or within twoo or three daies after, and at both the quarters the nepe streames. He doth knowe by experience, that the moone in suche a quarter of the skie maketh a full Sea, so that the raging wyndes doth neither hinder the same, nor cause it to be altered. He dothe knowe how that the streame or floudes or ebbes, doth set from place to place, from nas to nas, from poynt to poynt, in euery place about the Coaste that he hath occupied. And the chiefest thinge that belongeth to a Sea faring man, is to knowe the place that he shall happen to fall with, whiche thing he must knowe by the beholding of the countrey, by taking some principall marke thereof, & the chiefest thing is, to beholde the hilles and vales of the lande, that he may knowe them when he shall happen to see them againe, vpon euery side he must take heade. And as for wooddes and hedges with suche like, are not to be marked, because suche thinges may be cut or felled downe, and so youre marke is lost: Wherfore hylles, vales, cliftes, and Castels, with stepels and Churches, are the beste and most surest markes, that may or can be taken, and are better than hedgrowes, wooddes or trees. Furthermore, the good and wittie Nauigator, by the marke of the shore doth knowe whether there doth lie any daungers, as sandes, rockes, bankes, or shelffes: he knoweth by their soundinge howe neare that he is vnto them, he doth knowe when he hath any occasion to put into any herborowe, whether he haue water enough, yea or nay. Furthermore, he doth knowe by his soundinge, of the depth, and by the grounde that sticketh vpon the tallowe of the lead howe farre that he is shott, although he may see no lāde. Furthermore, he doth knowe when he is in the Occident Sea, by keeping of his course, with what place he shall fall or come to first, he knoweth howe farre that the ship shippe hath roone or gone, and howe farre that she hath to goe, by keping of accompt or pricking of his carde: he knoweth whether that the shippe goeth to leewardes, or maketh hir waye good: he knoweth whether about the head of the land there runneth any curraunt or not: He knoweth that the indraftes of the lande, causeth the tydes for to runne: he knoweth that the whyte waters or land waters, causeth the ebbes to runne swifter then the flouddes: he knoweth that in runninge up in to a Ryuer, the further he runneth in a floudde, the longer tyde or flowing with him: he knoweth that the salt water or sea water, is more stronger to beare a shippe then the freshe waters or ryuer waters be. Furthermore, he knoweth by the taking of the altitude or height of the Sunne, in what Parrell or Latitude he is in, considering the Sunnes declination, howe farre that he is aboue or beneath the Equinoctial lyne. And furthermore they doe knowe their Latitude or parrel by diuers starres fixed in the firmamēt, as by Oculus tauri, as Orione, his girdell, his shoulders, his feete, and also Alober, or otherwyse called the great Dogge and little dogge, with diuers other starres whiche I passe ouer. And some doe obserue the North star being the tippe of the tayle of Ursa minor, or little beare, but many mē be deceiued in taking of him. But for them whiche doe occupie the South, he is very good, so that they doe knowe the whole compasse of remouing. And for them that doe occupie the North, he will not serue their tournes, because his altitude is so great that your balastela wyll not take him perfecte.

The .vii. rule

Nowe in this table you must consider that the xi. daie of Marche, the Sunne is Equinoctiall, entring then the first point of Aries, called the Equinoctiall of Spring time, then hauing no declination, then the .x. daie of Aprill the Sunne entreth into the first minute of Taurus, thē hauing declination to the northwardes .xi. degrees .30. minutes, then the .xii. daie of May, the Sunne entreth the first point of Gemini, hauing then declination .20. degrees. .12. minut. the .12. daie of June, the Sunne entreth into Cancer, then the Sunne maketh his greatest prograce to the northwards, hauing .23. degrees .28. minutes of declinatiō now in this our time. But some doe affirme to be .23. degrees and a halfe, but there lacketh twoo min. Then the .14. daie of July, the Sunne entreth into Leo, coming downwardes to the equinoctiall, hauing .20. degrees .12. minutes. The .14. daie of August, the Sunne entreth into virgo, the declinatiō 11. degrees .30. minutes. Then

the .14. of September, the Sunne entreth into Libra, then being Equinoctiall, hauing no declination called the equinoctiall of Autumne or haruest,
then beginninge her south declination: then the .14. of October the Sunne
entreth into Scorpio, the declination .11. degrees .30. minutes. Then the
.12. of Nouember, the Sunne entreth into Sagittarius, the declination 20.
degrees .12. minutes. Then the .12. daie of December, the Sunne entreth
the first minute of Capricorne, then maketh the Sunne her greatest prograce to the southwardes and his declination .23. degrees .28. minutes, and
then retourneth to the Equinoctiall againe. Then the .11. of January the
Sunne entreth into Aquarius, and the declinatiō 20. degrees .12. minutes.
Then the .10. daie of February, the Sunne entreth into the first minute of
Pisces, and the declination .11. degrees .30. minut. then the .11. daie of
March, the Sunne returneth to the self same place that it did depart from
before. Wherfore the Egiptians did paint the yeare like to an Ader lightning her taile, hauing not the vse of letters, they made a ringe and named
it Anulus as it weare annus, that is a yeare, because a ring doth turne
round in it self as doth the yeare. Now you hauing your Astralabe, if you
doe require to know how many degrees the pole artick is aboue your horizon, take your Astralabe and hange it vpon one of your fingers, and lift vp
or put down the athladaie or rule with yᵉ sightes till yᵉ beames of the Sunne
doth pearse both the sightes of the rule or atheladaye, the Sunne beames
geuing shadowe through both the sightes, then looke in the table what
declination the sunne hath, whether that the declination be towardes the
south of the equinoctiall or towardes the north of it, then loke vpon your
astralabe what altitude the Sūne hath vpon the meridian vpon that daye
of the moneth in your table, then if that it hath north declinatiō subtract,
or pull away your declination, if south declination, adde or put to your
declination to the altitude of height of the Sunne, then that doth shew you
the true equinoctiall: then when you haue the true height of the equinoctiall, looke how many degrees that cometh vnto, subtract or pull that sum
out of .90. degrees, which degrees & minutes, then that doth remaine shalbe
the height of the pole aboue your horizon. For this you must cōsider, that
from the zeneth or prick ouer the crowne of your head to be .90. degrees
downe to the horizon, then looke what height the equinoctiall is from the
horizon, so much is the zeneth from the pole, thē must it nedes be said
that that is the altitude of the pole to be iust the distaunce of the zeneth
downe to the equinoctiall. As for an exāple, this at Grauesende the yeare
.1566. I take the Sunne vpon the meridian the .10. daie of Aprill, & founde
the altitude of the Sunne lifted aboue the horizon .49. degrees 49. minutes,
then I toke the Sunne the next daie and founde the Sunne vpon
the meridian. 50. degrees .9. min. so I found the declinatiō of the sunne
more the .11. daie then it was the .10. daie by .21. mi. which signifieth
to me that the Sunne entred into the first min. of Taurus, at one of the
clocke after midnight. The .x. daie of Aprill, the sunne hauing north

declination of the equinoctiall, and the declination .11. degrees .40. mi. Upon the .11. daie of the moneth, then I pulled yᵉ declinatiō .11. degrees 40. mi. out of .50. degrees .9. mi. the remainder was 38. degrees .29. mi. The altitude or height of the equinoctiall aboue the horizon: nowe I doe take or subtract .38. degrees .29. mi. out of .90. degrees, the remainder is the height of yᵉ pole, being .51. degrees. 31. mi. So in like case it is from the zeneth to the pole .38. degrees .29. min. & from the pole downe to the horizon .51. degrees .31. mi. as by the example of this figure.

The eleuenth Rule.

Nowe some there be that be very inquisatiue to haue a waye to gette the Longitude, but that is to tedious, for this they must consider, that the whole frame of the firmament is caried roūde from the east into the weast, in .24. houres, so there remayneth no light nor marke, but goeth rounde, sauing only the two poles of the world, and these twoo standeth faste. But as I saide before in the ix. Rule, he that goeth south or north, doth rayse or laye the pole, and in the like case of the Equinoctiall altering his parell, causing the lightes of the firmament to alter the time of their shining or byding aboue our horizon, and he that goeth directly east or weast, doth neither rayse nor delay the pole but still the lightes of the firmament, doth make one maner of arche, according to their latitude or declination, but the going east or weast, doth alter the meridiane, causinge the Planes to haue their aspectes at an other hour or tyme, altering the tyme of the chaunges of the moone, and also the tyme of the Eclipses, whiche is necessary for all trauaylers by Sea or by Lande. Therefore I thought it neadefull to bee spoken of. For as countries haue Latitude from the poles, so in like manner they haue appointed longitude. But nowe you may get the Latitude with instrumentes, but the Longitude you must bringe from an other place, whiche you can not doe but with a Globe, or els a Mappe or Carde, & then you must measure from the meridian landes of the Canary Ilandes, or otherwyse called the fortunate Ilāds: and in our Latitude of London euery .555. myles, whiche containeth .15. degrees, will aunswere to one houre of tyme, and vnder the Equinoctial .900 myles to fiftene degrees, the degrees be as longe as the degrees of Latitude, but towardes the pole fewer and fewer, till they come to nothing, vnder the twoo poles. And nowe .37. myles with vs at London, will aunswere to one degree to our Latitude at .51. or .52. degrees of eleuation of the pole. But the cause why the Longitude was fechte frō the Canary Ilandes, I knowe not: Yet as I suppose, because that it was thē the westermost place then knowen. For Ptholomeus was the first that ordayned that rule. Nowe furthermore, because that you shall knowe the better, I wyll drawe out certayne of the chiefest places about this Realme of Englande, both their Longitude and Latitude, by whiche you shall knowe what manner of arche the Sunne with the other lightes doth make, and also by the Longitude to knowe at what tyme the Moone with any of the Planetes doth make any aspecte, & also the Eclipses of the Sūne or moone, with the chaunge, quarters, and full moone, by a true and epact Ephimerides through all Englād, to knowe the very true houre and minute of the tyme of the deametre, considering for what Longitude or place your Almanack was made for, & nowe to gette the Longitude, you may at the time of the Eclipse of the Moone, for the Eclipses of yᵉ Moone be generall, so that she is aboue your Horizon in any place vpon the supersticial partes of the earth or sea, considering as I saide before, by your Almanack, at what tyme the Eclipse should happen the very houre & minute, knowinge the place that your Almanack was made for, and then according to this rule with a precise instrument the alteration of the time and houre and minute of the Eclipse. And furthermore, you may knowe your longitude by the Ephimerides by the coniunction of the moone with the other fixed starres, and by the distaunce betwene them with a precise instrument, considering the moones course with degrees and minutes, but I am of opinion that it is to tedious for to be done vpon the Sea, but it may be done vpon the land, for the Sea doth always lift the shippe vp and down, and the least chop of a sea causeth a man to committe errour. Therfore lette no Sea men trouble them selfe with this rule, but according to their accustomed manner, lette them kepe a perfect accoumpt and reckening of the way of his shippe, whether the ship goeth to lewardes or maketh her way good, cōsidering what things be against him or with him: as tydes, corrantes, wyndes, or suche like.

What Nauigation is

Nauigation is this, how to direct his course in the Sea to any place assigned, and to consider in that direction what things may stande with him, & what things may stand against him, hauing consideration how to preserue the ship in all stormes and chaunges of weather that may happen by the way, to bring the ship safe vnto the port assigned, and in the shortest time.

The vse of Nauigation

The vse thereof is this, fyrste to knowe how that the place dothe beare from him, by what winde or poynte of the compasse, and also how farre that the place is from hym, and also to consider the streame, or tide gates, Currents, which way that they do set or driue the ship, and also to consider what daungers is by the waye, as rockes and sandes, and suche other lyke impedimentes, and also if that the wynde chaunge or shifte by the waye, to consider which way to stand, and direct his course vnto the most aduantage to attayne vnto the port in shortest time: and also if anye stormes doe happen by the way, to consider how far to preserue the shippe and the goodes, and too bring hir safe vnto the porte assigned. And also it is moste principally to be considered and foreseene, that if they haue hadde by occasion of a contrarye tempest, for too goe very muche out of the course or way, too knowe then howe that the place dothe then beare, that is to say, by what poynte of the compasse the place dothe stande from you: and also how farre it may be from you. Whyche way to bee knowne is this: firste to consider by what poynte that the shippe hath made hir way by, and how fast and swiftly that the shippe hathe gone, and to consider how often that the shippe hath altered hir course, and how muche that she hathe gone at euery tyme, and then to consider all thys in youre platte or carde, and so you may gyue an neere gesse, by what poynte or wynde it beareth from you, and also howe farre it is thither. And also you may haue a greate helpe by the Sunne or Starres, to take the heigthe of the Pole aboue the horizon, and also in some place you may gesse by the sounding, bothe by the depth, and also by the grounde. And also it is very meete and necessarye to knowe any place, when that hee dothe see it.

What maner of persons be meetest to take charge of Shippes in Nauigation

As touching those persons that are meete to take charge, that is to say, to be as maister of ships in Nauigation, he ought to be sober and wise, and not to be light or rash headed, nor to be to fumish or hasty, but such a one as can wel gouern himselfe, for else it is not possible for him to gouerne his cōpany well: he ought not to be to simple, but he must be suche a one as must keepe his companie in awe of him (by discretion) doing his companie no iuiurie or wrong, but to let thē haue that which men ought to haue, and then to see vnto them that they doe their laboure as men ought to doe in all points. And the principall point in gouernment is, to cause himself both to be feared & loued, & that groweth principally by this meanes, to cherishe men in well doing, and those men that be honestly addicted, to let them haue reasonable preheminence, so that it be not hurtfull vnto the Marchaunt nor to himselfe, and to punishe those that be malefactors and disturbers of their company, and for smal faults, to giue them gentle admonition to amende them: and principally these two pointes arte to be foreseene by the maisters. (that is) to serue God himselfe, and to see that the whole companie do so in like maner, at suche conuenient time as it is meete to be done: the second point is, that the master vse no play at the dise or cards, neither (as near as he cā) to suffer any, for yᵉ sufferance thereof may do very much hurt in diuers respects: And furthermore, the maister ought to be suche a one, as dothe knowe the Moones course, whereby he doth knowe at what time it is a full Sea, or a lowe water, knowing in what quarter or part of the skye, that the Moone doth make a full Sea at that place, and also the master ought to bee acquainted, or knowe that place well, that he doth take charge to goe vnto (except that he haue a Pilot) and also he that taketh charge vpon him, ought to be expert, how the tydegates or currentes doe set from place vnto place: and also not to bee ignorant of such daungers as lyeth by the way, as rocks, sandes, or bankes, and also most principally he ought to bee such a one, as can very well directe his courses vnto any place assigned, and to haue capacitie howe for to handle or shift himselfe in foule weather or stormes. And also it behoueth him too be a good coaster. that is to say, to knowe euery place by the sight thereof. And also he that taketh charge for long voyages, ought to haue knowledge in plats or cardes, and also in such instrumentes as be meet to take the heigth of the Sunne or any Starre, and to haue capacitie to correcte those instrumentes, and also he ought to be such a one, that can calculate the Sunnes declination, or else to haue some true regiment, and also he ought to knowe howe to handle the Sunnes declination, when that he hath taken the heigth of the Sunne.

89. Captain James Cook: secret orders from the Admiralty and his description of New South Wales

Captain James Cook, R.N. (1728–1779), was the outstanding figure in eighteenth-century maritime exploration. Considered to be one of the finest navigators and marine surveyors, he was put in command, in 1768, of the *Endeavour*, bound for the South Pacific to observe the transit of Venus. The first excerpt in this segment contains the secret instructions issued by the Admiralty for that voyage; the second, observations on New South Wales taken from Cook's journal. On his return from that voyage, in 1771, Cook was entrusted with command of a squadron consisting of the *Resolution* and the *Adventure*; his mission was to search for the unknown continent in the southern oceans, *terra australis nondum cognita* as it had been called on maps before his time. Although he did not find the southern continent, Cook added a great deal to the knowledge of those seas. In 1776 he sailed on his third voyage in command of the *Resolution* and the *Discovery*. His was the first English voyage to explore the northern Pacific; Cook also discovered the Hawaiian Islands, where he met his death at the hands of the natives in 1779.

From *The Explorations of Captain James Cook,* ed. A. G. Price (New York: Heritage Press, 1958), pp. 18–20, 81–84.

Secret   By the Commissioners for executing the office of Lord High Admiral of Great Britain &c^a.

Additional Instructions for Lt James Cook, Appointed to Command His Majesty's Bark the Endeavour

Whereas the making Discoverys of Countries hitherto unknown, and the Attaining a Knowledge of distant Parts which though formerly discover'd have yet been but imperfectly explored, will redound greatly to the Honour of this Nation as a Maritime Power, as well as to the Dignity of the Crown of Great Britain, and may trend greatly to the advancement of the Trade and Navigation thereof; and Whereas there is reason to imagine that a Continent or Land of great extent, may be found to the Southward of the Tract lately made by Captⁿ Wallis in His Majesty's Ship the Dolphin (of which you will herewith receive a Copy) or of the Tract of any former Navigators in Pursuits of the like kind; You are therefore in Pursuance of His Majesty's Pleasure hereby requir'd and directed to put to Sea with the Bark you Command so soon as the Observation of the Transit of the Planet Venus shall be finished and observe the following Instructions.

You are to proceed to the southward in order to make discovery of the Continent above-mentioned until you arrive in the Latitude of 40°, unless you sooner fall in with it. But not having discover'd it or any Evident signs of it in that Run, you are to proceed in search of it to the Westward between the Latitude before mentioned and the Latitude of 35° until you discover it, or fall in with the Eastern side of the Land discover'd by Tasman and now called New Zeland.

If you discover the Continent above-mentioned either in your Run to the Southward or to the Westward as above directed, You are to employ youself diligently in exploring as great an Extent of the Coast as you can; carefully observing the true situation thereof both in Latitude and Longitude, the Variation of the Needle, bearings of Head Lands, Height, direction and Course of the Tides and Currents, Depths and Soundings of the Sea, Shoals, Rocks, &c^a. and also surveying and making Charts, and taking Views of such Bays, Harbours and Parts of the Coast as may be useful to Navigation.

You are also carefully to observe the Nature of the Soil, and the Products thereof; the Beasts and Fowls that inhabit or frequent it, the fishes that are to be found in the Rivers or upon the Coast and in what Plenty; and in case you find any Mines, Minerals or valuable stones you are to bring home Specimens of each, as also such Specimens of the Seeds of the Trees, Fruits and Grains as you may be able to collect, and Transmit them to our Secretary that We may cause proper Examination and Experiments to be made of them.

You are likewise to observe the Genius, Temper, Disposition and Number of the Natives, if there be any, and endeavour by all proper means to cultivate a Friendship and Alliance with them, making them presents of such Trifles as they may Value, inviting them to Traffick, and Shewing them every kind of Civility and Regard; taking Care however not to suffer yourself to be surprized by them, but to be always upon your guard against any Accident.

You are also with the Consent of the Natives to take possession of Convenient Situations in the Country in the Name of the King of Great Britain; or, if you find the Country uninhabited take Possession for His Majesty by setting up Proper Marks and Inscriptions, as first discoverers and possessors.

But if you should fail of discovering the Continent before-mention'd, you will upon falling in with New Zeland carefully observe the Latitude and Longitude in which that Land is situated, and explore as much of the Coast as the Condition of the Bark, the health of her Crew, and the State of your Provisions will admit of, having always great Attention to reserve as much of the latter as will enable you to reach some known Port where you may procure a Sufficiency to carry you to England, either round the Cape of Good Hope, or Cape Horn, as from Circumstances you may judge the Most Eligible way of returning home.

You will also observe with accuracy the Situation of such Islands as you may discover in the Course of your Voyage that have not hitherto been discover'd by any Europeans, and take possession for His Majesty and make Surveys and Draughts of such of them as may appear to be of Consequence, without Suffering yourself however to be thereby diverted from the Object which you are always to have in View, the Discovery of the Southern Continent so often Mentioned.

But for as much as in an undertaking of this nature several Emergencies may Arise not to be foreseen, and therefore not particularly to be provided for by Instruction before hand, you are in all such Cases, to proceed, as upon advice with your Officers you shall judge most advantageous to the Service on which you are employed.

You are to send by all proper Conveyances to the Secretary of the Royal Society Copys of the Observations you shall have made of the Transit of Venus; and you are at the same time to send to our Secretary, for our information, accounts of your Proceedings, and Copys of the Surveys and drawings you shall have made. And upon your Arrival in England you are immediately to repair to this Office in order to lay before us a full account of your Proceedings in the whole Course of your Voyage, taking care before you leave the Vessel to demand from your Officers and Petty Officers the Log Books and Journals they may have Kept, and to seal them up for our inspection, and enjoyning them, and the whole Crew, not to divulge where they have been until they shall have Permission so to do.

Given under our hands the 30th of July 1768.

Ed. HAWKE
Piercy BRETT
C. SPENCER

By Command of their Lordships
PH. STEPHENS

New South Wales

Thursday 23rd August. In the Course of this Journal I have at different times made mention of the appeerence or Aspect of the face of the Country, the nature of the Soil, its produce &ca. By the first it will appear that to the Southward of 33° or 34° the Land in general is low and level with very few Hills or Mountains, further to the northward it may in some places be called a Hilly, but hardly any where can be call'd a Mountainous Country, for the Hills and Mountains put together take up but a small part of the Surface in comparison to what the Planes and Vallies do which intersect or divide these Hills and Mountains: It is indefferently well watered, even in the dry Seasons, with small Brooks and springs, but no great Rivers, unless it be in the wet Season when the low lands and Vallies near the Sea I do suppose are mostly laid under water; the small brooks may then become large Rivers but this can only happen with the Tropick. It was only in *Thirsty Sound* where we could find no fresh Water, excepting one small pool or two which Gore saw in the woods, which no doubt was owing to the Country being there very much intersected with Salt creeks and Mangrove land.

The low Land by the Sea and even as far in land as we were, is for the most part friable, loose, sandy Soil; yet indefferently fertile and cloathed with woods, long grass, shrubs, Plants &ca. The Mountains or Hills are Chequered with woods and Lawns. Some of the Hills are wholy covered with flourishing Trees; others but thinly, and the few that are on them are small and the spots of Lawns or Savannahs are Rocky and barren, especially to the northward where the country did not afford or produce near the Vegetation that it does to the southward, nor were the Trees in the woods half so tall and stout.

The Woods do not produce any great variety of Trees, there are only 2 or 3 sorts that can be call'd Timber; the largest is the Gum Tree which growes all over the Country, the Wood of this Tree is too hard and ponderous for most common uses. The Tree which resembles our Pines, I saw no where in perfection but in Botany Bay, this wood as I have before observed is some thing of the same nature as America Live Oak; in short most of the large Trees in this Country are of a hard and ponderous nature and could not be applied to many purposes. Here are several sorts of the Palm kind, Mangro[v]es and several other sorts of small Trees and shrubs quite unknown to me besides a very great Variety of Plants hetherto unknown, but these things are wholy out of my way to describe, nor will this be of any loss sence not only Plants but everything that can be of use to the Learn'd World will be very accuratly described by Mr Banks and Dr Solander. The Land naturly produces hardly any thing fit for man to eat and the Natives know nothing of Cultivation. There are indeed found growing wild in the woods a few sorts of fruits (the most of them unknown to us) which when ripe do not eat a miss, one sort especially which we call'd Apples, being about the size of a Crab-Apple, it is black and pulpy when ripe and tastes like a Damson, it hath a large hard stone or kernel and grows on Trees or Shrubs.

In the Northern parts of the Country as about *Endeavour River,* and probably in many other places, the Boggy or watery Lands produce Taara or Cocos which when properly cultivated are very good roots, without which they are hardly eatable, the tops however make very good greens.

Land Animals are scarce, as so far as we know confined to a very few species; all that we saw I have before mentioned, the sort that is in the greatest plenty is the Kangooroo, or Kanguru so call'd by the Natives; we saw a good many of them about Endeavour River, but kill'd only Three which we found very good eating. Here are like wise Batts, Lizards, Snakes, Scorpions, Centumpees &c^a but not in any plenty. Tame Animals they have none but Dogs, and of these we saw but one and therefore must be very scarce, probably they eat them faster than they breed them, we should not have seen this one had he not made us frequent Visets while we lay in Endeavour River.

The Land Fowles are Bustards, Eagles, Hawks, Crows such as we have in England, Cockatoes of two sorts, white and brown, very beautifull Birds of the Parrot kind such as Lorryquets &c^a, Pidgeons, Doves, Quales, and several sorts of smaller birds. The Sea and Water Fowls are Herons, Boobies, Nodies, Guls, Curlews, Ducks, Pelicans &c^a and when Mr Banks and Mr Gore were in the Country at the head of Endeavour River they saw and heard in the night great numbers of Geese. The sea is indifferently well stock'd with Fish of various sorts, such as Sharks, Dog-fish, Rock-fish, Mullets, Breames, Cavallies, Mackarel, old wives, Leather-Jackets, Five-fingers, Sting-Rays, Whip-rays &c^a—all excellent in their kind. The Shell-fish are Oysters of 3 or 4 sorts, viz Rock oysters and Mangrove Oysters which are small, Pearl Oysters, and Mud Oysters, these last are the best and largest; Cockles and Clams of Several sorts, many of these that are found upon the Reefs are of a Prodigious size; Craw-fish, Crabs, Musles, and a variety of other sorts. Here are also among and upon the Shoals & reefs great numbers of the finest Gree[n] Turtle in the world and in the Rivers and salt Creeks are some Aligators.

The Natives of this Country are of a middle Stature straight bodied and slender-limbd, their skins the Colour of Wood soot or of a dark Chocolate, their hair mostly black, some lank and others curled, they all wear it crop'd short, their Beards which are generaly black they like wise crop short or singe off. Their features are far from being disagreeable and their Voices are soft and tunable. They go quite naked both Men and women without any manner of Cloathing whatever, even the Women do not so much as Cover their privities. Altho none of us were ever very near any of their women, one gentleman excepted, yet we are all as well satisfied of this as if we had lived among them. Notwithstanding we had several interviews with the Men while we lay in Endeavour River, yet whether through Jealousy or disrigard they never brought any of their women along with them to the Ship, but always left them on the opposite side of the River where we had frequent oppertunities [of] Viewing them through our glasses. They wear as Ornaments Necklaces made of shells, Bracelets or hoops about their arms, made mostly of hair twisted and made like a cord hoop, these they wear teight about the uper parts of their Arms, and some have girdles made in the same manner. The men wear a bone about 3 or 4 Inches long and a fingers thick, run through the Bridge of the nose, which the Seamen call'd a sprit sail yard; they like wise have holes in their ears for Earrings but we never saw them wear any, neither are all the other oraments wore in common for we have seen as many without as with them. Some of those we saw on Posession Island wore Breast Plates which we suppose'd were made of Mother of Pearl shells. Many of them paint their bodies and faces with a sort of White paist or Pigment, this they apply different ways each according to his fancy. Their Offensive weaphons are Darts, some are only pointed at one end others are barb'd, some with wood others with the Stings of Rays and some with Sharks teeth &c[a], these last are stuck fast on with gum. They throw the dart with only one hand, in the doing of which they make use of a piece of wood about 3 feet long made thin like the blade of a Cutlass, with a little hook at one end to take hold of the end of the Dart, and at the other end is fix'd a thin peice of bone about 3 or 4 Inches long; the use of this is, I beleive, to keep the dart steady and to make it quit the hand in a proper direction; by the help of these throwing sticks, as we call them, they will hit a Mark at the distance of 40 or 50 Yards, with almost, if not as much certainty as we can do with a Musquet, and much more so than with a ball. These throwing sticks we at first took for wooden swords, and perhaps on some occasions they may use them as such, that is when all their darts are expended, be this as it may they never travel without both them and their darts, not altogether for fear of enimies but for killing of Game &c[a] as I shall shew hereafter. Their defensive weapons are Shields made of wood but these we never saw use'd but once in Botany Bay. I do not look upon them to be a warlike People, on the Contrary I think them a timorous and inoffensive race, no ways inclinable to cruelty, as appear'd from their behaviour to one of our people in Endeavour River which I have before mentioned. Neither are they very numerous, they live in small parties along by the Sea Coast, the banks of Lakes, River creeks &c^a. They seem to have no fix'd habitation but move about from place to place like wild Beasts in search of food, and I beleive depend wholy upon the success of the present day for their subsistance. They have wooden fish gigs with 2, 3 or 4 prongs each very ingeniously made with which they strike fish; we have also seen them strike both fish and birds with their darts. With these they like wise kill other Animals; they have also wooden Harpoons for striking Turtle, but of these I beleive they got but few, except at the Season they come a shore to lay. In short these people live wholy by fishing and hunting, but mostly by the former, for we never saw one Inch of Cultivated land in the whole Country; they know however the use of Taara and sometimes eat them. We do not know that they eat any thing raw but roast or broil all they eat on slow small fires.

Their Houses are mean small hovels not much bigger than an oven, made of peices of Sticks, Bark, Grass &c^a, and even these are seldom used but in the wet seasons for in the dry times we know that they as often sleep in the open air as any where else. We have seen many of their Sleeping places where there has been only some branches, or peices of bark ris about a foot from the ground on the windward side. Their Canoes are as mean as can be conceived, especially to the southward where all we saw were made of one peice of the bark of Trees, about 12 or 14 feet long, drawn or tied together at one end as I have before made mention. These Canoes will not carry above 2 people, in general their is never more than one in them, but bad as they are they do very well for the purpose they apply them to, better then if they were larger, for as they draw but little water they go in them upon the Mud banks and pick up shell fish &c^a without going out of the Canoe. The few Canoes we saw to the northward were made of a log of wood hollow'd out, about 14 feet long and very narrow with out-riggers, these will carry 4 people. During our whole stay in Endevour River we saw but one Canoe and had great reason to think that the few people that resided about this place had no more; this one served them to cross the River and to go a fishing in &c^a.

German Geographers of the Sixteenth Century

During the sixty years between the first Portuguese voyage to the Gulf of Guinea, in 1462, and the return of Magellan's surviving ship from the first circumnavigation of the earth, in 1522, the European view of the world underwent a complete change. The men who sailed on the voyages of discovery were Portuguese and Spaniards, Frenchmen and Englishmen. The best maps showing the results of these discoveries were drawn, engraved, and published in Italy. But the credit for reforming geography, for giving up—at least in part— the exclusive reliance on classical and biblical authorities, and for introducing personal observation belongs to the German humanists.

The first of the excerpts in this section is part of the inaugural lecture of Barthel Stein, the first geographer to occupy a university post specifically created for geography. The other selections include the first formulation of the modern method of determining longitude; representative passages from geographical encyclopedias of the sixteenth century; and descriptions of natural phenomena and of distant lands and peoples based on personal observation.

90. Barthel Stein gives an inaugural lecture on geography

Barthel Stein (1476/77–1521/22) was the first regularly appointed— and paid—representative of geography in a university, in this case

From Barthel Stein, *Praelectio Bartholomei Steni in libellum Pomponii;* German translation by H. Markgraf (Breslau, 1902), quoted in H. Beck, *Geographie* (Freiburg-München: Karl Alber Verlag, 1973), p. 105, by permission. Editor's translation.

the University of Wittenberg. The following passage is taken from his inaugural lecture on the work of Pomponius Mela, delivered at Wittenberg in 1509. He represents the new humanistic geographer's approach, accepting as the basis for his presentations a text by a Roman writer, yet emphasizing to his audience the importance of geographical knowledge.

Who can understand the third book of the *Aeneid* of Virgil, or that same book and several others of Lucan's *Pharsalia*, without knowledge of Geography? . . . It would be indeed difficult for you to understand how Aeneas, having left Phrygia, sailed through the Hellespont to the Thracian peninsula, thence to Delos and Crete, carried later from the Tyrrhenian Sea to Africa, returned to Sicily, and proceeded from there to the mouth of the Tiber and indeed to the founding of the city of Rome, unless you had acquired knowledge of the location of these places. How could you, without knowing the region, have a clear understanding of Cato's march to Numidia across the Libyan Syrte and deserts? . . . Rather than saying too little, I shall not tell you anything about history, since it is well known that it requires a truly accurate knowledge of places. One must not only know the location of mountains, passes, streams, forests, but also that of hills, small streams, paths, slopes, bridges and fords, to fully understand how a particular event took place.

91. Gemma Frisius describes a new method to determine longitude

Roger Gemma Frisius (1508–1555), a Dutch humanist, taught medicine and mathematics at Louvain university, made globes and instruments, and was the first person to formulate the principles of triangulation. In 1530 he set forth the method of determining longitude that was to be employed throughout the western world, once the time pieces of required accuracy had been made. The following statement on longitude is a translation, with modernized spelling, of Frisius' method, taken from the first English book on the new discoveries, *The Decades of the New World.*

We see that in these our days certain little clocks are very artificially made, the which for their small quantities are not cumbrous to be carried

From Gemma Frisius, *Principles of Astronomy and Cosmography* (Louvain, 1530), as translated by Richard Eden in *The Decades of the New World* . . . (London, 1555), p. 361.

about in all voyages. These often-times move continually for the space of 24 hours, and may with help continue their moving perpetually. By the help therefore of these the longitude may be found after this manner. Before we enter into any voyage, we must first foresee that the said clock exactly observe the hours of the place from whence we depart. And again that on the way it never cease. Accomplishing therefore 15 or 20 miles of the voyage, if we desire to know how much in longitude we are distant from the place of our departure, we must tarry until the point of the style of the clock do exactly come to the point of some hour and at the same moment, by our Astrolabe or globe, ought we to seek the hour of the place where we be. The which, if it agree in minute with the hour which the Horoscopium or ascendent does show, then it is certain that we are yet under the same meridian or the same longitude, and that our course has been toward the south or north. But if it differ one hour or any minute, then are the same to be reduced by degrees or minutes of degrees, as we have taught in the chapter herebefore.

And so shall the longitude be found. And by this art can I find the longitude of regions, although I were a thousand miles out of my attempted course and in an unknown distance, but the latitude must first be perfectly known.

92. Peter Apianus on Asia and America

Peter Apianus (1495–1552) is known equally well for his work in astronomy and in cosmography (the term commonly used in the early 1500s to describe geography). His *Cosmography, that is a description of the entire world,* first published in 1524, remained popular throughout the sixteenth century. Of the two chapters presented here the first, on Asia, typifies the traditional approach, relying on classic and medieval sources and even using ancient place names; the second, on America, is based on explorers' reports then current throughout Europe.

———

The third part of the earth, called Asia by Asius Maneus, son of Libius, is touched in three places by the Ocean, at the south by the southern Indian Ocean, at the North by the Scythian Ocean, and at the east by the Eastern Ocean, and at the opposite part has Europe and Africa, and that sea which is driven between them. In this region are found diverse and extraordinary forms of men, and various customs of the peoples:

———

From Peter Apianus, "Cosmographicus Liber" (Antwerp, 1524), an unpublished translation by Curtis A. Manchester. By permission of Mrs. Curtis A. Manchester.

besides the land is fertile and temperate, and in the variety of all animals is very well taken care of. The peoples of this land, living in splendid cities, occupy themselves in this manner. Firstly, the main division of Asia is Pontus[1] and Bithynia,[2] afterwards called by the name of nearer Asia, Phrygia,[3] Capadocia, Lycia,[4] Carya,[5] Pamphylia,[6] Mysia,[7] and Armenia, through the middle of which the gliding Tigris and Euphrates rivers flow. After the Capadocians come those who are among the foremost of men, who are called by the extraordinary name of Pontici, next are the Amazons, near the Sarmatian river Tanais,[8] the farthest Scythians, and the Caspian peoples encircling the Caspian bay. Then come the Medes,[9] the Hyrcanians,[10] below them the Parthians, the Carmanians,[11] and the Persians[12] next to the Persian Gulf, the Babylonians, Mesopotamians, and the Syrians. The Arabians are situated at the middle and hold the Arabian Gulf. Beyond Parthia is Aria,[13] Paropamisus,[14] Drangiana,[15] and Gedrosia,[16] and beyond these is India, Northern and Southern, beyond and within this side of the Ganges, where Pliny in his seventh book, *Natural History,* reports that there are many races of men who feed upon human bodies. And in Scythia are produced the Arimaspi,[17] who are distinguished by having one eye in the middle of their foreheads, and by whom war is carried on continually with the Gryphies[18] over their mines: and in a

---

1. A district in Asia Minor between Bithynia and Armenia. It was the kingdom of Mithridates, and afterwards a Roman province.

2. A very rich province in Asia Minor, between the Propontis and the Black Sea. The Romans carried on a considerable trade there.

3. In Asia Minor. The country of the Phrygians, who were noted to the ancients for their laziness and stupidity as well as their skillful embroidery in gold.

4. A country in Asia Minor, between Caria and Pamphylia; the place where the volcano Chimaera was located.

5. A province south of Lydia, now the provinces of Aidin and Mentesche in Ejalet Anandoli.

6. A country on the sea coast of Asia Minor, between Lycia and Cilicia.

7. A country in Asia Minor, divided into Lesser Mysia on the Hellespont and Greater Mysia on the Aegean.

8. Now called the Don.

9. Here Apian could have meant the Assyrians also, or the Assyrians exclusively.

10. Also on the Caspian Sea.

11. Their country, now Kermanshah and Laristan, is on the Persian Gulf.

12. This term could have meant specifically the inhabitants of the country of Persis, which is between Caramania, Media and Susiana, and is now called Fars or Farsistan.

13. A Persian province between Hyrcania, Gedrosia and India, now western Khorasan.

14. A high mountain beyond the Caspian.

15. The country of the Drangae, a Persian or Bactrian tribe.

16. Within the borders of the Mekran.

17. A Scythian people in the north of Europe.

18. Fabulous four-legged, griffins.

certain valley of the Imaus[19] mountains are men of the forests who, having turned their feet in flight, are extremely swift of foot. Furthermore, in many mountains are found a race of men with dog's heads, and they knock off their teeth for the sake of having a voice that barks. Also, there is a race of men called the Monosceli, who by means of their remarkable legs are of miraculous agility in leaping, and it is stated that they are called Sciapodes,[20] and whenever prostrated by the great heat of their region, they protect themselves by the shade of their feet. These men, restricted by the limited movements of their necks, have eyes in their shoulders. In the region around the source of the Ganges of India, he (Pliny) reports a living race called Astomi who have no mouths, and are extremely odoriferous of breath, beyond these are reported the Pigmaei[21] and various wonderful living things, and strange races of men, the same that Pliny has related.

America

America, which is what the fourth part of the earth is now called, has its name by chance from Americus Vespucci, its discoveror. And not undeservedly, since it is surrounded on all sides by the sea, is it called an Island. Moreover, it remained unknown to Ptolemy and the other ancients because of its extreme remoteness. Indeed, it was discovered in the year of our Lord 1497 by the commission of the King of Castile: and also, because of the greatness of his generosity the new world has been reached. In this region, the inhabitants go about entirely naked, and the Cannibals are exceedingly cruel. In the skill of shooting arrows they are most proficient: but they obey no one, and they have neither Gods nor Kings. Strong swimmers are to be found among both sexes. They do not have iron, or other metals, but they arm their arrows with the teeth of fish and animals. Also in this region there is an animal having below its breast a certain natural pouch, by which it carried its young, except when it is wont to bring them out for the sake of giving suck. In running they (the savages) are extremely nimble and swift. Their treasures consists of feathers from variously colored birds and of a multitude of certain stones, which are suspended from the ears and the lip as ornaments. Gigantic pearls and gold and other similar valuables they regard as being of no value. They are said to be exceedingly generous in giving and extremely greedy in accepting. Also, they bleed the blood from their loins and the flesh around the tibia.

Some bury their dead with water and food: others truly suspend those

---

19. A great mountain chain in Asia between the Caspian Sea and the Ganges.

20. A fabulous Libyan people said to have monstrous soles on their feet, which they turned up for use as umbrellas.

21. Pigmies.

struggling in death in silken nets between two trees in a huge forest, and after the food has been hung beside them, they spend the whole day in dancing about the suspended ones. They worship the Heavens, the Sun, Moon, and the Stars. Their homes are constructed in the likeness of a bell, being covered overhead by the leaves of palms. They lack grain, but they grind the roots of trees into meal, and knead them into loaves. This island is situated at a distance in this part of the world, in which to us the German Sun is plunged. Although it (the island) appears in the east on our chart, indeed it is necessary that the Map (which is what it is called) be curved, until the equator (at which the land, together with the water, is at its greatest extent and is round) is doubled back on itself in a perfect circle. Then it will appear to us in the west. Moreover America has many neighboring islands, such as Parianus island,[22] Isabella, which is also called Cuba, and Hispaniola, in which is found Guaiacum wood,[23] which our countrymen use to combat the French disease.[24] In truth, the inhabitants of Hispaniola island feed upon huge snakes and roots in place of bread. The rites and the culture of these particular adjacent Islands are the same as those of the inhabitants of America.

93. From the Cosmography of Sebastian Münster

Sebastian Münster, a Minorite later turned Protestant, spent many years as Professor of Hebrew at the University of Basel, Switzerland. His *Cosmography* was the best-seller of all geographical works of the sixteenth century: forty-six editions, in six languages, were published prior to 1650. Münster's training in classical learning is evident in his work, yet his spirit was that of the "new geography," and his insistence that his readers use maps to locate the relative positions of places and regions is thoroughly modern in flavor.

How God in the beginning created land and sea and how he put them together

We know through divine revelation in the Holy Scriptures how in the beginning the land was completely, from above and below, embraced by the ocean. Then God commanded the water to recede from the upper

---

From Sebastian Münster, "Cosmographie" (Basel, 1592), an unpublished translation by Anneliese Moore. By permission of Mrs. Curtis A. Manchester.

22. Paria peninsula, on the northeast coast of Venezuela. It encloses the Gulf of Paria on the north.

23. *Guaiacum officinale* or *Guaiacum sanctum*. The resin from the trunks of the trees was used as a remedy for gout, rheumatism and skin diseases.

24. Syphilis.

side of the land in order to make room for those beings who were to live on the land and to find there their dwelling and livelihood. From these facts we can gather that at the present time the sea is not in its original and natural place or proportion: for to the same degree as the land has been exposed, the water has been displaced from that area and has been added to the other ocean areas, so that those latter parts became twice as deep as before. The Greeks and the Romans call this part *Oceanum*, which is generally understood as the very big and bottomless ocean which stretches from beyond Spain, Ireland and Scotland southward over to Africa and Asia. This does not mean that you should consider this ocean in the mentioned areas as bottomless, but its depth is so enormous that man has no means to fathom it. Some natural scientists claim that the one element surpasses the other ten times not only in transparency but also in expansion. Now it has been established that the circumference of the land equals 5400 German miles. From that you might gather the greatness of the ocean and its depth in locations where it is still in its natural place. The Holy Scriptures tell further that the Great Flood rose fifteen . . . above the highest mountains, because at that time the ocean was ordered by God to move back into its original place, until the sinful men and all living beings on earth had perished. A number of very high mountains can be found on earth, mountains high enough that they are covered with eternal snow. But during the Great Flood the ocean rose yet higher so that no human being could save himself from the water—not even on the top of the mountains.

About the distribution of the ocean

It has to be considered that during the Great Flood the furious ocean ate many holes, caves and gaps into the land, and new seas came into existence where there had none been before, just as by the same cause mountains and valleys were created through the back and forth moving ocean in places where previously fields used to be. That is why so many weeds are found on the land, a fact about which I do not want to speak now but rather save it for a more appropriate place. The big rivers as the Rhine, the Danube, the Rhone came into existence during the Great Flood, but not without special order by God and that for man's good use. Now bear in mind that the great water which surrounds the land is called, as mentioned before, *Oceanum;* but the big caverns which it has eaten into the land are called *Sinus* in Latin, which means golf, and they carry the name of the country they are located at, as e.g. *Sinus Persicus, Sinus Arabicus, Sinus Indicus* etc. But the big sea which separates Africa from Europe is called the Mediterranean Sea: it intrudes into the land through a narrow entrance in the west, at Spain, but then it stretches out—I shall talk about that later at a more appropriate place. There is also a sea in Asia, and it is all around enclosed by land; it has neither entrance nor exit other than that many rivers flow into it. It is called the Hyrcanian or Caspian Sea. About the Red Sea I shall report when I get to that area. The sea around the North Pole and the South Pole are called *Mare congelatum* which means Frozen Sea: at those places far away from the sunshine the sea is always covered with ice on account of the great cold. That is the reason why man never has sailed thereto and he would not find a good place for living there.

In order to understand this chapter well, it is a good idea to look at the map "General description of the whole globe" which is found below, at the beginning of the map section. On that map you will see all the things which were mentioned above: how the water enters through a narrow passage between Spain and Africa into the land, how it stretches out and how it takes over a great part of the land, how it expands quite closely to the Red Sea, and how many hundreds of small and large islands are within its reach. You will also see how large a piece of land has been taken away by the Red or Arabian Sea, and by the Hyrcanian Sea. Likewise quite some area is occupied by the great rivers, by the pools and ponds—land which otherwise could be utilized by man—not to mention all the high mountains with their woods, wilderness, and steep slopes, which cannot be occupied by man. And yet we are so blind that we crave for fancy titles, honors and treasures; we furiously fight wars and shed human blood (as if we were to live here for good)—just to have a wider possession here and to enlarge our domain. O blindness, a folly of Adam's children.

Explanation of the first and the second general map which represent the whole world

To learn something fruitful about the provinces of Europe, Africa and Asia, it is in the first place necessary to visualize the world as a whole—how land and sea are related to each other and how the land rises out of the sea. The first two maps serve this purpose. The first one is the printed form of all the land and the sea, as it is known in our time through exploration by water and by land, with mountains, islands, rivers, meridians and noon circles and parallels. Both are general maps, but the first comprises the whole globe and the second depicts that half of the globe which Ptolemy described as the world at his time. That means that this map is not complete: for Ptolemy has not completed the description towards the south to the end of Africa, as can be seen in the second general map. Now keep in mind what is here said in short words: The whole world is divided into three unequal parts: Europe, Asia, and Africa. Europe you see well on the second general map separated from Africa by the *Mare Mediterraneum*, that is the Mediterranean Sea, and from Asia by the river Tanais

(Don), which flows from the north into the *Pontus Euxinus*. Europe is divided and described by Ptolemy in ten special maps as will be shown later. All the land, however, that lies beyond the Mediterranean towards the south is attributed to Africa, and stretches toward the east to the river Nile, or as others regard it, to the *Mare Rubrum*, that is the Red Sea. In the first general map the longitudes are indicated on the tropic of the capricorn and the latitudes on the left margin of the map. The longitudes begin behind (west of) Spain and Mauretania and are counted toward the east, and since they are discontinued at the eastern margin, they are resumed at the western edge, as indicated by the numbers. The latitudes begin at the equator and are counted towards north and south. When you now examine the second (Ptolemeic) general map, which describes not more than a half of the globe, you will find the longitudes indicated in the upper and in the lower part, and the latitudes on the right hand side. At the right you'll also find indicated the seven climates. The parallels, however, are indicated by number and name on the left margin. If you compare this (Ptolemeic) general map with the first general map, you will see, how many areas have been discovered which were not known at Ptolemy's times. With the help of the first general map you can realize how the Spaniards have to navigate to reach the new islands, or how they have to go around Africa and set their sails in the east towards Callicut. Here you see how the land masses could be likened to a big island floating in water. You also realize that, compared with all the land masses, Europe is only a small piece, and that Europe just missed to be a big island surrounded everywhere by water. Only at the area of the river Tanais (Don) Europe is attached to the big continent Asia, otherwise it is surrounded by the sea. The countries north of the 63rd parallel were unknown to Ptolemy, but they have become known at our time. These countries will be described later on special maps.

About Europe and something about Turkey

Europe is the first third of the world, and though it is smaller in extension than Africa or Asia, it is quite a considerable area: according to Ptolemy's reckoning Europe's length from Spain to Constantinople, in the outer part of Greece, equals 550 German miles. The north-south extension is somewhat smaller—as you can gather from the second general map and from that new map which represents Europe specially. Yet, if one considers the large areas in the north as a part of Europe, then the north-south extension is bigger than the west-east extension. But as Ptolemy described Europe, the west-east extension was larger than the one from north to south. Europe is certainly a fertile and well cultivated area. It has a higher population than Africa, though Africa is bigger. In Europe you find neither big deserts, barren sands nor great heat as in Africa, and no place or land is of so little value that it were not inhabited and cultivated. For who would have originally thought that men would search for food and create a beloved fatherland in the high Alps, an area of rugged, snow-capped mountains like Switzerland? If snowcapped mountains can contribute to an area of fertility, how fertile must the other European countries be, which have no snowcapped mountains. And when you examine the many European islands, you'll find that they ornate our continent like gems in a golden crown—especially those in the south. It was not just by chance that the two powerful cities, Rome in Italy and Carthago in Africa, fought bloody wars on account of the islands of Sicily and Sardinia. At this point I shall not go into detail about Negropont, Peleponnes, Candia, and other islands, about which I shall talk farther down at a more appropriate place.

About the division of the big European land mass

As mentioned before the ancients thought Europe started in the west and ended in the east at the river Tanais (Don) which flows into the Meotic Sea. To the north Europe reaches as far as one finds land. Later on the ancients divided Europe into many separate lands, i.e. the islands Albion and Ibernia; Hispania; Gallia which later was called France; Germania—that is Germany; Italy; Sarmatia—that is Poland; Lithuania; Hungary and Walachia; Greece. Ptolemy made maps of various islands which will be seen below.

How to Study the European maps

It will be necessary that you study the European map frequently and thoroughly in order to obtain a picture of the various countries: for it is very useful to know where a country is located—whether to find it in the west or in the east or in the south. With the help of the map you'll see that a person could board a ship at Basel or at Mainz which would take him to S. Jacob, Spain. You asked me which route such a person would have to take—and my answer is: "Take the map of Europe which you find here in the book and you'll see that he can sail down the Rhine river into the ocean, then between Flanders and England towards Normandy, and then on to Spain. If you want to go from S. Jacob to Rome, Italy, you must sail from S. Jacob alongside Portugal and then turn at Granat from the ocean into the Mediterranean Sea. Thus you get, passing the islands, to Rome. By means of this trip you might learn how to sail to other places and cities, i.e. if somebody wants to make a voyage from Venice to Constantinople he has to sail southward on the Adriatic Sea, then turn north at the Peleponnes or Morea to get to Constantinople, for that city is located at the sea, as you can find out on the map of Europe.

We here in Germany have yet a closer way to get to Constantinople. From Ulm or Regensburg we could sail down the Danube to the sea and then on the sea to Constantinople. If you want to get somewhere by foot from Germany the map will likewise show your way. E.g. you have to go by foot on a business trip from Mainz to Sicily and you don't know the way. You look at the map of Europe which will inform you that Sicily is situated in the south and that you have to go through Italy down to Calabria; from there you sail across the channel to the mentioned island. The map informs you also that you cannot march directly from Germany to S. Jacob, but that you have to pass first through Gallia, or France.

The New Islands

The new Islands, also called NOVUS ORBIS, that is the New World, are situated beyond India. Therefore the New Islands are called by some people the Indian Islands. They are also located beyond Spain, just half the way between Spain and India. Therefore you find on two maps the land Cathay and the archipelagus, that is the Great Ocean with its 7448 and more islands. On the map you will recognize the island Zipangri (Japan) which the great Khan had subdued after he learned how rich that island was. Much gold is found there, and the island is governed by a sovereign king. They have red pearls there, superior to the white variety, and precious stones. The inhabitants worship idols; whenever they get hold of their enemies, they cook and devour them. Most of the many small islands around Zipangri are inhabited and everywhere grow spices and sweet smelling trees. In regard to the other New Islands it should be remembered that A.D. 1492 Christopherus, a Genoese from Italy, of the family Columba, asked the king of Spain, at whose court he had spent some time, for a well equipped fleet in order to discover unknown countries. In the beginning the king and his council did not take his request serious, but finally the king consented. Thus Christopherus departed in the early fall of the mentioned year for his voyage and came with two ships to the islands which in earlier times were named *Fortunatas,* but which now are called Canaries. Yet first there is an island called Madeiras which the Spaniards and especially the Portuguese have found in an uncultivated state partly covered with forest. Since they noticed the excellent soil, they burned the forests down, prepared the ground for agriculture and settled down in such a way that there is no other island in that ocean which surpasses Madeiras. One finds many springs and there are sawmills along the rivers. In the sawmills the red, redolent lumber of cedar and cypress is cut and used for the construction of furniture. The king of Portugal introduced sugar cane onto the island. The cane grows well, the crop is abundant, and the quality of the sugar surpasses that of Sicily and Cyprea.

They also transplanted grapes from Candia and they grow so enormously that the clusters have more grapes than leaves, and some grapes are four spans long. Partridges, doves, wild peacocks, wild pigs and many other not domesticated animals are found on this island. Not less fertile are the *Fortunatae,* i.e. the "Fortunate Islands", which are now called the *Canaries,* because so many dogs are found there. There are ten islands in the Canary group—seven are cultivated, three are not. The cultivated islands are called *Fracta Lancea, Magna Sors, Grancanaria, Taneriffa, Ginera, Palma, Ferrum.* When Columbus got there, the people ran about in the nude, shamelessly, just like the animals. They neither served nor feared God. Later on, however, they became Christians, especially on four islands. It is said that each of these islands has a different language, though they are not far from each other. *Tanriffa* and *Grancanaria*—which means Great Canaria—are the biggest among the ten islands. When the weather is fine, *Tanriffa* can be seen from a distance of fifty German miles, and that because a high rock rises fifteen Leucas, i.e. about twelve German miles, from the center of the island. This mountain is in perpetual eruption like Mount Etna on Sicily. The inhabitants of this island eat barley bread, meat and milk products. They have goats, mountain donkeys and fig trees. Grapes and wheat do not grow there.

From these islands the above named Christopherus Columbus sailed twenty three days westwards without finding any land, until the lookout saw some islands. They approached those islands and there were especially two which they meant to look over, and they gave names to the islands. One they called *Johanna* and the other *Hispana.* They sailed around Johanna island. They heard the birds singing and especially on St. Martin's day they heard the nightingales, saw the fresh water fall from the rivers into the ocean, sailed on for another 100 or 200 German miles, and since they did not see anybody they turned back toward the other island. There they noticed the inhabitants flee into the woods. They disembarked, grabbed a woman, led her onto the ship, gave her wine to drink, dressed her into nice clothes and sent her back to her people. When they saw the clothes on the woman (for they all went around in the nude) and when they heard what the woman had to eat and to drink aboard ship, they all ran down to the ocean, gave the Spaniards gold from earthen and glass vessels—there was nothing childish enough on the ship that they would not have given gold in exchange for. When mutual confidence was established, the Spaniards gathered information about their way of living. They noticed that there was a king by whom they were received in a friendly and honest manner. The inhabitants of that island prepare their bread from roots which taste like fresh chestnuts. They appreciate gold, though they do not trade with other islands in order to increase their wealth. They dig for gold in the sand of a big river which has its source way up in the mountains. They melt the gold into lamina. As to animals, there are big, but harmless, snakes; doves as big as our ducks; geese as white as swans and with red heads; parrots of many colors, which indicated to the Spaniards that they were not too far from India. They also found on this island mastic, aloe, red pepper, cinnamon and ginger.

While Columbus was with his crew on the island, the inhabitants complained about the cannibals who attack other islands, caught the people, killed and devoured them. They acted like tigers and lions who attack tame animals. Boys are castrated and fattened like capons; men are butchered on the spot, their intestinals discarded, the other organs, hands and feet being eaten right away while the rest is being salted and preserved. Women are not eaten but kept for breeding, just like hens are kept for their eggs. Old women have to render services. Therefore everybody flees when these cannibals visit the island, for they are strong and ferocious, and ten of them can master a hundred of other people.

After Columbus had stayed all through the winter until next spring on that island, he decided to sail home to bring the king the good news about the newly found land. He left thirty-eight of his men on the island with the order to explore the place during his absence and to gather information about the vegetation. He took ten of the inhabitants along in order to learn from them their language. When Columbus arrived in Spain, he was received with highest honors by the king and by the people, and plans were discussed how to bring Christianity to the pagans.

94. Josias Simler describes glaciers and avalanches

Josias Simler (1530–1576), a Swiss theologian, published his commentary on the Alps in 1574. The book is clearly based on observation rather than tradition, and Simler's description of the world of glaciers and avalanches is that of a modern natural scientist.

Old masses of snow are called here, in contrast to new snow, *Firn;* old wines are also called Firnwine, when they have been stored for a year. Hard and partially frozen snow retains the characteristics of snow, but once melted and frozen again, it is no longer snow but ice, which we call "Glacier," a term probably originating from *glacies,* ice. But this distinction is not always observed, often the two [words], firn and glacier, are used in addition to ice. One finds in the Alps ice hundreds of years old, and since new ice is constantly being added to the old, huge masses are formed which sometimes culminate in frightening, rock-like ice formations [séracs].

From Josias Simler, *De Alpibus commentarius;* German translation by A. Steinitzer (München, 1931), quoted in H. Beck, *Geographie* (Freiburg-München: Karl Alber Verlag, 1973), pp. 95–97, by permission. Editor's translation.

There is a mountain in Valais, called Sylvius by some, the Salassians call it Rosa. Surrounding it there is an immense deposit of eternal ice, four miles long; there is a path to the Salassians across it, and it has much higher, ice-covered peaks towering above it. The people of Valais call this entire area "the glacier," from the word *glacies*. This world of ice has long been clear of sand, earth, stone, or other impurities, so that it has an almost crystalline luster; for that reason, some people believe that this crystal was created out of century-old ice. . . . . . .

The most serious danger [in the Alps] is the sliding downslope of accumulated snow, called by us *Löuwine,* by the Rhaetians [in Canton Graubünden] *Labinae;* even the German term *Lawine* is but a distortion of *Labinae.* The smallest event suffices to set masses of snow in motion, thus snow on a steep and treeless slope can be shaken by a bird flying past, by another animal, by strong wind, or by the shouts of men passing by. In this last instance the air, by its reflection of the sound, which we call echo, provides the impulse that sets the snow in motion. Through such quakes, no matter how insignificant, snow set in motion forms a ball which gets bigger and bigger as it slides downslope and becomes so shapeless that it has to follow a straight course. It then slides down with extraordinary force, which keeps increasing, gaining in size, and it will carry rocks, trees, chamois and other beasts, men, houses, indeed all that happens to be in its path, to the foot of the mountain. Masses of snow that slid down in this manner often cover several across of ground; they move with such a roar that the earth seems to quake, and someone far away, unaware of what is happening, believes he can hear distant thunder.

95. Leonhart Rauwolff on the lands, peoples, and plants of the Near East

Leonhart Rauwolff (1535–1596), a native of Augsburg, Germany, and official physician of that city, traveled throughout the Near East between 1573 and 1576, visiting Syria, Iraq, Lebanon, and Palestine. He was very much interested in botany, and his observations on plants occupy an important place in the description of his travels.

___

Of the site of the mighty city Aleppo

The valley where Aleppo lies, which is not particularly wide, and for the most part surrounded by rough mountainsides, is rich in grain, barley, etc.,

___

From Leonhart Rauwolff, *Aigentliche Beschreibung der Raiss in die Morgenlaender* [True description of the journey to the lands of the Orient] (Lauingen, 1583), passim. Editor's translation.

which they harvest for the most part in April and May. They do have but little oats, and even less hay, since the drought is hard, the soils sandy, the mountains rough and covered with weed, that little hay is harvested. They feed their animals with barley and with straw that is threshed by carts drawn by oxen. There are quite a few handsome olive trees in the valley, and they harvest thousands of pounds of oil every year, used mostly to make soap. There are also both wild and domesticated almond trees, fig trees, and quince trees, white mulberry trees that are large and tall. The pistachio tree is rather common there; they have thick trunks, with widespread branches covered with ash-colored bark and having handsome green, round leaves that resemble those of the carob. In the spring, long shoots appear, and before these open into leaves, the Moors cut them off and use them in salads, the way we prepare asparagus.

A short and simple report on plants that I collected, not without considerable danger and labor, both in and around Aleppo, and pasted on paper

One also finds here the liquid apothecaries call *opium* . . . which the Turks, Moors, Persians and other peoples use, not only in wartime, when they need a strong heart and courage while fighting their enemies, but also in peacetime, to forget worries and imaginings or at least to reduce them. Their religious use it, especially the dervishes, who use it so much that they become sleepy, even to the extent of becoming possessed, and inflicting wounds on themselves or burning themselves, without feeling much pain. Once they start using opium, they cannot give it up (they usually take a quantity of the size of a pea), unless they become ill, and then they do feel very ill indeed. Opium is collected from the head of the white poppy, by cutting into it, so that the milk begins to run. They leave it on the plant until it solidifies, then roll it into small balls.

A simple description of the city of Jerusalem, as it appears in our days, and of the surroundings thereof

The magnificent and royal city Jerusalem, which the Saracens called *Kurzitadon,* and which its inhabitants call *Kuds,* still lies in its old place, in the midst of Judaea, on a high mountain . . . One can see Jericho, and the great plain surrounding it which is traversed by the river Jordan from north to south, together with the Dead Sea, where Sodom and Gomorra once stood, from a barren hilltop. It seems as if one could get there in two hours, yet it takes an entire day's travel to get there. Beyond the river (Jordan) that separates Arabia from Judaea are the high mountains Abarim and Nebo, whence Moses saw the land of Canaan, promised by Abraham, Isaac and Jacob. Those mountains, and the mountain Seir, which lies adjacent to them, appear so clear, from the same spot, as if they and the lands of Moab and the Ammonites, were very near indeed . . .

The Arabs called *Beduin* live in the desert. They do not have a steady habitation, but are always on the way, crisscrossing these lands to find good pastures for their cattle and camels. I have met many of them in my travels; like other Arabs, they carry bow and arrow and long spears made from bamboo, and because their lightning speed and courage are much preferred to others.

The Beginnings
of Modern Geography:
the Seventeenth Century

Geographical writing in the seventeenth century at first followed closely the established custom of presenting a general view of the earth, its shape, and dimensions, as is witnessed by the selection from Philip Cluver's *Introduction to Universal Geography,* a text that was popular during the first half of the century. There were also those who, not content with description only, searched for the causes of terrestrial phenomena: Conrad Gessner's prefatory letter to his work on mountains is a good example of this type of geographical writing. There is general agreement, however, as to which was the first work to introduce a more disciplined approach to the subject of geography: the *Geographia Generalis* of Varenius, published in 1650. It was reprinted in 1672 and again in 1681 in Cambridge, where Isaac Newton used it as a text for his students. William Warntz points out that Varenius' book was "the first textbook on record as being used in general geography in an American college."[1]

96. A geography textbook by Cluverius

Philip Cluverius, or Cluver, (1580–1622) was a German scholar whose *Introductio in Universam Geographiam* was first published in 1624. The following excerpts are typical both of his approach to the subject in general, and of his summary manner of describing regions of major importance.

From Philip Cluverius, *Introduction à la Géographie Universelle tant nouvelle qu'ancienne* (Paris, 1631), pp. 1–3. Editor's translation.

1. William Warntz, *Geography Now and Then* (New York, 1964), p. 109.

Geography is the description of the entire earth, insofar as it has been discovered to this date. This word geography is Greek, composed of the word *gea,* that is, the earth, and the verb *graphon,* that is, to write; hence the word *geographia,* the description of the earth.

Geography is as different from Cosmography as one part from the whole; it is also as different from chorography as the whole from the part. For cosmography is the description of the entire earth which comes from the Greek word *cosmos,* that is, the world, and the verb *graphon,* that is, to write. When you say the world you understand by that term this earth as well as the heavens.

Chorography is a detailed particular description of a region, or a district and it comes from the Greek word *choros,* that is, a region, and the verb *graphon,* which means to write. Thus we have a description of Spain, Italy, Germany, Germania, and of all Gaul. Topography is a detailed description of a particular area such as a district or territory, a town or a city, within which are included cultivated lands, meadows, trees, places and buildings described in detail; this also is a Greek word composed of the noun *topos,* that is, a place, and the verb *graphon,* meaning to write.

But rightly defined geography is a description of the earth only. We have to add that in these terms the word geography is not only applied to one of the four elements as in terms of physics. We define geography here as the earth and the waters that are expanded upon it, the two together making the center of the entire world which, because of its round shape, is called *Orbis,* that is, the terrestrial globe. For the globe is a solid body completely round, within it is a center and all lines drawn from that center to the surface are equal in length. It is therefore but a single globe composed of these two elements earth and water, hence it has a single convex surface, that is, it appears without any concavity.

North America, or Mexico or Mexicane

North America is called by some Mexicane after the principal city of that region, Mexico. Its length between the strait of Anian and the Isthmus that joins it to South America is 1500 leagues. Its width is the same as that of the entire region, that is 900 German leagues. Half of it that is westward, and the interior is still unknown and only the coast has been discovered, and the entire north coast extending toward the pole is also unknown.

The remaining part of it is divided into several regions called Canada or New France, Virginia, Florida, New Spain, New Granada, California and in the west near the strait of Anian two kingdoms, Quivira and Anian, which has given its name to the strait.

About Canada

Canada is thus called after a river and we do not know whether it is an island or a continent. That part of it which has been discovered is divided into Estotiland, Cortereal, the land of Labrador and the adjacent islands, which are very large. The most important islands are Goleme, Beaupais, the Mountain of Lions, and Newfoundland or Codfishland, thus called because of the large quantities of fish which are found in this part of the sea and which, at times, even slow down ships.

Canada, even though very cold, is very fertile and rich in metals and gold; its inhabitants are rather pleasant, very clever in the mechanical arts; they go about wearing skins and pay obeisance to the king of France.

Of New France

New France was discovered by Frenchmen in the days of King Francis I. It is poor in all respects though it does produce some wheat and some vegetables; its inhabitants are savages and in some places are cannibals who eat one each other. Nonetheless that part of it which is along the sea, called *Norumbega,* after a city, has a pleasant enough climate and fertile soils.

About Virginia

Virginia lies adjacent to new France, and has been named after a city on its shores, thus called after the king who was called Virguina, or after Elizabeth Queen of England, by whose command it was discovered. Earlier it was called Apalchen; its land is sterile and infertile, its shores unfriendly and without harbours. There is a city there called Medano.

97. Conrad Gessner contemplates the Alps

People of the Renaissance were fascinated by mountains. The wide views from mountain peaks and the variety of fauna and flora found at different elevations attracted people who left detailed descriptions of their mountain climbing. Thus Petrarch climbed Mont Ventoux in southern France in 1336, Dante made it to a mountain peak in Italy in 1311. Conrad Gessner (1516–1565), a physician from Zürich,

From Conrad Gessner, "On the Admiration of Mountains," the prefatory letter addressed to Jacob Avienus, physician, in Gessner's pamphlet *On Milk and Substances prepared from Milk,* trans. H. B. D. Soule (San Francisco: The Grabhorn Press, 1937), pp. 5–13.

Switzerland, climbed Mount Pilatus in the Alps near Bern in 1555, and wrote the first treatise on Alpine flora, published between 1551 and 1571. The following excerpt from his preface to *On the Admiration of Mountains* shows that he was not only interested in the present aspects of his beloved Swiss Alps but was concerned with their life history too.

I have determined for the future, most learned Avienus, so long as the life divinely granted to me shall continue, each year to ascend a few mountains, or at least one, when the vegetation is flourishing, partly for the sake of becoming acquainted with the latter, partly for the sake of suitable bodily exercise and the delight of the spirit. For how great the pleasure, how great, think you, are the joys of the spirit, touched as is fit it should be, in wondering at the mighty mass of mountains while gazing upon their immensity and, as it were, in lifting one's head among the clouds. In some way or other the mind is overturned by their dizzying height and is caught up in contemplation of the Supreme Architect.

I ask you, to use the words of our exceeding wise friend Grynaeus, what is there to prevent such a heap of rocks, a weight of such vast size, where the softness of the ground is forever yielding, from being buried by constantly sliding in the depths toward which even by its own very nature it tends; particularly nowhere more than at the bases of these mountains, where the earth is rather soft and marshy: so that it must needs be either that their foundations are of unmeasured depth; or that even as the roots of trees, they are very widely spread. How is it that often as of set purpose rocks will so overhang; in vain threatening ruin for such ages as to seem impossible by any force to hold back? And with what intent do such heights uplift themselves? Within these vaults forsooth is begotten that continual supply of water anon bursting forth in so great a course, while the air shut within the hollows of their sides with icy echo ceaselessly distils to drops: and these presently drawn together into one stream become forthwith, by their constancy, springs supplying lakes, ponds and rivers in abundance with so great an amount of water. Lo! such a blessing do we owe to those mighty masses, to so amazing a miracle of nature hidden within them. The rest of them indeed, even those of the smaller sort, either provide wider spaces, through the swelling of sunny hills, since the flat places of the earth have been appropriated to other uses; or like a rampart that is set up they shut out the hostile quarter of the world. For the most part they turn aside and draw off into the neighboring fields the courses of rivers; not seldom also they provide implements of iron, the stuff not only for cultivating the land and building fortifications, such as timber, stones and iron; but for conducting business, that is to say silver and gold . . .

I say therefore that he is an enemy of nature, whosoever has not deemed lofty mountains to be most worthy of great contemplation. Surely the height of the more elevated mountains seems already to have risen above a baser lot and to have escaped from our storms, as though lying within another world. There the force of the most powerful sun, of the air and of the winds is not the same. The snows linger perpetually, the softest of objects, which melts even at the touch of the fingers, recks not at all for any glow or violence of the sun's heat: nor vanishes with time but rather freezes into the most enduring ice and everlasting crystal. The summit of Olympus shows even after a year the letters which were written in the ashes (of the sacrificial fires lit the previous summer). Who might properly reckon the varieties of animals and the fodder of wild beasts aloft in the mountains? Whatever in other places nature offers here and there but sparingly, in the mountains she presents everywhere and in sufficiency as though in a heap so to speak, she spreads it out, unfolds it, and sets before the eye all her treasure, all her jewels. And so of all the elements and the variety of nature the supreme wonder resides in the mountains. In these it is possible to see "the burden of the mighty earth," just as if nature were vaunting herself & making trial of her strength, by lifting to such a height so great a weight, which still of its own accord and by reason of the most heavy pressure is ever ready to slip downward. Hence gush most copious springs of water sufficient for moistening the earth. Often there are lakes upon the summits, as though nature were sporting and exulting at drawing upwards the water from afar out of the deepest wells of her caverns. It is possible to gaze upon the air spreading far and wide nourished and increased by the thin mountain mists of the waters. At times shut within hollow caves, it rouses an earthquake and in some places those are continuous. Fire also dwells there, by the aid of which, as if by that of an artisan, metals are fashioned. Elsewhere the sources of warm healing baths give evidence of the presence of fire, which happens in a great many places in our Switzerland. There are places where the flames break forth, as on Aetna, Vesuvius and the mountains close by Grenoble. In other places moreover, though it may give no sign of itself, still fire is lurking in the bowels of the earth. For why do not the mountains in the long succession of the centuries sink down; why are they not consumed by any storms, to which they are continually exposed, nor by the rushing waters of the rains? Doubtless by reason of the fire, which is the cause both of the generation of the mountains and likewise of their continuance, as is demonstrated by the testimony of Philo the philosopher. For when indeed the glowing substance hidden within the earth is hurled up aloft by its native force, its directs itself toward the region proper to it; and if it has found a breathing-hole, however small, it carries on high with it much earthy substance (as is evident in the craters of Aetna), and just as much as it is able. But as soon as it has burst forth from the earth itself, it is borne along a short path. Furthermore this earthy matter compelled to follow the fire as it bursts forth, rising to a great height is compressed into narrow compass, and finally ends in a point like a sword, mimicking the nature of fire: inasmuch as at such a time it must needs be that the lightest matter and the heaviest, which are opposites, strive against each other, since each is impelled by its own tendency to its own proper place, and on the other hand is drawn asunder by the violence of its opponent. Accordingly the fiery matter, dragging with it on high the earth, is impelled to turn backward when the earthy overcomes it; while the earth pressed to the lowest depths by its own gravity, and contrariwise lifted up by the fire, which of its own volition is taken up on high, and finally vanquished, though with difficulty, by its all-prevailing and sustaining force, is propelled aloft into the abode of fire and there remains . . .

Further (to return to the point from which I departed), those persons who do not scruple to say that the nature of mountains differ not a whit from that of trees are clearly ignorant as to why the mountains suffer no loss from diminishing with age and as to what is the manner of their arising. For just as the latter with the changes of the seasons lose their leaves and by turns grow young again, in the same manner likewise, say they, certain portions of the mountains are broken off and some in turn are born later; even though this accretion does not become known save after a long lapse of time, for the reason that trees being endowed with a swifter nature get their growth rapidly, but mountains more slowly. Thence it comes about, say they, that the portions of the mountains which have arisen are scarcely observed by the perception of mankind, unless over a long period. Such persons, therefore, I leave with their opinion. But what of the fact that mountains are manifestly hollowed out into the great halls of caverns, like those of dwelling houses, which seem to have been reared not without painstaking care? Is it that torrents of water once upon a time there made a passage for themselves by violence, carrying earth and rocks with them? Or, in the beginning, when by the action of fire the ground was heaved up unequally, was a space perchance left empty betwixt the vaulted sides? Or, again, did the fire seeking the highest regions make thin that which stood in its way? Or, in truth, do portions when smitten by earthquakes, with which they are so often disturbed, gape and yawn? The following fact, too, is not devoid of marvelousness, that the bases of mountains are soft and smooth, if they be compared with their summits, the huge mass of which is for the most part of flint. Doubtless the water coursing down softens the foot of them, which feels to excess the force of neither sun nor winds. The top being exposed to sun and wind becomes dry, whatever there is of moisture runs off, leaving it parched. When, therefore, the lighter and watery portions have been filtered away, whatever is hard, dense and especially earthy alone remains, which thereupon turns to stone whether through the force of cold or through the violence of heat (which in hard, compact bodies is greater), doubtless from the innate fire which acts very little upon the lower portions and those which border on them, but in the highest degree upon the crest by reason of its pyramidal shape.

98. From the Geographia Generalis of Bernardus Varenius

Bernardus Varenius was born Bernhard Varen, in northern Germany, in 1622; he probably adopted the latinized form of his name when he first matriculated at the University of Königsberg. He moved from there to Leiden, where he obtained his degree in medicine, but his principal interest seems to have been geography. He first published, in 1649, a book on Japan and Siam which may well be considered his contribution to what he called "Special Geography," that is, the description of a single country, or a couple of countries, in considerable detail. One year later his Geographia Generalis was published in Latin. He died that same year, and as little is known of the cause of his death as of his life in general.

Geography in Varenius' time was closely linked to theology, since the works of man and the earth itself were considered to be the creation of divine providence. Varenius dissented from that view, and his book is the first scientific approach to the discipline. A single quote from Varenius' book represents his point of view: "When the ocean moves, everything moves"; there is a basic unity, a set of close linkages among all terrestrial phenomena, a single earth system.

The selections following that from the preface are taken from the first English translation, by Blome, published in 1693.

[General geography] studies the Earth in general, describing its various divisions and the phenomena which affect it as a whole. The latter kind, to wit special Geography, while keeping in mind the general laws, passes in review the position and divisions and boundaries of individual countries, and other noteworthy facts about them. And yet those who have hitherto

---

The first selection is from Bernardus Varenius, the "Espistola Dedicatoria" to *Geographia Generalis,* as translated by William Warntz in *Geography Now and Then: Some Notes on the History of Academic Geography in the United States,* American Geographical Society Research Series, no. 25 (New York, 1967), pp. 112–113. © by the American Geographical Society. Courtesy of the American Geographical Society. The second selection is from *Cosmography and Geography, in Two Parts, the First, Containing the General and Absolute Part ... being a Translation from that Eminent and much Esteemed Geographer Varenius* (London, 1693).

written on Geography have treated almost exclusively of Special Geography, and at tedious length. They have described very little which concerns General Geography, ignoring or completely omitting many necessary aspects of the subject. The result is that our young men while learning more and more Special Geography were largely ignorant of the foundations of the study, and thus Geography scarcely vindicated her claims to be called a Science.

I saw this state of affairs and to remedy what was amiss, I began to bend my thoughts towards supplying the deficiency by composing a General Geography. I did not desist from my labors until I had completed the task. According to such powers as God has granted me, cultivated by years spent in mathematics, my object was to render service to the World of Letters and to the studies of youth or at least leave no doubt of my good-will to do so.

The definition of geography

GEOGRAPHY is called a mixt *Mathematical Science,* which teacheth the affections or qualities of the *Earth,* and the parts thereof depending of quantity; that is to say, the *figure, place, magnitude,* and other like properties.

Geography by some (but too strictly) is taken for the only description and placing the Countrys of the Earth; And on the contrary, by others it is extended (but too largely) to the political description of every Country. But these Men are easily excused, seeing they do it to retain and stir up the Readers affections; who otherwise by a bare account, and naked description of those Countrys, would be made drowsie and heedless.

The division of geography

We will divide *Geography* into General and Special, or Universal and Particular. *General* or *Universal Geography* is that, which doth generally consider the *Earth,* and declare its properties without any respect of particular Countrys. *Special* or *Particular Geography* is that, which teacheth the constitution and placing of all single Countrys, or every Country by itself. And this *particular Geography* is twofold, to wit *Chorography* and *Topography. Chorography* proposeth the description of any Country, having at least a mean magnitude. *Topography* describeth any little tract of Land, or place.

In this Book we will present you with a *General Geography;* which we have distributed into Three parts, to wit, the Absolute part, the Respective part, and the Comparative part. In the *Absolute part* we will consider the very Body of the Earth, with its parts, and proper affections and qualities; as *figure, magnitude, motion, Lands, Seas, Rivers, &c.* In the *Re-*

The contents of the sections and chapters

I place hereunder a Table, which openeth the order in *Special Geography,* to the observing the Explication of single Countries.

Special Geography considereth in every Region,

Ten *Terrestrial,*

1. Limits and circumscription.
2. Longitude of place, and scituation.
3. Figure.
4. Magnitude.
5. Mountains, The Appellation, Scituation, and Altitude. Their properties, and things contained in them.
6. Mines.
7. Woods and Deserts.
8. Waters, The Sea, Lakes, Marshes, Rivers. Their Springs, Inlets, Tracts, and Latitude. The quantity of Water, the celerity, the quantity, the Cataracts.
9. Fertility, Sterility, and Fruits.
10. The Animals.

Eight *Celestial,*

1. The distance of place from the Aequator and pole.
2. The obliquity of Motion above the Horizon.
3. The Quantity of Dayes.
4. The Clime and Zone.
5. The Heat, the Seasons of the Year, the Winds, Rain, and other Meteors.
6. The rising and stay of the Stars above the Horizon.
7. The Stars passing through the Vertex of the place.
8. The celerity or quantity of their Motion according to the Hypothesis of *Copernicus.*

Ten *Human Things,*

1. The Stature, Life, Meat and Drink, and the Original of the Inhabitants.
2. The Income, Arts, Merchandize or Traffick.
3. Vertues and Vices, the Genius and Erudition.
4. Customs about Marriages, Children, and Funerals.
5. Speech and Language.
6. Politick Government.
7. Religion, and Ecclesiastical Affairs.
8. Cities.
9. Memorable Histories.
10. Famous Men and Women, Artificers, and Inventions.

spective part we will contemplate those properties and accidents which from *Celestial causes* happen to the Earth: And lastly, the *Comparative part* shall contain an explication of those properties, which arise from the comparing of divers places of the Earth.

The object of geography

The *Object* of *Geography*, or *Subject* about which it is employed, is the *Earth;* but principally its *Superficies* and *parts.*

The properties of geography

Those things which deserve to be considered in every Country, seem to be of a triple kind, to wit *Celestial, Terrestrial,* and *Human;* and there-fore may be declared in the particular Geography for every *Country,* with the profit of Learners and Readers.

I call those *Celestial properties* which depend on the apparent motion of the *Sun, Stars,* and other Planets: and seem to be Eight. 1. *The elevation of the Pole, the distance of the place from the Equator, and from the Pole.* 2. *The obliquity of wriness of the daily motion of the Stars above the Horizon of that place.* 3. *The Quantity of the longest and shortest day.* 4. *The Climate and Zone.* 5. *Heat and Cold, and the Seasons of the year: also Rain, Snow, Winds, and other Meteors:* for although these things may be referred to Terrestrial properties, yet because they have a great affinity with the four Seasons of the Year, and motions of the Sun; there-fore we have marshalled them in the order and rank of Celestials. 6. *The rising of the Stars, their appearance and continuance above the Horizon.* 7. *The Stars passing through the Vertical point of the place.* 8. *The quan-tity or swiftness of the Motion wherewith,* according to *Copernicus* his Hypothesis, *each one is every hour wheeled about.* According to Astrologers a Ninth property may be added; because they do appoint one of the Twelve Signs of the *Zodiack,* and the peculiar Planet of that *Sign,* to rule and govern every Country. But this Doctrine hath ever seemed to me frivolous, neither can I perceive any ground for it: nevertheless at the end of our Special or Particular *Geography,* we will reckon up this their distribution.

These may suffice for the Celestial affections or properties. I call those *Terrestrial properties,* which are considered in the place of every Country it self; of which I shall note Ten. 1. *The bounds and circumference of the Country.* 2. *Its Figure.* 3. *Its Magnitude.* 4. *Its mountains.* 5. *Its Waters, as Rivers, Springs, Bays of the Sea.* 6. *The Woods and Deserts.* 7. *The Fruitfulness and Barrenness, as also the kinds of Fruits.* 8. *The Minerals, or things dig'd out of the Earth.* 9. *The living Creatures.* 10. *The Longi-*

tude of the Place, which ought to be added to the first Terrestrial property, to wit the *Circumference.*

I make the third kind of Properties, which are to be considered in every Country, to be *Humane,* which do depend of the Men, or Natives and Inhabitants of the Countries: of which *Humane properties* about Ten also may be made. 1. *The stature of the Natives, as to their shape, colour, length of life, Original, Meat, Drink, &c.* 2. *Their Trafficks and Arts in which the Inhabitants are employed.* 3. *Their Vertues, Vices, Learning, Wit, &c.* 4. *Their Customs in Marriages, Christnings, Burials, &c.* 5. *Their Speech and Language.* 6. *Their State-Government.* 7. *Their Religion and Church-Government.* 8. *Their Cities, and most renowned Places.* 9. *Their memorable Histories.* And 10. *Their famous Men, Artificers, and Inventions of the Natives* of all Countries.

These are the three sorts of Properties to be declared in *Special Geography;* although those *Terrestrial properties,* which make up the third rank, are not so rightly referr'd to Geography: But we must yield somewhat to Custom and the Profit of Learners. We will besides these, joyn many Chapters to *Particular Geography,* concerning the practice of Geography.

But in *General Geography,* which we will unfold in this Book; first the *absolute* properties of the *Earth,* and its constitution, are considered. Lastly, in the *Comparative* part those things shall be proposed, which are offered unto us in the comparing one place with another.

The principles of geography

The *Principles* which *Geography* useth for the confirming the truth of her Propositions, are threefold: 1. *Geometrical, Arithmetical,* and *Trigonometrical* Propositions. 2. *Astronomical* Precepts and Theorems; although it may seem like a miracle for the knowledge of the Earth in which we dwell, to use the Celestial Bodies, which are so many thousand miles remote from us. 3. *Experience;* for indeed the greatest part of *Geography,* especially that which is Particular, is upheld by the only Experience and Observation of men who have described every Country.

The order of geography

Concerning the *Order* which I esteem fitting to observe in this Art of *Geography,* it hath been already spoken in the *Division* and *Explication* of the properties thereof; yet here meets us a certain difficulty concerning the Order to be observed in the explication of these Properties: Forsooth, whether to all Countries their own Properties are to be attributed, or whether the Countries themselves are to be ascribed to the Properties generally explicated? *Aristotle* in the first Book of the History of *Living Creatures,* as also in his first Book of the Parts of *Living Creatures,* moveth the like doubt, and disputes it at large; whether according to the single sorts of *Living Creatures,* their Properties are singly to be reckoned up; or else, whether these *Properties* are generally to be declared, and the *Living Creatures* in which they may be found are then to be subjoyned? The like difficulty occurs also in other parts of Philosophy. We in *General Geography* have generally unfolded some *Properties,* which in *Special Geography* we will apply to the application of single Countries.

The method of geography

As touching the *method* and manner of proving the truth of *Geographical* Tenents, very many are proved in *general Geography* by Demonstrations properly so called, especially *Celestial Properties:* but in *Special Geography* (the Celestial Properties only excepted, which may be demonstrated) are in a manner declared without demonstration, because experience and observation doth confirm them, neither can they be proved by any other means.

Also very many *Propositions* are proved, or rather demonstrated by the *Terrestrial Artificial Globe,* and also by *Geographical Maps;* and some of these Propositions which are thus explained upon the globe, &c. may be confirmed by lawful demonstrations. Again, some Propositions can in no wise be so proved, but are therefore received; because we suppose, that all places in the Globe and Maps are so disposed, even as they lie on the Earth. Yet in these things we will rather follow the Descriptions made by Authors of Geography. The *Globe* and *Maps* serve for the clearing and more easie comprehension thereof.

The original of geography

The *Original* of *Geography* is not New, nor brought into the world at one birth, neither came she to us from one Man: but her Principles and Foundations were laid long ago, yea many Ages since; although ancient Geographers were employed only in describing Countries, which is the part of *Chorography,* and *Topography.* The *Romans* were accustomed, when any Country by them was subdued, to shew in their Triumph the *Chorography* thereof lively pencilled, and drawn on a Table, and flourished with Pictures to the Beholders. There were besides at *Rome* in *Lucullus* his Porch, many Tables of Geography exposed to the view of all men. The Senate of *Rome* about a hundred years before Christs Birth, sent Surveyors and Geographers into divers parts of the World, that they might measure out the Earth; but they came far short thereof. *Neco* King of the *Egyptians,* many Ages before the Birth of Christ, commanded that the whole outerside of *Africa* should be discovered by the *Phoenicians* in three years space. King *Darius* commanded, that the Mouths of the River *Indus*, and the *Ethiopian* Eastern-Sea should be searched out. *Alexander* the Great in his Voyage to *Asia*, took with him *Diognetus* and *Beton* (as *Pliny* noteth) two Surveyors and Describers of his Journies; out of whose Annotations and Journals *Geographers* of succeeding Ages took many things.

But the *Geography* of the Ancients was very lame and imperfect; for first they knew not *America* in the least. 2. The *Northern-Lands*. 3. The *Southland* and *Magellan* were utterly unknown to them. 4. They knew not whether the *Earth* might be sailed about, or the Main Ocean with a continual trace did encompass it; but yet I deny not, but that some of the Ancients were of that opinion; yet I utterly deny they knew it certainly. 5. They knew not whether the *Torrid Zone* were habitable. 6. They were ignorant of the true *dimensions* of the *Earth*, although they wrote many things in this business.

The excellency of geography

First, the study of *Geography* is commended to us by the great worthiness thereof, because it most of all becometh Man, being an Inhabitant of the Earth, and endued with Reason above all Living Creatures. Secondly, it is also a pleasant thing, and indeed an honest recreation to contemplate the Kingdoms and Properties of the Earth. Thirdly, The commodity and necessity of it is notable, insomuch as neither Divines, Physitians, Lawyers, Historians, nor other Professors can want the knowledge thereof. But the Excellency of *Geography* hath been sufficiently handled.

Eighteenth Century
Concepts of Geography

In the eighteenth century newly-formed scientific societies in England, France, Sweden, and other European countries brought new life to the ancient science of geography by providing leadership for scientific work and by encouraging travel for scientific purposes. This, too, was the century of the great naturalist-taxonomists, represented in this section by Buffon contemplating the general characteristics of the earth and its population, and by Linnaeus writing about his tour of northern Sweden. Some of the truly original work in eighteenth century geography was done by men who worked in other fields, especially in cartography, as is shown by an excerpt from one of Philippe Buache's essays.

Germany, long a center of geographical studies, continued to be a place where the tasks of geography were defined in the light of current developments; this tendency is reflected here in the writings of Leyser, Franz, Herder, and Büsching. An early example of botanical geography is given in a selection from von Haller.

99. Buffon on the history of the earth, on earthquakes, and on the different races

Georges-Louis Leclerc, Count de Buffon (1707–1788), was Superintendent of the King's Garden, the precursor of the Jardin des Plantes of Paris. The following excerpt is from the *Natural History,* the great taxonomist's major work.

From Count de Buffon, *Natural History, General and Particular,* trans. William Smellie (London, 1791), I, 15–17, 432–436.

On the history of the earth

It is certain, that the waters of the sea have, at some period or other, remained for a succession of ages upon what we now know to be dry land;
and, consequently, that the vast continents of Asia, Europe, Africa, and
America, were then the bottom of an immense ocean, replete with every
thing which the present ocean produces. It is likewise certain, that the different strata of the earth are horizontal, and parallel to each other. This
parallel situation must, therefore, be owing to the operation of the waters,
which have gradually accumulated the different materials, and given them
the same position that water itself invariably assumes. The horizontal position of strata is almost universal: In plains, the strata are exactly horizontal.
It is only in the mountains that they are inclined to the horizon; because
they have originally been formed by sediments deposited upon an inclined
base. Now, I maintain, that these strata must have been gradually formed,
and that they are not the effect of any sudden revolution; because nothing
is more frequent than strata composed of heavy materials placed above light
ones, which never could have happened, if, according to some authors, the
whole had been blended and dissolved by the deluge, and afterwards pre-
cipitated. On this supposition every thing should have had a different aspect
from what now appears. The heaviest bodies should have descended first,
and every stratum should have had a situation corresponding to its specific
gravity. In this case we should not have seen solid rocks or metals placed
above light sand, nor clay under coal.

Another circumstance demands our attention. No cause but the motion
and sediments of water could possibly produce the regular position of the
various strata of which the superficial part of this earth is composed. The
highest mountains consist of parallel strata, as well as the lowest valleys. Of
course, the formation of mountains cannot be imputed to the shocks of
earthquakes, or to the eruptions of volcanoes. Such small eminences as have
been raised by volcanoes or convulsions of the earth, instead of being composed of parallel strata, are mere masses of weighty materials, blended together in the utmost confusion. But this parallel and horizontal position of
strata must necessarily be the operation of a uniform and constant cause.

We are, therefore, authorised to conclude, from repeated and incontro-
vertible facts and observations, that the dry and habitable part of the earth
has for a long time remained under the waters of the sea, and must have
undergone the same changes which are at present going on at the bottom of
the ocean. To discover what has formerly happened to the dry land, let us
examine what passes in the bottom of the sea; and we shall soon be enabled
to make some rational conclusions with regard to the external figure and
internal constitution of the earth.

On earthquakes

Earthquakes are of two kinds: Those occasioned by the action of subterraneous fires, and by the explosions of volcanoes are only felt at small distances, previous to, or during the time of eruptions. When the inflammable matters in the bowels of the earth begin to ferment and to burn, the fire makes an effort to escape in every direction; and if it finds no natural vents, it forces a passage, by elevating and throwing off the incumbent earth. In this manner volcanoes commence, and their effects continue in proportion to the quantity of inflammable matter they contain. When the quantity of inflamed matter is inconsiderable, it produces only an earthquake, and exhibits no marks of a volcano: The air generated by subterraneous fire may also escape through small fissures; and, in this case, likewise, it will be attended with a succussion of the earth; but no volcano will appear. But when the quantity of inflamed matter is great, and when it is confined on all sides by solid and compact bodies, an earthquake and a volcano are the necessary consequences. All these commotions, however, constitute only the first species of earthquakes, which are not felt but in the neighbourhood of the places where they happen. A violent eruption of Aetna, for example, will shake all the island of Sicily; but it will never extend to the distance of three or four hundred leagues. When Vesuvius bursts open a new mouth, it excites an earthquake in Naples and in the neighbourhood of the volcano; but these earthquakes never shake the Alps, nor do they extend to France or other countries distant from Vesuvius. Thus, earthquakes produced by volcanoes are limited to a small space; they are nothing but effects of the reaction of the fire, and they shake the earth in the same manner as the explosion of a powder-magazine occasions an agitation to the distance of several leagues.

But there is another species of earthquakes which are very different in their effects, and perhaps also in their cause. These earthquakes are felt at great distances, and shake a long tract of ground without the intervention either of a new volcano, or of eruptions in those which already exist. There are instances of earthquakes which have been felt at the same time in Britain, in France, in Germany, and in Hungary. These earthquakes always extend more in length than in breadth. They shake a zone or belt of earth with greater or less violence in different places; and they are generally accompanied with a hollow noise, like that of a heavy carriage rolling with rapidity.

Of the varieties of the human species

The heat of the climate is the chief cause of blackness among the human species. When this heat is excessive, as in Senegal and Guiney, the men are perfectly black; when it is a little less violent, the blackness is not so deep;

when it becomes somewhat temperate, as in Barbary, Mogul, Arabia, &c. the men are only brown; and, lastly, when it is altogether temperate, as in Europe and Asia, the men are white. Some varieties, indeed, are produced by the mode of living. All the Tartars, for example, are tawny, while the Europeans, who live under the same latitude, are white. This difference may safely be ascribed to the Tartars being always exposed to the air; to their having no cities or fixed habitations; to their sleeping constantly on the ground; and to their rough and savage manner of living. These circumstances are sufficient to render the Tartars more swarthy than the Europeans, who want nothing to make life easy and comfortable. Why are the Chinese fairer than the Tartars, though they resemble them in every feature? Because they are more polished; because they live in towns, and practice every art to guard themselves against the injuries of the weather; while the Tartars are perpetually exposed to the action of the sun and air.

But, when the cold becomes extreme, it produces effects similar to those of violent heat. The Samoiedes, the Laplanders, and the natives of Greenland, are very tawny. We are even assured, that some of the Greenlanders are as black as the Africans. Here the two extremes approach each other: Great cold and great heat produce the same effect upon the skin, because each of these causes acts by a quality common to both; and this quality is the dryness of the air, which, perhaps, is equally great in extreme cold as in extreme heat. Both cold and heat dry the skin, and give it that tawny hue which we find among the Laplanders. Cold contracts all the productions of nature. The Laplanders, accordingly, who are perpetually exposed to the rigours of frost, are the smallest of the human species. Nothing can afford a stronger example of the influence of climate than this race of Laplanders, who are situated, along the whole polar circle, in an extensive zone, the breadth of which is limited by nothing but excessive cold; for that race totally disappears, whenever the climate becomes a little temperate.

The most temperate climate lies between the 40th and 50th degree of latitude, and it produces the most handsome and beautiful men. It is from this climate that the ideas of the genuine colour of mankind, and of the various degrees of beauty, ought to be derived. The two extremes are equally remote from truth and from beauty. The civilized countries, situated under this zone, are Georgia, Circassia, the Ukraine, Turkey in Europe, Hungary, the south of Germany, Italy, Switzerland, France, and the northern part of Spain. The natives of these territories are the most handsome and most beautiful people in the world.

100. The Lapland journey of Linnaeus

In May 1732 Carl Linnaeus, then a student at Uppsala University, left on horseback for a journey to Lapland, the northernmost part of Sweden. It was his first field trip, and the impressions he gathered during its course were to exert a lasting influence on his career as a naturalist.

We pursued our journey by water with considerable labour and difficulty all night long, if it might be called night, which was as light as the day, the sun disappearing for about half an hour only, and the temperature of the air being rather cold. The colonist who was my companion was obliged sometimes to wade along in the river, dragging the boat after him, for half a mile together. His feet and legs were protected by shoes made of birch bark. In the morning we went on shore, in order to inquire for a native Laplander, who would undertake to be my guide further on. Finding only an empty hut at the spot where we landed, we proceeded as fast as we could to the next hut, a quarter of a mile distant, which likewise proved unoccupied. At length we arrived at a third hut, half a mile further, but met with as little success as at the two former, it being quite empty. Upon which I dispatched my fellow-traveller to a fourth hut, at some distance, to see if he could find any person fit for my purpose, and I betook myself to the contemplation of the wild scenes of Nature around me.

The soil here was extremely sterile, consisting of barren sand (*Arena Glarea*) without any large stones or rocks, which are only seen near the shores of the waters. Fir trees were rather thinly scattered, but they were extremely lofty, towering up to the clouds. Here were spacious tracts producing the finest timber I ever beheld. The ground was clothed with Ling, Red Whortle-berries (*Vaccinium Vitis Idæa*), and mosses. In such parts as were rather low grew smaller firs, amongst abundance of birch, the ground there also producing Red Whortle-berries, as well as the common black kind (*Vaccinium Myrtillus*), with *Polytrichum* (*commune*). On the dry hills, which most abounded with large pines, the finest timber was strewed around, felled by the force of the tempests, lying in all directions, so as to render the country in some places almost impenetrable ...

I wondered that the Laplanders hereabouts had not built a score of small houses, lofty enough at least to be entered in an upright posture, as they have such abundance of wood at hand. On my expressing my surprise at this, they answered: "In summer we are in one spot, in winter at another,

From Carl Linnaeus, *A Tour in Lapland*, ed. Robert Goldwyn, M.D. (reprint ed., New York: Arno Press, Inc., 1971) I, pp. 122–124, 167–169, 283–284, 321–325; II, pp. 130–132.

perhaps twenty miles distant, where we can find moss for our reindeer." I asked "why they did not collect this moss in the summer, that they might have a supply of it during the winter' frosts?" They replied, that they give their whole attention to fishing in summer time, far from the places where this moss abounds and where they reside in winter.

These people eat a great deal of flesh meat. A family of four persons consumes at least one reindeer every week, from the time when the preserved fish becomes too stale to be eatable, till the return of the fishing season. Surely they might manage better in this respect than they do. When the Laplander in summer catches no fish, he must either starve, or kill some of his reindeer. He has no other cattle or domestic animals than the reindeer and the dog: the latter cannot serve him for food in his rambling excursions; but whenever he can kill Gluttons (*Mustela Gulo*), Squirrels, Martins, Bears or Beavers, in short any thing except Foxes and Wolves, he devours them. His whole sustenance is derived from the flesh of these animals, wild fowl, and the reindeer, with fish and water. A Laplander, therefore, whose family consists of four persons, including himself, when he has no other meat, kills a reindeer every week, three of which are equal to an ox; he consequently consumes about thirty of those animals in the course of the winter, which are equal to ten oxen, whereas a single ox is sufficient for a Swedish peasant.

The peasants settled in this neighbourhood, in time of scarcity eat chaff, as well as the inner bark of pine trees separated from the scaly cuticle. They grind and then bake it in order to render it fit for food. A part is reserved for their cattle, being cut obliquely into pieces of two fingers' breadth, by which the fodder of the cows, goats, and sheep is very much spared. The bark is collected at the time when the sap rises in the tree, and, after being dried in the sun, is kept for winter use. They grind it into meal, bake bread of it, and make grains to feed swine upon, which render those animals extremely fat, and save a great deal of corn.

The Lapland Alps

July 6

In the afternoon I took leave of Hyttan, and, at the distance of a mile from thence, arrived at the mountain of *Wallavari* (or *Hwallawari*), a quarter of a mile in height. When I reached this mountain, I seemed entering on a new world; and when I had ascended it, I scarcely knew whether I was in Asia or Africa, the soil, situation, and every one of the plants, being equally strange to me. Indeed I was now, for the first time, upon the Alps! Snowy mountains encompassed me on every side. I walked in snow, as if it had been the severest winter. All the rare plants that I had previously met with, and which had from time to time afforded me so much pleasure, were here as in miniature, and new ones in such profusion, that I was overcome with astonishment, thinking I had now found more than I should know what to do with . . .

July 11

We rose early this morning, and after walking a quarter of a mile arrived at the lofty icy mountain. This is indeed of a very great elevation, and covered with perpetual snow, the surface of which was, for the most part, frozen quite hard. Sometimes we walked firmly over it, but it occasionally gave way, crumbling under our feet like sand. Every now and then we came to a river taking its course under the snowy crust, which in some parts had yielded to the force of the currents, and the sides of each chasm exhibited many snowy strata one above another. Here the mountain streams began to take their course westward, a sign of our having reached Norwegian Lapland. The delightful tracts of vegetation, which had hitherto been so agreeably interspersed among the alpine snows, were now no longer to be seen. No charming flowers were here scattered under our feet. The whole country was one dazzling snowy waste. The cold east wind quickened our steps, and obliged us to protect our hands that we might escape chilblains. I was glad to put on an additional coat. As we proceeded across the north side of this mountain, we were often so violently driven along by the force of the wind, that we were taken off our feet, and rolled a considerable way down the hill. This once happened to me in so dangerous a place, that, after rolling to the distance of a gunshot, I arrived near the brink of a precipice, and thus my part in the drama had very nearly come to an end. The rain, which fell in torrents on all sides, froze on our shoes and backs into a crust of ice. This journey would have been long and tiresome enough without any such additional inconvenience. At length, after having travelled betwixt three and four miles, the mountains appeared before us, bare of snow though only sterile rocks, and between them we caught a view of the western ocean. The only bird I had seen in this icy tract, was what the Laplanders call *Pago* (*Charadrius Hiaticula*). Its breast is black, throat white, feet orange.

Having thus traversed the alps, we arrived about noon upon their bold and precipitous limits to the westward. The ample forests spread out beneath us, looked like fine green fields, the loftiest trees appearing no more than herbs of the humblest growth. About these mountains grew the same species of plants that I had observed on the other side of the alps. We now descended into a lower country. It seems, as I write this, that I am still walking down the mountain, so long and steep was the descent, but the alpine plants no longer made their appearance after we had reached the more humble hills. When we arrived at the plains below, how grateful was the transition from a chill and frozen mountain to a warm balmy valley!

I sat down to regale myself with strawberries. Instead of ice and snow, I was surrounded with vegetation in all its prime. Such tall grass I had never before beheld in any country. Instead of the blustering wind so lately experienced, soft gales wafted around us the grateful scent of flowery clover and various other plants. In the earlier part of my journey, I had for some time experienced a long-continued spring (whose steps I pursued as I ascended the Lapland hills); then unremitting winter and eternal snow surrounded me; summer at length was truly welcome. Oh how most lovely of all is summer!

Torneå

August 4

A reindeer may be driven in a carriage twelve (Swedish) miles in a day, or, at the utmost, fifteen; but the animal is generally killed immediately afterwards with the Laplanders' spears, and eaten.

The Laplanders are perhaps so called from the (Swedish) word *lappa*, to sew or patch together, because their garments usually answer to that description.

The degree of cold is certainly greater on the alps than in less elevated regions; for instance, it is colder in Jämtland than at Torneå, though the former lies about one hundred miles further south. Thus it appears that the cause of the severity of the cold does not depend so much on the approach towards the pole, as on the elevation of the ground, which ought to be carefully observed. Hence the plants of the north of Lapland are such only as are capable of resisting the most severe and long-continued cold, and hence snow lies on the alps of Italy.

Alpine plants are, for the most part, perennial, except the Little Blue Centaury (*Gentiana nivalis*), and perhaps the Single-flowered *Lychnis* (*apetala*). I wish botanists would endeavor to discover and make known any new kinds of pulse or grain, especially such as are of a hardy constitution, and not likely to suffer from the severity of winter. Some are perhaps to be found among the grasses. It is necessary to ascertain what degree of northern latitude they would bear, and whether they are capable of growing within the frigid zone. It is worthy of inquiry in what respects the Alps, properly so called (of Switzerland,) agree with ours of Lapland, as well as how far, and by what means, they are susceptible of culture. The descriptions that have been given of them may be consulted. The Greenland alps would also be worth examining, to see how far these different countries agree in their native plants, in varieties of situation, and in plenty of soil, compared with the proportion of rocks or large stones. Do they all agree in the diminutive stature of their plants?

101. Buache's "Framework of the Earth"

Philippe Buache (1700–1773) was one of the leading cartographers of the eighteenth century, with the title *premier géographe du Roi*. He was a member of the French Academy of Sciences and at a meeting of that body on November 15, 1752, Buache presented a novel scheme for describing the physical characteristics of the earth, together with a map illustrating his scheme.

Buache introduced his essay by reminding his audience that geography had had a utilitarian origin, and that it was not until the time of Augustus that it had become an organized form of knowledge which would be extended by the Arabs and, in particular, by the discovery of America and the navigations to the Indies.

But can we be satisfied now, knowing that we are ignorant of nearly everything beyond latitude 50° South, without even mentioning the fact that we are not even certain of the exact location of cities, and so forth?

Having stated his belief in the three branches of geography— namely, physical geography, historical geography (corresponding to regional-political-economic geography in our time) and mathematical/theoretical geography—Buache proceeds to set forth his views on physical geography.

I see a kind of framework [*charpente*] that supports the several parts of our globe, formed by high mountain systems that surround it and traverse it in a proportion that will appear even more admirable when we examine it closely . . . I was influenced in setting forth these views, first, by the sources of the great rivers which serve as natural indicators of the highest mountains and terrain where these sources lie, and second by the islands, rocks, reefs, etc., in the sea which we are familiar with, and which allow us to have an idea of the mountain systems that I shall call marine . . .

Having established a classification of mountains according to high, medium, or low elevation—a classification based upon these mountains being the source of large streams, medium-sized rivers, or short or coastal rivers—Buache proceeds to divide the world ocean into its several parts. These include first what Buache calls "the Ocean" (the Atlantic); second, the Indian Ocean; and third, the

From Philippe Buache, *An Essay on Physical Geography, wherein we present general views on what may be called the Framework of the Earth composed of mountain systems that traverse seas as well as continents, together with some specific remarks on the various basins of the sea, and on its interior configuration* (Paris, 1752), pp. 399–416 passim. Editor's translation.

Pacific or Southern Ocean, each consisting of several subdivisions. Buache's conception of submarine topography, illustrated by his map, is set forth in considerable detail; the description of the Indian Ocean is a good example.

The second great subdivision of the sea is the Indian Ocean, which lies between Africa and Australia and laps at the southern coasts of Asia. It extends all the way to the land masses of Antarctica (*sic*) . . .

We may distinguish in this body of water three basins, separated by a marine [submarine] mountain system that begins with Madagascar, continues to Sumatra, thence to Van Diemen's Land and New Guinea. The first of these basins of the Indian Ocean, lying in the west, consists of the gulfs that used to be called Red Sea and Persian Gulf, and I propose to use the single term Arabian Gulf for them, since there are ancient Arab settlements on its shores. The second is none other than the Gulf of Bengal, separating the two great Indian peninsulas. The third is the great archipelago, consisting of the Sunda Islands [Indonesia], the Moluccas, and the Philippines; this is like a massif of mountains connecting Asia and the Southern Continent . . . It is separated from the Pacific by a small basin, limited to the east by the marine mountain system formed by the Marianas . . .

What is most remarkable about the continuous nature of this framework that traverses the continents and supports the several parts of our globe is the fact that these mountain systems seem to extend, to radiate, from certain places that must be the highest on earth, and from Plateaus formed by mountains, grouped and piled one upon the other . . .

The following excerpt is typical of Buache's views of great mountain systems.

Five great mountain systems have their origins in the African plateau. The first, having extended one of its branches towards the Straits of Bab-el-Mandeb and Ormuz, follows the shores of the Red Sea, reaches the isthmus of Suez, and then joins the mountain system of Europe and Asia. The second reaches the Atlas Mountains in the vicinity of Tripoli. The third, having extended a branch that further on becomes the marine mountain range that passes across the Canaries and the Azores [and forms a connection with North America], reaches Cape Tagrin in Guinea, and continues beyond as the great marine mountain range that connects Africa with South America. The fourth system extends as far as the Cape of Good Hope, and continues as the marine mountain range that connects Africa with the Antarctic lands. Finally, the fifth system extends to the coast across from Madagascar, and continues as the marine range across the Indian Ocean.

Having discussed in considerable detail the structure of these several mountain systems and presented a detailed view of submarine topography (based upon his pioneer map of the English Channel complete with bathymetric lines), Buache completed his lecture with a specific suggestion regarding a new kind of globe.

A way to improve upon this system would be to make a relief globe, one made up of movable parts, that is, continents and seas could be removed and replaced at will. Thus, having observed higher and lower elevations above sea level, and the direction taken by streams and rivers, together with the location of cities, and even facts concerning Natural History [vegetation], one could lift up the surface of the sea, and view its depths, the layout of marine mountain systems, and the general nature of the sea itself.

102. Polycarp Leyser on geography and history

Polycarp Leyser (1690–1728) was a professor at the University of Helmstedt in Prussia. In this selection he praises geography as a companion to historical studies and attempts to define the task of physical geography.

He who wishes to read the works of historians, or desires to hold forth in the proper manner about history, must know all disciplines, arts and sciences. Yet there are certain disciplines which are of such a nature that they relate more closely to history than others: chronology, archaeology, the study of coinage, and geography.

Geography, which I list in the last place, surpasses the others both in dignity and excellence. For it assists in a wonderful way the study of history, making it easier to remember historical events when relying on geography.

I would even add that geography is the touchstone both of history and historians, which reveals the errors of historians with ease . . .

Geographia naturalis . . . observes natural boundaries and indicators that place a city in a definite and unchangeable location. Indicators of this type include mountains, valleys, rivers, lakes and seas, in a word, all those variations of the settled parts of the earth which are part of nature. These indicators are extraordinarily stable. Mountains cannot easily be worn down, rivers cannot have their courses changed with ease, nor can seas be changed to dry land as part of a continent . . .

Since mountains and valleys cannot be found everywhere, it seems to

---

From Polycarp Leyser, *Commentatio de vera geographiae methodo* (Helmstedt, 1726), quoted from the German translation in H. Beck, *Geographie* (Freiburg-München: Karl Alber Verlag, 1973), pp. 136–139, by permission. Editor's translation.

me that the best way to establish topographic divisions is to consider water. One finds water in most places, since these settlements were founded by some sea, river, or small stream.

If you add to these indicators the ocean, and the rivers draining to it, in a definite sequence, all areas can be given a topographic definition with ease and in a sequence that is constant, and that leads to the desired location, through a history of untold centuries.

103. Johann Michael Franz defines the state geographer

Geography in the eighteenth century was often considered one of the essential branches of knowledge necessary for government. Johann Michael Franz (1700–1761), professor at the University of Göttingen, describes the ideal geographer as one fully prepared for the service of the state.

Our geographer [*Staatsgeographus*] is not a mapmaker who buys a sketch from an engineer, or puts a sort of map together from scraps that he had begged for, then gives it to the engraver, and when it is finished, has it sold through merchants of pictures in the world at large. He is not an author of compendia, who uses information that he pasted together in his study out of books and letters, and thus describes regions and even continents that he has never seen, and thus depends entirely on other peoples' observations. In a word, he is not a copyist, a compiler, a purveyor of secondhand knowledge.

Our geographer is a much more useful and needed member of the community. He is a real geographer, who does not stay at home, buried in a small corner; for what is more remarkable in this world than a man who desires to present the world to others, a world that he does not know for he is a stay-at-home! Our geographer travels all the time, his greatest pleasure is to be a constant traveler; he will only stop traveling when he dies, in a stage coach . . .

He must be fully familiar with mathematics, history, and natural science. He must be qualified to carry out surveys and establish boundary lines. He must be able to provide full descriptions of the state and its cities, towns, and villages, putting together a complete repertory of information of all sorts. On the basis of that information, he must prepare a description of the state, illustrate it with maps, and see to its publication. He has the obligation to propose schemes for reclamation and ameliora-

From Johann Michael Franz, *Der Deutsche Staatsgeographus* (Frankfurt & Leipzig, 1753), quoted in H. Beck, *Geographie* (Frankfurt-München: Karl Alber Verlag, 1973), pp. 201–204, by permission. Editor's translation.

tion of the soil, and be well informed on agriculture, architecture, and commerce in order to suggest improvements. He must prepare lists of all roads and of the postal system, properly illustrated with maps, for the use of travelers, as well as for purposes of improving communications. He must keep the library of his sovereign up to date as regards books, maps, charts, and all manner of geographical publications regarding new knowledge and discoveries. In the manner customary in France, where the first Geographer Royal also provides instruction in geography to the heir to the throne, the geographer must provide this service to the offspring of his sovereign, and prepare handbooks for schools for the nobility and schools in general. In time of war, he serves as the military geographer. And when the sovereign or his envoys travel abroad, the geographer will keep a travel diary, entering information gathered during the travel as regards matters terrestrial and celestial.

104. Johann Gottfried von Herder on the charm and necessity of the study of geography

Johann Gottfried von Herder (1744–1803), distinguished writer and philosopher, was very much taken with geography as an indispensable part of knowledge to be acquired by people of intelligence.

––––––––

Of course, if you consider geography merely as a dry catalogue of names of lands, rivers, frontiers, and cities, it is a knowledge of mere words, dry and yet unjustly treated and misunderstood, just as it would be the case if history consisted only of a list of names of worthless kings and of dates. Such study is not only without educational value, but it is to a high degree repulsive, without flavor, and without value.

Much of political geography and political history, is without interest to young people. Indeed, were one to tell the truth, such study is not even understandable, since youth has so few correct notions of most events of wars and of statecraft, as does many an adult. But is this true geography, true history? Is a miserable list of words a language? Does learning a dictionary by heart make one a good writer? Would we not find it foolish for a man who wishes to learn Latin and Greek to limit himself to studying nothing but a dictionary?

And yet this is precisely geography and history, if we use it only as a

––––––––

From Johann Gottfried Herder, *"On the Charm, Usefulness, and Need for the Study of Geography"* (1784), from part 16 of Herder's Works, ed. Heinrich Düntzer (Berlin, 1879), quoted in H. Beck, *Geographie* (Frankfurt-München: Karl Alber Verlag, 1973), pp. 176–178, by permission. Editor's translation.

list of rivers, lands, cities, kings, battles, and peace treaties. All this is the raw material necessary to construct a building; otherwise it remains only stone and lime, only rubble that is not prized by anyone, nor inhabited by anyone. A painter needs color, but only to create a painting; then and only then does the eye rejoice in it, only then does it provide enjoyment for the soul. Let us see what the word *geography* tells us, in its very name.

Geography means a description of the earth; therefore a knowledge of the earth, especially of physical geography, is of the utmost importance. It is knowledge that is as important as it is easy to understand, and entertaining. Who would not wish to know the marvelous edifice that we inhabit, the changing scene where the goodness and wisdom of creation have found it proper for us to settle? This means that we must know about the earth, that is, a globe, a planet; we must familiarise ourselves with the general laws ruling its motion upon its own axis and around the sun; and we must know how these motions create days and years, climates and regions. And we must present this in the professional and dignified manner that this great subject demands. If all this does not quicken and dignify the spirit, what will? . . .

With the help of geography history becomes like a richly illuminated map, helping imagination and judgment. Only with its help does it become clear why these peoples and no others played particular parts on the scene that is our earth; why these rulers ruled in one place, others in different ones; why one realm lasted longer than another . . . why science and civilization, inventions and the arts followed a particular course . . .

In short, geography is the foundation of history, and history is but the geography of eras and peoples set in motion. He who cultivates one without the other understands neither; he who despises both should, like a mole, live under the earth rather than on its surface. All of the sciences that our century loves, treasures, promotes, and rewards are based mainly on philosophy and history; trade and statecraft, economics and jurisprudence, medicine, and all practical knowledge and manipulation of men is based on geography and history. These are the setting and the text of God's creation in our world: history is the text, geography the scene.

105. Anton Friedrich Büsching on geography

Anton Friedrich Büsching (1724–1793) represents a point of view typical of his time. He defines geography's field of study in his *Description of the Earth.*

From Anton Friedrich Büsching, *Erdbeschreibung* (Hamburg, 1787), quoted in H. Beck, *Geographie* (Frankfurt-München: Karl Alber Verlag, 1973), pp. 174–175, 197–199, by permission. Editor's translation.

Geography can be defined as a thorough report on the natural and political characteristics of the known earth. This definition consists of two principal parts.

The first of these is the description of the earth, that is, of the natural and political characteristics of its known parts. Since our earth is but a part of the universe, so geography is part of cosmography, a subject with which it is closely connected, and which provides it with numerous explanations.

Geography deals with our earth to the extent that it is known to us; there are unknown regions both toward the north and the south pole about which we know little, except the fact that they exist; in fact there are some whose existence we only suspect, since we have no information about them. But that part of our earth that is known must be considered both in terms of its natural characteristics and its political organization. Among the natural characteristics we list those mathematical considerations regarding the earth as a celestial body, and compare its shape, size, and position with that of other celestial bodies. But natural characteristics also include a knowledge of all things mobile and immobile both on the surface of the earth and underneath it, that is, true physical geography.

Considering the political characteristics, we view the several states, not restricting ourselves to their position, which provides us with correct information as to their size, strength, institutions, form of government, inhabitants, etc., but we also include a description of their constitutional systems, their religious organizations, their fortresses, castles, villages, and other important settlements and foundations.

The second characteristic of geography is that it provides reliable information on all things. Such information may be brief or extensive, according to the purpose of the author; but in every instance useless and irrelevant matters must be kept out of it, to prevent such books from assuming an unnecessary and uncomfortable size.

106. Albrecht von Haller on the vertical zoning of vegetation

Albrecht von Haller (1708–1777), physician and naturalist, contemporary of Linnaeus and Buffon, was both a pioneer in general geographical studies and a distinguished scholar of plant geography. His *Introduction to the History of Swiss Plants* establishes clear verti-

From Albrecht von Haller, *Vorrede zur Geschichte schweizerischer Pflanzen* (Bern, 1772), quoted in H. Beck, *Geographie* (Frankfurt-München, Karl Alber Verlag, 1973), pp. 142–145, by permission. Editor's translation.

cal limits of vegetation zones, and compares them to latitudinal zones of similar floristic character.

———————

Switzerland is representative of nearly all European countries, from remote Spitzbergen to Spain. On the glaciers and in the highest valleys of the Alps the air is like Spitzbergen. Summer lasts forty days at most, and even that period may be interrupted by snowfall. The rest of the year rough wintry weather prevails. Thus the plants growing around the glaciers are for the most part those that Friedrich Martens found in Spitzbergen. Since these plants grow along the seashore in Spitzbergen and in Greenland, it is evident that the reason for the presence of these plants in the Alps is not the air, but the cold . . .

Once the eternal ice is left behind, one encounters rough pastures, accessible only to sheep, dominated by low plants . . . Next are more durable meadows, where cattle find enough feed during those forty days when these are not snow-covered. It is here that numerous plants are found that are called Alpine plants, many of these can be found in Lapland, Siberia, Kamchatka, some even on the highest mountains of Asia . . .

Somewhat further down one encounters stands of fir on the Alpine slopes. In some of these forests that face to the north one finds northern plants characteristic of Lapland and Siberia . . . In the other forests of this type plants typical of the Harz Mountains and of Sweden can be found . . . Between these forests are thick meadows, created through the burning of the woods . . . Further down, at the foot of the Alps, one encounters a mixture of fields, meadows, and woods . . . Finally, there are the plains, with, on the hills surrounding them, vineyards . . .

It is truly remarkable: this great variety of plants within so small an area. When one climbs from the city of Sitten (Sion), in the Valais, to the mountain called Sanetsch, a climb of about seven hours, one leaves behind at Sitten ephedra and pomegranate which thrive on the rocks of a nearby fortress; chestnut trees, walnut trees, and vineyards producing the best of wines; and soon thereafter the fields of wheat. Gradually beech and oak disappear; then one leaves behind fir and pine, and soon all trees; and one can take his ease among the saxifrage and other plants typical of Spitzbergen. In half a day, one can collect plants that thrive at latitude 80° and others that can be found at latitude 40°.

Measuring the Earth

In the sixteenth century Gemma Frisius suggested that longitude could be measured accurately by using a reliable—and transportable—timepiece to relate local time, as measured for instance by the sun, with time at the place of departure of the observer. Since twenty-four hours of time equals the circumference of the earth, 360 degrees, a difference of one hour between local time and that of the standard—the prime meridian—would equal a difference of fifteen degrees of longitude. Frisius' suggestion was first put to practical use by the seventeenth century astronomer J. D. Cassini.

Cassini determined the exact time of the appearance and disappearance of the satellites of Jupiter using a pendulum clock, an invention of the seventeenth century, and published these data in an astronomical almanac, the *Ephemerides* of Jupiter. French scientists working under Cassini's direction compared the time given by Cassini for the motions of the moons of Jupiter as observed at the University of Bologna with local time, and thus were able to develop an accurate method of determining longitude. The first result of their work was a world map, published in 1696, bearing Cassini's name; it was the first map giving the longitude of a number of places around the world based on this new and exact method.

At the same time, the introduction of optical instruments into the field of geodesy enabled scientists to accurately measure an arc of the meridian. Under the impulse of Newton's theory of the shape of the earth, the French followed up that first measurement, undertaken by Jean Picard near Paris in 1671, with the measurement of two more meridian arcs.

The first of these measurements (1735–1743) was carried out by an expedition led by Bouguer and La Condamine. They measured an arc of the meridian on the equator, in present-day Ecuador. The second measurement, made by an expedition led by Maupertuis in 1735, took place in Lapland on the present Swedish-Finnish frontier. By comparing the length of the degree of the meridian in France, in a mid-latitude area, with an arc on the Arctic Circle and one on the equator, Maupertuis made a first contribution to the actual measurement of the earth's circumference. The final computation of that figure led, in 1791, to the establishment of the metric system. The French surveys had also confirmed Newton's definition of the earth as a rotational spheroid.

107. Maupertuis on the dimensions of the earth

Determination of [the length of] a degree of the Meridian comprised between these two churches [Notre Dame of Paris and Notre Dame of Amiens]

We shall therefore take as the true amplitude of the arc of the meridian defined by the parallels passing through Paris and Amiens 1°1′12″, which is the median amplitude and compare this amplitude with the distance of 58327 toises [1 toise equals 1.949 meters]. This figure is obtained by deducting from 59530.5 toises, the length of the arc of the meridian comprised between the two churches of Paris and Amiens, the distance of 1105.5 toises, that is, the northerly distance of the measuring instrument [sector] from the tower of Notre Dame of Paris, and the distance of 98.5 toises, the distance between the royal residence where observations were taken in Amiens, and the bell tower of the cathedral. Thus, the length of a degree of the meridian between Paris and Amiens is 57183 toises.

Comparison of a degree of the meridian in France with the degree measured at the Arctic Circle, whence the proportion of the earth's axis to the equatorial diameter is deduced

Now that we have [the length of] two degrees of the meridian, measured at considerable distance, that which lies across the Arctic Circle and that between Paris and Amiens, it is possible to determine with considerable

From "Measuring a degree of the meridian between Paris and Amiens, determined on the basis of M. Picard's measurement and by the observations of MM. Maupertuis, Clairaut, Camus, Le Monnier . . . Whence one obtains the shape of the earth, by comparing this degree with the one measured at the Arctic Circle," (Paris, 1740), pp. liv–lvi. Editor's translation.

accuracy the shape of the earth, that is, the proportion of its axis to its equatorial diameter.

For this, we must recall our statement that the earth's meridian may be taken as an ellipse, very close to a circle, of which one-half of its greater axis, corresponding to the radius of the Equator, equals l, and half of the smaller axis, which corresponds to half of the earth's axis, equals m. Two degrees of the meridian, one measured nearer, the other farther, from the Pole are E and F. The sines of the latitudes corresponding to these degrees are S and s for the radius.

We must now recall the formula

$$1 - mm = \frac{2}{3} \quad \frac{E - F}{SSE \quad ssF}$$

which includes the relationships between the two degrees, the sines of their latitudes, the axis and the diameter of the earth. If E equals 57438 (the distance between Kittis and Torneå, in Lapland), and F equals 57183 (the degree between Paris and Amiens); and using the sines of 66°20' for S and of 49°22' for s, we shall find that m equals $\dfrac{177}{178}$ that is, the proportion of the earth's diameter to its axis is 178 to 177.

Immanuel Kant, Geographer

Immanuel Kant (1724–1804), philosopher, held a chair at the University of Königsberg, in Prussia, during the second half of the eighteenth century. It is a matter of record that he offered a course of lectures on physical geography over a period of forty years, between 1757 and 1797. His lectures, which were published during his lifetime, and his other writings on the subject of geography establish him as one of the founders of modern geography, a discipline concerned with spatial phenomena. The first of the excerpts in this section is a brief statement by a twentieth-century scholar on Kant's significance in the history of geography; it is followed by selections from Kant. The first of these is from the announcement of Kant's course at Königsberg during the winter semester of 1765–1766. The second and third selections are taken from his lectures. The third selection is believed to be the first scientific explanation of the phenomenon of monsoon winds.

108. From the geographical writings of Kant

Kant's view of geography

In [Kant's] view the human element was an integral part of the subject matter of geography. Kant divided the communication of experience between persons into two branches, narrative or historical, and descriptive or geographical; and he regarded both history and geography as descriptions, the former in time and the latter in space. He claimed physical geog-

The first selection is from Robert E. Dickinson, *The Makers of Modern Geography* (London: Routledge & Kegan Paul, Ltd., 1969), pp. 10–11, by permission of the publisher. The second selection is from Kant's *Works* (Berlin: Prussian Academy of

raphy to be "a summary of nature," the basis not only of history but also of "all the other possible geographies." Five of the other geographies he identified as part of physical geography are mathematical (the form, size, and movement of the earth and its place in the solar system), moral (the customs and character of man in relation to environment), political, mercantile (commercial), and theological (the distribution of religions). Physical geography thus embraced the outer physical world, the earth's surface, and its cover of life forms of plants, animals, and man and his works.

Kant followed essentially the Ptolemaic definitions of geography. However, there was an essential difference between his work and that of the classical scholars. The latter laid primary emphasis on the areal divisions of the earth and the business of systematically describing their distinctive content. Kant, on the other hand, was not so much concerned with composite terrestrial units of different orders, but with the orderly investigation of particular areally differentiated phenomena. Both are integral aspects of the areal differentiations of the earth's surface, but out of the difference of emphasis that is reflected in the works of Ptolemy and Strabo on the one hand and Kant and Varenius on the other emerge differences of approach that run throughout the development of modern geography.

These differences of approach have been recently described as the theoretical (deductive) or nomothetic approach and the empirical (descriptive) approach. The theoretical approach seeks to establish theories relevant to the location and interrelations of places and to establish laws and make deductions on the basis of the laws. The empirical or ideographic approach places primary emphasis on the description of particular groupings of nations and people in terms of lands, seas, countries, and places. It does not seek to develop laws but to find out how phenomena account for the *genius loci,* the character of a place and its relations with other places. These are the two basic approaches and traditions in all geographical inquiry and their contrast and conflict have become more marked and difficult to bridge as knowledge of the surface of the earth has increased.

Kant's course announcement

I recognized when I first started my university teaching that one of the serious deficiencies of our students consists of their learning to speculate

Sciences, 1901), I, 312–313; editor's translation. The first part of the third selection is from Kant, *Physische Geographie,* vol. I, pt. 4, as translated by Richard Hartshorne in *The Nature of Geography, Annals of the Association of American Geographers* 29 (1939); reprinted by permission. The second part of the third selection is also from *Physische Geographie* but is taken from the *Works,* IX, 160–161; editor's translation. The fourth selection is from the *Works,* I, 492–501; editor's translation.

early on, without historical knowledge sufficient to replace experience. I then resolved to present the history of the present state of the earth, that is, geography in its widest sense, as a pleasant and easy summary of what is needed to prepare for common sense; and thus to create the desire to ever-widen knowledge first presented in this way.

I called such a discipline physical geography, from that part of it that first caught my attention. I have since gradually broadened this outline, and I now am thinking of narrowing that section dealing with the physical characteristics of the earth, thus gaining time to extend my lectures to those other sections which are of even greater use. This discipline will thus be divided into physical, moral, and political geography. Within it will first be discussed the characteristics of nature in its three realms; but through a selection of those which, among innumerable others—because of their rarity or because of the influence they have through commerce and industry upon states—attract seekers of knowledge. It is this part, which also deals with the natural relationships of all lands and seas and their interconnections, that is the true foundation of all history, which otherwise would hardly differ from fairy tales.

The second segment considers man on earth, in terms of the variety of his natural characteristics and the moral differences that exist; this is a very important and equally fascinating set of considerations; without it it would be difficult to make general judgments on man. Here a comparison of various races and of moral conditions of the present and of older times create for us a huge map of the human race.

Finally the conditions of states, which we can view as a result of the mutual influences of the forces discussed previously, and of the peoples of the earth will be considered. This will be viewed not only insofar as it is based upon the random causes of man's actions and his fate, that is, upon the sequence of governments, of conquests, and of the ranking of states, but also insofar as it is based upon that which is more permanent, and which consists of the fundamental of these: the position of countries, their products, customs, industry, commerce, and population. The very rejuvenation, if I may call it that, of a science of such wide-ranging views, even to a small extent, has great utility, for it is only in this manner that the unity of recognition, without which all knowledge is but fragmentary, can be achieved.

On the nature of geography and its distinction from history

We may classify our empirical knowledge in either of two ways: either according to conceptions, or according to time and space in which they are actually found. The classification of perceptions according to concepts is the *logical* classification, that according to time and space the *physical* classification. Through the former we obtain a system of nature, such as that of Linnaeus, through the latter, a geographical description of nature.

Thus, if I say that cattle are to be included under the class of quadrupeds, and under the group of this class that have cloven hooves, that is a classification that I make in my head—that is, a logical classification. The system of nature is likewise a register of the whole: there I place each thing in its class, even though they are to be found in different, widely separated places.

According to the physical classification, on the other hand, things are regarded according to the places than include them on the earth ... In contrast to the system of nature with its division by classes, the geographical description of nature shows the places in which things are actually to be found on the earth. For example, the lizard and the crocodile are fundamentally the same animal: the crocodile is merely an extremely large lizard. But they exist in different places: the crocodile in the Nile, the lizard on the land. In general, we consider here the scene (*Schauplatz*) of nature, the earth itself, and the regions where the things are actually found, in contrast with the system of nature which is concerned with the similarity of forms.

Description according to time is history, that according to space is geography ... Geography and history fill up the entire circumference of our perceptions: geography that of space, history that of time ...

History, therefore, and geography broaden our knowledge in regard to time and space. History is about events that, with regard to time, have occurred one following the other. Geography is about phenomena that, as regards space, happen at the same time. According to the several subjects that geography concerns itself with, it may be referred to by different terms. Therefore it is called physical, mathematical, political, moral, theological, literary, or mercantile geography ...

History differs from geography in terms of space and time. The first ... is a report on events that follow one another and are related to time. The second however is a report on events that take place simultaneously in space. History tells a story, geography, a description.

On monsoon winds

First remark
Greater heat, which influences the air in one region more than in another, creates a wind toward the region of heated air; this lasts as long as the increased heat of the region.

Second remark
A region in which air is cooling to a greater degree than in the neighboring one creates a wind in the neighboring region that blows toward the place where the cooling is taking place.

Third remark

A wind that blows from the Equator toward the Pole will become gradually more westerly as it proceeds further, while wind that moves from the Pole towards the Equator changes its direction in a collateral movement from the east.

Fourth remark

The prevailing easterly wind that dominates the entire ocean between the Tropics is to be ascribed to no other cause than that resulting from the combination of the first and third remarks.

Fifth remark

The monsoons or periodic winds which dominate in the Arabian, Persian, and Indian oceans may be explained in the easiest manner by the law proved in the Third Remark.

In these seas southwesterly winds prevail from April into September, followed by calms, and then by opposite, northeasterly winds from October into March. Alerted by the preceding [remarks], the reason for this becomes immediately clear. The sun reaches our northern hemisphere in March, and heats up Persia, Arabia, India, the neighboring peninsulas, and in the same way China and Japan to a greater degree than the seas located between these lands and the Equator. Air stationary over these seas is forced, through the thinning of northern air, to expand in that direction, and we know that wind blowing from the Equator toward the North Pole has to deflect southwestward.

On the contrary, as soon as the sun is past the fall equinox and thins the air of the southern hemisphere, air from the northern areas moves toward the Equator. A wind blowing from northerly areas toward the Equator, left alone, will become a northeasterly wind; it is quite clear why it is this wind which follows the previous southwesterly wind.

It is simple to observe the casual connections, insofar as these fit the creation of periodic winds. There has to exist near the tropic, a large landmass which receives more heat due to the sun than the seas that are located between it and the Equator. Then, air from these seas will be forced to move over the landmass and create a westerly collateral wind, and alternately will move from the landmass over the seas.

The Founders of Modern Geography: Humboldt and Ritter

Alexander von Humboldt (1769–1859) and Carl Ritter (1779–1859) are unique figures in the long history of geography. Both attempted a synthesis of knowledge about the earth. Both made a deep impression on their contemporaries and their successors. Humboldt mainly through his writings, Ritter more through his university teaching. Both men reflect the viewpoints of the eighteenth century and Humboldt in particular has been described as the last of the great polymaths. The distinguished German geographer Carl Troll, speaking in 1959, characterized Humboldt as "the creator of modern physical and biological geography," Ritter as "the creator of historical and cultural geography."[1] In their lifetimes and, even more so after their deaths, geography's main stream divided into specialized fields of knowledge. Humboldt and Ritter are thus the last representatives of classical geography, and they stand at the beginnings of geography as we know it today.

Alexander von Humboldt came from the Prussian aristocracy, and was trained in technology and science in preparation for a career as a public official. He did serve as an official of the Prussian mining administration, but his heart was already set on travel. In 1799, having inherited part of his family's fortune, he left Europe in the company of a French botanist, Aimé Bonpland, and spent the next five years traveling in tropical America, from the llanos of the Orinoco to the high Andes, from Ecuador to Mexico. The two men returned to Europe via the United States where Humboldt visited Washington, the new capital.

---

1. Carl Troll, "The Work of Alexander von Humboldt and Carl Ritter: A Centenary Address," in *The Advancement of Science* 64 (1960), 441.

From 1805 to 1827 Humboldt lived in Paris, writing and publishing the results of his travels. These are pioneer works in geography and in natural history: monumental in their scope, innovative in their concepts, original in their organization, impressive in their magnificent illustrations. Humboldt spent a large part of his fortune financing his travels and the publication of his books, and in 1827 returned to Germany to spend the rest of his life as a high official of the Prussian court. He continued to write and, occasionally, to lecture. In 1829 he undertook one more long voyage at the invitation of the Russian government—to the Urals and to the Asian borderlands of the Russian empire.

The first selection from Humboldt's works is a passage from one of his earliest writings, on the nature of geography. This is followed by selections from his book on travels in equatorial America. A letter from Thomas Jefferson to Humboldt follows, illustrating the esteem in which Humboldt was held and the value that the President of the United States placed on the information Humboldt possessed.

When Humboldt returned to Berlin, famous both as scientist and traveler, he offered a series of lectures at the local university. In subsequent years those lectures served as an outline for his *Kosmos;* a statement describing the plan of the lectures is included here, followed by excerpts from the great work, which was published in five volumes between 1845 and 1862. The final selection, taken from his *Aspects of Nature,* illustrates Humboldt's work as naturalist and underlines his lasting concern with the unity of nature.

109. Humboldt on "geognosy"

Geognosy (*Erdkunde*) studies animate and inaminate nature . . . both organic and inorganic bodies. It is divided into three parts: solid rock geography, which Werner has industriously studied; zoological geography, whose foundations have been laid by Zimmerman; and the geography of plants, which our colleagues have left untouched. Observation of individual parts of trees or grass is by no means to be considered plant geography; rather plant geography traces the connections and relations by which all plants are bound together among themselves, designates in what lands they are found, in what atmospheric conditions they live, and tells of the destruction of

---

From Alexander von Humboldt, *Florae Fribergensis Specimen,* as translated by Richard Hartshorne in *The Concept of Geography as a Science of Space, from Kant and Humboldt to Hettner.* Reprinted, by permission, from the *Annals of the Association of American Geographers* 48 (1958), p. 100.

rocks and stones by what primitive forms of the most powerful algae, by what roots of trees, and describes the surface of the earth in which humus is prepared. This is what distinguishes geography from nature study, falsely called nature history; zoology (*zoognosia*) botany (*phytognosia*) and geology (*oryctognosia*) all form parts of the study of nature, but they study only the forms, anatomy, processes, etc. of individual animals, plants, metallic things or fossils. Earth history, more closely affiliated with geography than with nature study, but as yet not attempted by any, studies the kinds of plants and animals that inhabited the primeval earth, their migrations and disappearance of most of them, the genesis of mountains, valleys, rock formations and ore veins . . .

110. From Humboldt's Personal Narrative of Travels to the Equinoctial Regions of America

Traveling on the Orinoco

During the whole of my voyage from San Fernando to San Carlos del Rio Negro, and thence to the town of Angostura, I noted down day by day, either in the boat or where we disembarked at night, all that appeared to me worthy of observation. Violent rains, and the prodigious quantity of mosquitos with which the air is filled on the banks of the Orinoco and the Cassiquiare, necessarily occasioned some interruptions; but I supplied the omission by notes taken a few days after. I here subjoin some extracts from my journal. Whatever is written while the objects we describe are before our eyes bears a character of truth and individuality which gives attraction to things the least important.

On the 31st March a contrary wind obliged us to remain on shore till noon. We saw a part of some cane-fields laid waste by the effect of a conflagration which had spread from a neighbouring forest. The wandering Indians everywhere set fire to the forest where they have encamped at night; and during the season of drought, vast provinces would be the prey of these conflagrations if the extreme hardness of the wood did not prevent the trees from being entirely consumed. We found trunks of desmanthus and mahogany which were scarcely charred two inches deep.

Having passed the Diamante we entered a land inhabited only by tigers, crocodiles, and chiguires; the latter are a large species of the genus Cavia of Linnæus. We saw flocks of birds, crowded so closely together as to appear

From Alexander von Humboldt, *Personal Narrative of Travels to the Equinoctial Regions of America during the Years 1799–1804,* trans. and ed. Thomasina Ross (London, 1852), II, 152–154, 494–495.

against the sky like a dark cloud which every instant changed its form. The river widens by degrees. One of its banks is generally barren and sandy from the effect of inundations; the other is higher, and covered with lofty trees. In some parts the river is bordered by forests on each side, and forms a straight canal a hundred and fifty toises broad. The manner in which the trees are disposed is very remarkable. We first find bushes of *sauso*, forming a kind of hedge four feet high, and appearing as if they had been clipped by the hand of man. A copse of cedar, brazilletto, and lignum-vitæ, rises behind this hedge. Palm-trees are rare; we saw only a few scattered trunks of the thorny peritu and corozo. The large quadrupeds of those regions, the jaguars, tapirs, and peccaries, have made openings in the hedge of *sauso* which we have just described. Through these the wild animals pass when they come to drink at the river. As they fear but little the approach of a boat, we had the pleasure of viewing them as they paced slowly along the shore till they disappeared in the forest, which they entered by one of the narrow passes left at intervals between the bushes. These scenes, which were often repeated, had ever for me a peculiar attraction. The pleasure they excite is not owing solely to the interest which the naturalist takes in the objects of his study, it is connected with a feeling common to all men who have been brought up in the habits of civilization. You find yourself in a new world, in the midst of untamed and savage nature. Now the jaguar,—the beautiful panther of America,—appears upon the shore; and now the hocco, with its black plumage and tufted head, moves slowly along the sausos. Animals of the most different classes succeed each other. *"Esse como en el Paradiso,"* "It is just as it was in Paradise," said our pilot, an old Indian of the Missions. Everything, indeed, in these regions recalls to mind the state of the primitive world with its innocence and felicity. But in carefully observing the manners of animals among themselves, we see that they mutually avoid and fear each other. The golden age has ceased; and in this Paradise of the American forests, as well as everywhere else, sad and long experience has taught all beings that benignity is seldom found in alliance with strength.

When the shore is of considerable breadth, the hedge of sauso remains at a distance from the river. In the intermediate space we see crocodiles, sometimes to the number of eight or ten, stretched on the sand. Motionless, with their jaws wide open, they repose by each other, without displaying any of those marks of affection observed in other animals living in society. The troop separates as soon as they quit the shore. It is, however, probably composed of one male only, and many females; for as M. Descourtils, who has so much studied the crocodiles of St. Domingo, observed to me, the males are rare, because they kill one another in fighting during the season of their loves. These monstrous creatures are so numerous, that throughout the whole course of the river we had almost at every instant five or six in view. Yet at this period the swelling of the Rio Apure was scarcely perceived; and consequently hundreds of crocodiles were still buried in the mud of the savannahs. About four in the afternoon we stopped to measure a dead crocodile which had been cast ashore. It was only sixteen feet eight inches long; some days after M. Bonpland found another, a male, twenty-two feet three inches long.

Geophagy among South American Indians

The situation of the mission of Uruana is extremely picturesque. The little Indian village stands at the foot of a lofty granitic mountain. Rocks everywhere appear in the form of pillars above the forest, rising higher than the tops of the tallest trees. The aspect of the Orinoco is nowhere more majestic, than when viewed from the hut of the missionary, Fray Ramon Bueno. It is more than two thousand six hundred toises broad, and it runs without any winding, like a vast canal, straight toward the east. Two long and narrow islands (*Isla de Uruana* and *Isla vieja de la Manteca*) contribute to give extent to the bed of the river; the two banks are parallel, and we cannot call it divided into different branches. The mission is inhabited by the Ottomacs, a tribe in the rudest state, and presenting one of the most extraordinary physiological phenomena. They eat earth; that is, they swallow every day, during several months, very considerable quantities, to appease hunger, and this practice does not appear to have an injurious effect on their health. Though we could stay only one day at Uruana, this short space of time sufficed to make us acquainted with the preparation of the *poya,* or balls of earth. I also found some traces of this vitiated appetite among the Guamos; and between the confluence of the Meta and the Apure, where everybody speaks of dirt-eating as of a thing anciently known. I shall here confine myself to an account of what we ourselves saw or heard from the missionary, who had been doomed to live for twelve years among the savage and turbulent tribe of the Ottomacs.

The inhabitants of Uruana belong to those nations of the savannahs called wandering Indians (Indios andantes), who, more difficult to civilize than the nations of the forest (Indios del monte), have a decided aversion to cultivate the land, and live almost exclusively by hunting and fishing. They are men of very robust constitution; but ill-looking, savage, vindictive, and passionately fond of fermented liquors. They are omnivorous *animals* in the highest degree; and therefore the other Indians, who consider them as barbarians, have a common saying, "nothing is so loathsome but that an Ottomac will eat it." While the waters of the Orinoco and its tributary streams are low, the Ottomacs subsist on fish and turtles. The former they kill with surprising dexterity, by shooting them with an arrow when they appear at the surface of the water. When the rivers swell fishing almost entirely ceases. It is then very difficult to procure fish, which often fails the poor missionaries, on fast-days as well as flesh-days, though all the young Indians are under the obligation of "fishing for the convent." During the period of these inundations, which last two or three months, the Ottomacs swallow a prodigious quantity of earth. We found heaps of earth-balls in their huts, piled up in pyramids three or four feet high. These balls were five or six inches in diameter. The earth which the Ottomacs eat, is a very fine and unctuous clay, of a yellowish grey colour; and, when being slightly baked at the fire, the hardened crust has a tint inclining to red, owing to the oxide of iron which is mingled with it. We brought away some of this earth, which we took from the winter-provision of the Indians; and it is a mistake to suppose that it is steatitic, and that it contains magnesia. Vauquelin did not discover any traces of that substance in it: but he found that it contained more silex than alumina, and three or four per cent of lime.

111. Jefferson asks for Humboldt's views on the American West

On his return journey from South America Humboldt spent two weeks in Washington, in June, 1804. For part of that time he was a guest of President Jefferson at the White House. Jefferson, greatly interested in all scientific matters, was especially eager to find out as much as he could from the distinguished explorer-scientist about the western parts of North America. The United States had purchased the territory west of the Mississippi from France the year before Humboldt's visit, and the question of American-Spanish boundaries in the West was very much on Jefferson's mind. On June 9 he wrote a memorandum to his house guest. It is interesting to consider Jefferson's request to Humboldt in terms of Johann Michael Franz's description of the role of the state geographer (selection 103).

Jefferson asks leave to observe to Baron de Humboldt that the question of limits of Louisiana between Spain and the United States is this: they claim to hold to the river Mexicana or Sabine and from the head of that Northwardly along the heads of the waters of the Mississippi to the source either of its Eastern or Western branch, thence to the head of the Red River & so on. Can the Baron inform me what population may be between those lines of white, red or black people? and whether any & what mines are within them? The information will be thankfully received.

From *The Writings of Thomas Jefferson* (New York, 1859), IV, 544.

112. From Humboldt's Kosmos

The numerous listeners who had followed my lectures at the University with such good will may well be pleased if, as a remembrance of days long ago, and also as an expression of my appreciation, I include here the distribution of subjects in my lecture series. These lectures, held from November 3, 1827 to April 26, 1828, numbered 61 in all. Of these, five dealt with general aspects and limits of physical geography, and general aspects of nature; three with the history of geography; two with stimuli for the study of nature; sixteen with the heavens; five with the shape, density, interior heat, and magnetism of the earth, and with the aurora borealis; four with the characteristics of the solid crust of the earth, with hot springs, earthquakes, and volcanism; two with types and formation of mountains; two with the general aspects of the earth's surface, the distribution of continents, and the rising of the crust along synclines; three with the humid envelope of the earth, the seas; ten with the atmosphere and the distribution of temperatures; one with the general distribution of organic life; three with the geography of plants; three with the geography of animals; and two with the distribution of the races of mankind.

On physical geography

The physical description of the world, considering the universe as an object of the external senses, does undoubtedly require the aid of general physics and of descriptive natural history, but the contemplation of all created things, which are linked together, and form one *whole*, animated by internal forces, gives to the science we are considering a peculiar character. Physical science considers only the general properties of bodies; it is the product of abstraction,—generalization of perceptible phenomena; and even in the work in which were laid the first foundations of general physics, in the eight books on physics of Aristotle, all the phenomena of nature are considered as depending upon the primitive and vital action of one sole force, from which emanate all the movements of the universe. The terrestrial portion of physical cosmography, for which I would willingly retain the expressive designation of *physical geography*, treats of the distribution of magnetism in our planet with relation to its intensity and direction, but does not enter into a consideration of the laws of attraction or repulsion of the poles, or the means of eliciting either permanent or transitory electro-magnetic currents. Physical geography

---

The first part of this selection is the editor's translation of a passage from Alexander von Humboldt's *Kosmos* (Stuttgart and Tübingen, 1845), vol. I, pp. xi–xii. The second and third parts are from the edition of *Cosmos* translated by E. C. Otté (London, 1849), I, 40–43, 355–364.

depicts in broad outlines the even or irregular configuration of continents, the relations of superficial area, and the distribution of continental masses in the two hemispheres, a distribution which exercises a powerful influence on the diversity of climate and the meteorological modifications of the atmosphere; this science defines the character of mountain-chains, which, having been elevated at different epochs, constitute distinct systems, whether they run in parallel lines, or intersect one another; determines the mean heights of continents above the level of the sea, the position of the centre of gravity of their volume, and the relation of the highest summits of mountain-chains to the mean elevation of their crests, or to their proximity with the sea-shore. It depicts the eruptive rocks as principles of movement, acting upon the sedimentary rocks by traversing, uplifting, and inclining them at various angles; it considers volcanoes either as isolated or ranged in single or in double series, and extending their sphere of action to various distances, either by raising long and narrow lines of rocks, or by means of circles of commotion, which expand or diminish in diameter in the course of ages. This terrestrial portion of the science of the Cosmos describes the strife of the liquid element with the solid land; it indicates the features possessed in common by all great rivers in the upper and lower portion of their course, and in their mode of bifurcation when their basins are unclosed; and shows us rivers breaking through the highest mountain-chains, or following for a long time a course parallel to them, either at their base, or at a considerable distance, where the elevation of the strata of the mountain system and the direction of their inclination correspond to the configuration of the table-land. It is only the general results of comparative orography and hydrography that belong to the science whose true limits I am desirous of determining, and not the special enumeration of the greatest elevations of our globe, of active volcanoes, of rivers, and the number of their tributaries; these details falling rather within the domain of geography properly so called. We would here only consider phenomena in their mutual connection, and in their relations to different zones of our planet, and to its physical constitution generally. The specialities both of inorganic and organised matter, classed according to analogy of form and composition, undoubtedly constitute a most interesting branch of study, but they appertain to a sphere of ideas having no affinity with the subject of this work.

The description of different countries certainly furnishes us with the most important materials for the composition of a physical geography; but the combination of these different descriptions, ranged in series, would as little give us a true image of the general conformation of the irregular surface of our globe, as a succession of all the floras of different regions would constitute that which I designate as a *Geography of Plants*. It is by subjecting isolated observations to the process of thought, and by combining and comparing them, that we are enabled to discover the relations existing in common between the climatic distribution of beings and the individuality of organic forms (in the morphology or descriptive natural history of plants and animals); and it is by induction that we are led to comprehend numerical laws, the proportion of natural families to the whole number of species, and to designate the latitude or geographical position of the zones in whose plains each organic form attains the maximum of its development. Considerations of this nature, by their tendency to generalization, impress a nobler character on the physical description of the globe; and enable us to understand how the aspect of the scenery, that is to say, the impression produced upon the mind by the physiognomy of the vegetation depends upon the local distribution, the number, and the luxuriance of growth of the vegetable forms predominating in the general mass. The catalogues of organized beings, to which was formerly given the pompous title of *Systems of Nature,* present us with an admirably connected arrangement by analogies of structure, either in the perfected development of these beings, or in the different phases which, in accordance with the views of a spiral evolution, affect in vegetables the leaves, bracts, calyx, corolla, and fructifying organs; and in animals, with more or less symmetrical regularity, the cellular and fibrous tissues, and their perfect or but obscurely developed articulations. But these pretended systems of nature, however ingenious their mode of classification may be, do not show us organic beings, as they are distributed in groups throughout our planet, according to their different relations of latitude and elevation above the level of the sea, and to climatic influences, which are owing to general and often very remote causes. The ultimate aim of physical geography is, however, as we have already said, to recognise unity in the vast diversity of phenomena, and by the exercise of thought and the combination of observations, to discern the constancy of phenomena in the midst of apparent changes. In the exposition of the terrestrial portion of the Cosmos, it will occasionally be necessary to descend to very special facts; but this will only be in order to recall the connection existing between the actual distribution of organic beings over the globe, and the laws of the ideal classification by natural families, analogy of internal organization, and progressive evolution.

It follows from these discussions on the limits of the various sciences, and more particularly from the distinction which must necessarily be made between descriptive botany (morphology of vegetables) and the geography of plants, that in the physical history of the globe, the innumerable multitude of organised bodies which embellish creation are considered rather according to *zones of habitation* or *stations* and to differently inflected *isothermal bands,* than with reference to the principles of gradation in the development of internal organism. Notwithstanding this, botany and zoology, which constitute the descriptive natural history of all organised beings, are the fruitful sources whence we draw the materials necessary to give a solid basis to the study of the mutual relations and connection of phenomena . . .

On the geography of plants and animals and the distribution of the races

Vital organisms, whose relations in space are comprised under the head of the geography of plants and animals, may be considered, either according to the difference and relative numbers of the types, (their arrangement into genera and species,) or according to the number of individuals of each species on a given area. In the mode of life of plants, as in that of animals, an important difference is noticed; they either exist in an isolated state, or live in a social condition. Those species of plants which I have termed *social,* uniformly cover vast extents of land. Among these we may reckon many of the marine algæ—cladoniæ and mosses which extend over the desert steppes of Northern Asia—grasses, and cacti growing together like the pipes of an organ—avicenniæ and mangroves in the tropics—and forests of coniferæ and of birches in the plains of the Baltic and in Siberia. This mode of geographical distribution determines, together with the individual form of the vegetable world, the size and type of leaves and flowers, in fact, the principal physiognomy of the district; its character being but little, if at all, influenced by the ever-moving forms of animal life, which, by their beauty and diversity, so powerfully affect the feelings of man, whether by exciting the sensations of admiration or horror. Agricultural nations increase artificially the predominance of social plants, and thus augment, in many parts of the temperate and northern zones, the natural aspect of uniformity; and whilst their labours tend to the extirpation of some wild plants, they likewise lead to the cultivation of others which follow the colonist in his most distant migration. The luxuriant zone of the tropics offers the strongest resistance to these changes in the natural distribution of vegetable forms.

Observers who in short periods of time have passed over vast tracts of land, and ascended lofty mountains, in which climates were ranged, as it were, in strata one above another, must have been clearly impressed by the regularity with which vegetable forms are distributed. The results yielded by their observations furnished the rough materials for a science, to which no name had as yet been given. The same zones or regions of vegetation which, in the sixteenth century, Cardinal Bembo, when a youth, described on the declivity of Etna, were observed on Mount Ararat by Tournefort. He ingeniously compared the Alpine flora with the flora of plains situated in different latitudes, and was the first to observe the influence exercised in mountainous regions, on the distribution of plants by the elevation of the ground level of the sea, and by the distance from the poles in flat countries. Menzel, in an inedited work on the flora of Japan, accidentally made use of the term *geography of plants;* and the same expression occurs in the fanciful but graceful work of Bernardin de St. Pierre, *Etudes de la Nature.* A scientific treatment of the subject began, however, only when the geography of plants was intimately associated with the study of the distribution of heat over the surface of the earth, and when the arrangement of vegetable forms in natural families admitted of a numerical estimate being made of the different forms which increase or decrease as we recede from the equator towards the poles, and of the relations in which, in different parts of the earth, each family stood with reference to the whole mass of phanerogamic indigenous plants of the same region. I consider it a happy circumstance, that at the time during which I devoted my attention almost exclusively to botanical pursuits, I was led by the aspect of the grand and strongly characterised features of tropical scenery, to direct my investigations towards these subjects.

The study of the geographical distribution of animals, regarding which Buffon first advanced general, and in most instances very correct views, has been considerably aided in its advance by the progress made in modern times in the geography of plants. The curves of the isothermal lines, and more especially those of the isochimenal lines, correspond with the limits which are seldom passed by certain species of plants, and of animals which do not wander far from their fixed habitation, either with respect to elevation or latitude. The elk, for instance, lives in the Scandinavian peninsula, almost ten degrees further north than in the interior of Siberia, where the line of equal winter temperature is so remarkably concave. Plants migrate in the germ; and, in the case of many species, the seeds are furnished with organs adapting them to be conveyed to a distance through the air. When once they have taken root, they become dependent on the soil and on the strata of air surrounding them. Animals, on the contrary, can at pleasure migrate from the equator towards the poles; and this they can more especially do where the isothermal lines are much inflected, and where hot summers succeed a great degree of winter cold. The royal tiger, which in no respect differs from the Bengal species, penetrates every summer into the north of Asia as far as the latitudes of Berlin and Hamburg—a fact of which Ehrenberg and myself have spoken in other works.

The grouping or association of different vegetable species, to which we are accustomed to appy the term *Floras,* do not appear to me, from what I have observed in different portions of the earth's surface, to manifest such a predominance of individual families as to justify us in marking the geographical distinctions between the regions of the Umbellatæ, of the Solidaginæ, of the Labiatæ, or the Scitamineæ. With reference to this subject, my views differ from those of several of my friends, who rank among the most distinguished of the botanists of Germany. The character of the floras of the elevated plateaux of Mexico, New Granada, and Quito, of European Russia, and of Northern Asia, consists, in my opinion, not so much in the relatively larger number of the species presented by one or two natural families, as in the more complicated relations of the co-existence of many families, and in the relative numerical value of their species. The Gramineæ and the Cyperaceæ undoubtedly predominate in the meadow lands and steppes, as do Coniferæ, Cupuliferæ, and Betulineæ, in our northern woods; but this predominance of certain forms is only apparent, and owing to the aspect imparted by the social plants. The north of Europe, and that portion of Siberia which is situated to the north of the Altai mountains, have no greater right to the appellation of a region of Gramineæ and Coniferæ, than have the boundless llanos between the Orinoco and the mountain chain of Caracas, or the pine forests of Mexico. It is the co-existence of forms which may partially replace each other, and their relative numbers and association, which give rise either to the general impression of luxuriance and diversity, or of poverty and uniformity in the contemplation of the vegetable world.

In this fragmentary sketch of the phenomena of organisation, I have ascended from the simplest cell—the first manifestation of life—progressively to higher structures. "The association of mucous granules constitutes a definitely-formed cytoblast, around which a vesicular membrane forms a closed cell," this cell being either produced from another pre-existing cell, or being due to a cellular formation, which, as in the case of the fermentation-fungus, is concealed in the obscurity of some unknown chemical process. But in a work like the present we can venture on no more than an allusion to the mysteries that involve the question of modes of origin—the geography of animal and vegetable organisms must limit itself to the consideration of germs, already developed, of their habitation and transplantation, either by voluntary or involuntary migrations, their numerical relation, and their distribution over the surface of the earth.

The general picture of nature which I have endeavoured to delineate, would be incomplete, if I did not venture to trace a few of the most marked features of the human race, considered with reference to physical gradations—to the geographical distribution of contemporaneous types—to the influence exercised upon man by the forces of nature, and the reciprocal, although weaker, action which he in his turn exercises on these natural forces. Dependent, although in a lesser degree than plants and animals, on the soil, and on the meteorological processes of the atmosphere with which he is surrounded—escaping more readily from the control of natural forces, by activity of mind and the advance of intellectual cultivation, no less than by his wonderful capacity of adapting himself to all climates—man everywhere becomes most essentially associated with terrestrial life. It is by these relations that the obscure and much contested problem of the possibility of one common descent, enters into the sphere embraced by a general physical cosmography. The investigation of this problem will impart a nobler and, if I may so express myself, more purely human interest to the closing pages of this section of my work.

The vast domain of language, in whose varied structure we see mysteriously reflected the destinies of nations, is most intimately associated with the affinity of races; and what even slight differences of races may effect, is strikingly manifested in the history of the Hellenic nations in the zenith of their intellectual cultivation. The most important questions of the civilisation of mankind, are connected with the ideas of races, community of language, and adherence to one original direction of the intellectual and moral faculties.

As long as attention was directed solely to the extremes in varieties of colour and of form, and to the vividness of the first impression of the senses, the observer was naturally disposed to regard races rather as originally different species than as mere varieties. The permanence of certain types in the midst of the most hostile influences, especially of climate, appeared to favour such a view, notwithstanding the shortness of the interval of time from which the historical evidence was derived. In my opinion, however, more powerful reasons can be advanced in support of the theory of the unity of the human race, as, for instance, in the many intermediate gradations in the colour of the skin and in the form of the skull, which have been made known to us in recent times by the rapid progress of geographical knowledge—the analogies presented by the varieties in the species of many wild and domesticated animals—and the more correct observations collected regarding the limits of fecundity in hybrids. The greater number of the contrasts which were formerly supposed to exist, have disappeared before the laborious researches of Tiedemann, on the brain of negroes and of Europeans, and the anatomical investigations of Vrolik and Weber, on the form of the pelvis. On comparing the dark-coloured African nations, on whose physical history the admirable work of Prichard has thrown so much light, with the races inhabiting the islands of the South-Indian and West-Australian archipelago, and with the Papuas and Alfourous (Haroforas, Endamenes), we see that a black skin, woolly hair, and a negro-like cast of countenance are not necessarily connected together. So long as only a small portion of the earth was known to the Western nations, partial heat and a black skin consequently appeared inseparable. "The Ethiopians," said the ancient tragic poet, Theodectes of Phaselis, "are coloured by the near Sun-god in his course, with a sooty lustre, and their hair is dried and crisped with the heat of his rays." The campaigns of Alexander, which gave rise to so many new ideas regarding physical geography, likewise first excited a discussion on the problematical influence of climate on races. "Families of animals and plants," writes one of the greatest anatomists of the day, Johannes Müller, in his noble and comprehensive work, *Physiologie des Menschen,* "undergo, within certain limitations peculiar to the different races and species, various modifications in their distribution over the surface of the earth, propagating these variations as organic types of species. The present races of animals have been produced by the combined action of many different internal, as well as external conditions, the nature of which cannot in all cases be defined, the most striking varieties being found in those families which are capable of the greatest distribution over the surface of the earth. The different races of mankind are forms of one sole species, by the union of two of whose members descendants are propagated. They are not different species of a genus, since in that case their hybrid descendants would remain unfruitful. But whether the human races have descended from several primitive races of men, or from one alone, is a question that cannot be determined from experience."

113. From Humboldt's Aspects of Nature

The carpet of flowers and of verdure spread over the naked crust of our planet is unequally woven; it is thicker where the sun rises high in the ever cloudless heavens, and thinner towards the poles, in the less happy climes where returning frosts often destroy the opening buds of spring, or the ripening fruits of autumn. Everywhere, however, man finds some plants to minister to his support and enjoyment. If new lands are formed, the organic forces are ever ready to cover the naked rock with life. Sometimes, as at an early period among the Greek Islands, volcanic forces suddenly elevate above the surface of the boiling waves a rock covered with Scoriæ: sometimes, by a long-continued and more tranquil series of phenomena, the collective labors of united Lithophytes raise their cellular dwellings on the crusts of submarine mountains, until, after thousands of years, the structure reaches the level of the ocean, when the creatures which have formed it die, leaving a low flat coral island. How are the seeds of plants brought so immediately to these new shores? by wandering birds, or by the winds and waves of the ocean? The distance from other coasts makes it difficult to determine this question; but, no sooner is the rock of the newly raised islands in direct contact with the atmosphere, than there is formed on its surface, in our northern countries, a soft silky network, appearing to the naked eye as coloured spots and patches. Some of these patches are bordered by single or double raised lines running round their margins; other patches are crossed by similar lines traversing them in various directions. Gradually the light colour of the patches becomes darker, the bright yellow which was visible at a distance changes to brown, and the bluish gray of the Leprarias becomes a dusty black. The edges of neighboring patches approach and run into each other; and on the dark ground thus formed there appear other lichens, of a circular shape and dazzling whiteness. Thus an organic film or covering establishes itself by

From Alexander von Humboldt, *Aspects of Nature,* trans. Mrs. Sabine (Philadelphia, 1850), pp. 230–234.

successive layers; and as mankind, in forming settled communities, pass through different stages of civilization, so is the gradual propagation and extension of plants connected with determinate physical laws. Lichens form the first covering of the naked rock, where afterwards lofty forest trees rear their airy summits. The successive growth of mosses, grasses, herbaceous plants, and shrubs or bushes, occupies the intervening period of long but undetermined duration. The part which lichens and mosses perform in the northern countries is affected within the tropics by Portulacas, Gomphrenas, and other low and succulent shore plants. The history of the vegetable covering of our planet, and its gradual propagation over the desert crust of the earth, has its epochs, as well as that of the migrations of the animal world.

Yet although organic life is everywhere diffused, and the organic powers are incessantly at work in reconnecting with each other the elements set free by death or dissolution, the abundance and variety of organized beings, and the rapidity with which they are renewed, differ in different climates. In the cold zones, the activity of organic life undergoes a temporary suspension during a portion of the year by frost; fluidity is an essential condition of life or vital action, and animals and plants, with the exception of mosses and other cryptogamia, are in those regions buried for several months of each year in winter sleep. Over a large part of the earth, therefore, there could only be developed organic forms capable of supporting either a considerable diminution of heat, or, being without leaves, a long interruption of the vital functions. Thus we see variety and grace of form, mixture of colours, and generally the perpetually youthful energy and vigour of organic life, increase as we approach the tropics. This increase can be denied only by those who have never quitted Europe, or who have neglected the study of physical geography. When, leaving our oak forests, we traverse the Alps or the Pyrenees, and enter Italy or Spain, or when we direct our attention to some of the African shores of the Mediterranean, we might easily be led to draw the erroneous inference that hot countries are marked by the absence of trees. But those who do so, forget that the South of Europe wore a different aspect on the first arrival of Pelasgian or Carthaginian colonies; they forget that an ancient civilization causes the forests to recede more and more, and that the wants and restless activity of large communities of men gradually despoil the face of the earth of the refreshing shades which still rejoice the eye in Northern and Middle Europe, and which, even more than any historic documents, prove the recent date and youthful age of our civilization. The great catastrophe which occasioned the formation of the Mediterranean, when the swollen waters of what was previously an immense lake burst through the barriers of the Dardanelles and of the Pillars of Hercules, appears to have stripped the adjacent countries of a large portion of their coating of vegetable mould. The traditions of Samothrace, handed down to us by Grecian writers, appear to indicate the recentness of the epoch of the ravages caused by this great change. In all the countries which surround the Mediterranean, and which are characterized by beds of the tertiary and cretaceous periods (nummulitic limestone and neocomian rocks), great part of the surface of the earth consists of naked rock. One especial cause of the picturesque beauty of Italian scenery is the contrast thus afforded between the bare rock and the islands, if I may so call them, of luxuriant vegetation scattered over its surface. Wherever the rock is less intersected with fissures, so that it retains water at the surface, and where it is covered with vegetable mould, there, as on the enchanting shores of the Lake of Albano, Italy has her oak forests, with glades as deeply embowered, and verdure as fresh as those which we admire in the North of Europe.

The deserts to the south of the Atlas, and the immense plains or steppes of South America, must be regarded as only local phenomena. The latter, the South American steppes, are clothed, in the rainy season at least, with grass, and with low-growing almost herbaceous mimosas. The African deserts are, indeed, at all seasons devoid of vegetation; seas of sand, surrounded by forest shores clothed with perpetual verdure. A few scattered fan palms alone recall to the wanderer's recollection that these awful solitudes belong to the domain of the same animated terrestial creation which is elsewhere so rich and so varied. The fantastic play of the mirage, occasioned by the effects of radiant heat, sometimes causes these palm trees to appear divided from the ground and hovering above its surface, and sometimes shows their inverted image reflected in strata of air undulating like the waves of the sea. On the west of the great Peruvian chain of the Andes, on the coasts of the Pacific, I have passed entire weeks in traversing similar deserts destitute of water.

The origin of extensive arid tracts destitute of plants, in the midst of countries rich in luxuriant vegetation, is a geognostical problem which has hitherto been but little considered, but which has doubtless depended on ancient revolutions of nature, such as inundations or great volcanic changes. When once a region has lost the covering of plants with which it was invested, if the sands are loose and mobile, and are destitute of springs, and if the heated atmosphere, forming constantly ascending currents, prevents precipitation taking place from clouds, thousands of years may elapse ere organic life can pass from the verdant shores to the interior of the sandy sea, and repossess itself of the domain from which it had been banished.

Those, therefore, who can view nature with a comprehensive glance and apart from local phenomena, may see from the Poles to the Equator organic life and vigor gradually augment with the augmentation of vivifying heat. But, in the course of this progressive increase, there are reserved to each zone its own peculiar beauties: to the tropics, variety and grandeur of vegetable forms; to the north, the aspect of its meadows and green pastures, and the periodic reawakening of nature at the first breath of the mild air of spring. Each zone, besides its own peculiar advantages, has its own distinctive character. Primeval laws of organization, notwithstanding a certain degree of freedom in the abnormal development of single parts bind all animal and vegetable forms to fixed ever-recurring types. As we recognise in distinct organic beings a determinate physiognomy, and as descriptive botany and zoology, in the restricted sense of the terms, consist in a detailed analysis of animal and vegetable forms, so each region of the earth has a natural physiognomy peculiar to itself. The idea indicated by the painter by expressions such as "Swiss nature," "Italian sky," &c., rests on a partial perception of this local character in the aspect of nature. The azure of the sky, the lights and shadows, the haze resting on the distance, the forms of animals, the succulency of the plants and herbage, the brightness of the foliage, the outline of the mountains, are all elements which determine the total impression characteristic of each district or region.

Humboldt's great contemporary, Carl Ritter, exerted his influence on the development of geography first, as a teacher and second, as a writer. After his extensive travels in Europe, and with the recognition of his gifts as a writer, Ritter was appointed Professor of Geography at Berlin in 1820, becoming the holder of the first university chair of geography in modern times. His greatest work, *Earth Science in Relation to Nature and the History of Man, that is, a General Comparative Geography as a Secure Foundation for the Study and the Teaching of the Physical and Historical Sciences,* was to have been a geographical survey of the world. Although nineteen volumes appeared in Ritter's lifetime, the work was not complete, for the published volumes dealt only with Africa and Asia. In his *Earth Science (Erdkunde),* and in other, shorter works Ritter was as concerned as Humboldt with the interdependence of terrestrial phenomena, but his outlook was dominated by historical and even theological considerations in contrast to Humboldt's main concern with natural science.

Humboldt's influence was greatest perhaps in physical geography, in which his concepts are an integral part of today's outlook. Ritter, through his position at the University of Berlin, contributed much to the methodology of geography and firmly established it as a university discipline.

The first selection from Ritter's work is from his introduction to *Earth Science;* it is followed by a statement of his views of the earth as a whole, and one illustrating his constant concern with the relationship of geography to history. The next two selections are taken from *Earth Science* and from a lecture delivered in Berlin on "Forms and Numbers as Auxiliary in Representing the Relations of Geographical Spaces." The final selection is a statement by a contemporary historian of geography, Robert Dickinson, on Ritter's geographical concepts.

114. Ritter's method of organization in geography

The title of the present work indicates that it belongs in the realm of historical or experiential sciences. Their completion can only take place as a function of the sum of important experiences, and thus [they are] handed over to coming generations in ever improved form . . .

The method used to organize this special part of observational natural science is called, characteristically, that of reduction or the objective method, one that endeavors to identify the principal types of natural formations and thus to construct a natural system by investigating those links that are basic to the existence of nature itself.

Thus the entire organization differs completely from those earlier and outstanding works, which presented this same science under the name of geography or physical description of the earth according to the subjective method, one of classification, which fits the needs of other sciences and of particular ends . . .

An attempt is made here to organize physical geography in this manner; in order to reserve every effort for characteristics of its own, all those cosmic, statistical, and political conditions of the globe which have been heretofore closely connected with it will be omitted, since these have already been explored in separate works by outstanding writers.

We call this science "physical" because it deals with the forces of nature to the extent that they operate in space, demand clearly defined forms, and create changes. But we are speaking here not only of the effects of mechanical and chemical forces but also of those created by organic and even less significant forces . . . effective over time and influencing matters intellectual and moral. It is for this reason that we omit the usual term, "physical geography," as foreign and having many meanings, as too restricted a definition . . . and replace it by two distinct terms.

This description of the earth is called "general," not because it attempts to present everything, but because it tries to explore, without regard for any particular end, every part of the earth and all of its forms with equal

From Carl Ritter's introduction to his *Earth Science,* reprinted in *Introduction to General Comparative Geography and Studies for the Foundation of a more Scientific Treatment of Geography* (Berlin, 1852), pp. 23–25. Editor's translation.

attention, be they on the liquid or solid surface, in a distant continent or right here in our own land, the domain of a civilized people or a desert. A natural system can only be the result of a study of the basic types of all natural formations.

We attempt to call it "comparative" in the same sense in which other disciplines are being developed, especially comparative anatomy.

115. Ritter on the contrasts between the land and water hemispheres

On December 14, 1826, Ritter spoke to the Royal Academy of Sciences in Berlin "On the Geographical Position and Horizontal Development of the Continents." The following excerpt is taken from that lecture.

––––––––

Several observers have called attention to the unequal distribution of land in contrast to water, pointing out that land masses surround the North Pole of the earth more than the South Pole, and are more concentrated in what we consider the eastern than the western part.

The result of this distribution, keeping in mind solid and liquid forms, is a natural organization of a northeastern hemisphere of lands and an even greater southwestern hemisphere of waters, and these certainly present the greatest contrast on the globe. We can characterize this relationship, widening the meaning of the term earth circle [*Erdkreis*] as it was once used, as a contract between a great circle of land and a great circle of water, thus speaking of a continental and a pelagic side of our planet. These two circles are not separated by a simple line but by a wide belt which extends over continents and oceans and which is easily identified: it circles the entire globe and crosses the equator to the northeast of the Mozambique Strait and along the coastal waters of Peru at an angle of approximately 45 degrees. Europe rules all continents because its advanced civilization lies in the middle of this northern land hemisphere ... It has the greatest number of contacts with the great continents of the planet, is in the center of the greatest creative activity, and has come to rule over the widest part of this area through its historic impact and development.

In the midst of the water hemisphere are located the shores of Australia and its archipelagos, fully outside natural, early contacts with the great

––––––––

From Carl Ritter, *Introduction to General Comparative Geography* ... (Berlin, 1852), pp. 106–108. Editor's translation.

continents. As a result, these people, our antipodes, could not be drawn into the circle of general civilization until the perfectioning of means of oceanic navigation, that is, after a history of thousands of years. This relationship was created by their position on the planet, just as the remarkable capacity for assimilation of the various and diverse characteristics of the rest of the continents, and the early ripening of civilization with its universal distribution displayed by Europe, was at least in part the result of its central position as regards the earth; in other words, [the result of] the world position of this part in relation to the whole.

116. From Ritter's introduction to general comparative geography

Ritter's concern with the historical dimension in geography was perhaps best demonstrated in his lecture presented to the Royal Academy of Sciences in Berlin, on January 10, 1833, from which this excerpt is taken.

The geographical sciences are mainly concerned with spaces on the earth's surface ... that is, with the descriptions and relationships of the coexistence of areas as such, both in terms of their most unique characteristics and in their general earthly phenomena. In this way they differ from the historical sciences, whose task it is to investigate and present the sequence of events, or the chronology and development of things both in particulars and as a whole ...

Areas, times, shapes, and forms, the occupational phenomena of space in terms of their construction and organization on the earth ... do not remain the same in their relation to the globe, viewed as the dwelling place of mankind, rather they truly change their relative values with the progress of centuries and millenia. The way in which space is occupied will thus be differently viewed from decade to decade, from century to century.

The greatest changes, more significant than those (be they ever so grandiose) wrought by volcanoes, earthquakes, tidal waves, or other destructive natural phenomena which for a time claim everyone's attention, took place gradually on this globe, definitely under observation by history, influencing the nature of the globe as the abode of mankind, yet

---

From Carl Ritter, *Introduction to General Comparative Geography* ... (Berlin, 1852), p. 152. Editor's translation.

almost escaping observation. They made the globe, as compared to what it had been in earlier millenia, entirely different, and created a completely different set of relationships between its occupied spaces.

117. From Ritter's Earth Science

What geography as a science has yet to accomplish

Geography can just as little be contented with being a mere description of the Earth, and a catalogue of its divisions, as a detailed account of the objects in nature can take the place of a thorough and real natural history. The very word Geography, meaning a description of the Earth, has unfortunately been at fault, and has misled the world: to us it merely hints at the elements, the factors of what is the true science of Geography. That science aims at nothing less than to embrace the most complete and the most cosmical view of the Earth; to sum up and organize into a beautiful unity *all* that we know of the globe. The whole body of facts revealed by past and present discovery must be marshaled into harmony, before we gain the high pinnacle of Geographical Science. The Earth, in all its parts, must be known in all its relations, before we can speak of it as the scholars of our day ought to speak of the world they inhabit.

Moreover, the Earth is to be considered in two main relations—a relation, and an absolute relation; that is, we are to regard it in its connection with the greater whole, of which it forms a part, the Universe, and as a body standing alone, existing, as it were, for itself. It is the latter view which falls within the strict province of Geography. The very prominence of the old Greek word γη indicates the pre-eminence which, in this science, our own planet, rather than others, receives. *Ge*-ography confines our attention to the Earth, and concentrates it upon the globe, regarded *per se,* rather than in its relations to the Universe. Taken, therefore, strictly, as already hinted at in the foregoing remark, Geography is the department of science that deals with the globe in all its features, phenomena, and relations, as an independent unit, and shows the connection of this unified whole with man and with man's Creator. Should we go beyond this, and discuss the relations of the Earth to the Universe, (as is often done in our geographical treaties, in a singularly imperfect and unfruitful manner,) we should outrun the strict bounds of a single science, and should be encroaching on the domain of the sister science of astronomy. This we have no right to do. Yet, from time to time, we must borrow the results of other departments of learning to confirm our own. The field which we have to till has been immensely reduced in its proportions by the publication of

From Carl Ritter, *Comparative Geography,* trans. William L. Gage, (Philadelphia, 1865), xix–xxvii.

"Kosmos," which great work has almost exhausted the subject of the earth in its external relations. The limiting of our own department may, perhaps, give more opportunity for thorough investigation within itself.

The Earth, if discussed exhaustorily, must be spoken of in its relations to Time as well as to Space. The word by which we characterize it, in this regard, is History. The duration of the Earth outruns all measurement. By thinking of its beginning, is the only way we have of gaining a conception of Time. We cannot conceive of the universe as antedating the creation of our earth. By this indefinite, not to say infinite, duration of time, the Earth is discriminated from all that it contains; it is older than any of its parts; it antedates all its kingdoms. The nature of the whole is, therefore, radically different from that of any of its divisions. The Earth has had a development of its own; hence the too common error of treating it as passive and inorganic. The history of the Earth displays, in all the monuments of the past, that it has been subjected in every feature, in every division of itself, to ceaseless transformation, in order to show that, as a whole, it is capable of that organic development on which I lay so much stress. The natural powers which the earth includes are constantly obedient to the mechanical laws of chemistry and physics. The animate creation, plants, animals, man, come and go, in accordance with the laws of their being, and as subordinate dependents on the great forces which the earth holds locked up within her bosom. The earth, the mother of them all, has her own special advance, her own development, to use that overburdened German word. She has relations to herself alone; not simply to organized forms, plants, and animals; just as little to organic things; not simply to her own countries, her rocks, and her crystals. These are but isolated parts; or, if not isolated, yet bound *together* by a common tie. There is another tie above this; it is that which binds the earth to itself alone; that subordinates its parts to such an extent that they almost disappear. There is, above all this thought of parts, of features, of phenomena, the conception of the Earth as a whole, existing in itself, and for itself, an organic thing, advancing by growth, and becoming more and more perfect and beautiful. Without trying to impose on you anything vague and transcendental, I wish to lead you to view the globe as almost a living thing,—not a crystal, assuming new grace by virtue of an external law,—but a world, taking on grandeur and worth, by virtue of an inward necessity. The individuality of the earth must be the watchword of re-created Geography. To think of the Earth, as a seed sown from the hand of God himself on the great fields of space, and filled with a germinant power of life, which will transform it more and more, and make it more and more worthy of its noblest inhabitant, is the first, as it is the last, idea which we must take and keep in these inquiries.

Formerly, Geography was regarded as a mere auxiliary of History, Politics, Military Science, Natural History, and Industrial Arts, and Commerce. And in truth it does reach out and teach all these departments of knowledge and action; but only in the most recent times has it assumed the place of an independent branch of study. Only through the widening of the whole circle of sciences has room been made for this.

Geography used, for the sake of commerce, to be divided into three divisions: mathematical, physical, and political. This was at the time when it was thought that the whole frame-work of the sciences was a disjointed and sundered thing; before that minor principle of unity which binds them all together was recognized as one of the noblest conceptions that the mind can cherish. In the first two of these arbitrary divisions into which Geography was severed, the relations of astronomy, mathematics, and physics were studied, and their applications to the confused phenomena of the globe investigated. Yet the most important thing of all escaped notice; students overlooked their chief task, the tracing of causation and interdependence in the phenomena, and the relation of every one to the country which supplies its conditions of being. It was not suspected that each phenomenon was one link of a great chain of phenomena, the whole revealing a comprehensive law. Men discussed porphyritic formations, basaltic columns, hot springs, and a thousand features which dot the earth, and a thousand kinds of rock which rift the surface of the globe, and treated them singly as if each was a spore and the whole combination only a sporadic group. They did not discover that in the one feature was to be found the reason of the existence of its neighbor; that all the layers of stone owe their singularities of structure to one another rather than to themselves; that each one stands in the closest connection with the up-heaval of the loftiest mountains, with the formation of great volcanic islands, and, in truth, with the building up of entire continents. And, in like manner, plants were discussed as if they were obedient to no law of grouping, as if they were scattered broadcast over the earth, having no relation to zones of vegetation, to isothermal and isochimenal lines; as if, in fact, there was no suspicion of any principle underlying the very existence of the whole vegetable kingdom. And so, too, with such phenomena as the Aurora Borealis; they were treated as isolated features, rather than in their relations to the globe; the connection was not seen between the maritime discoveries of voyagers and the great system of oceanic currents, on which voyagers are so dependent; in fact, the whole influence of the world of matter on the world of mind was unexplored.

And in order to study what was called Political Geography, a vast mass of materials was converted into a stiff, ritualistic frame-work, in the effort to impose some system and imaginal completeness on it, and not in order to grasp facts and truths in their mutual relations and inward life; they were merely arranged for convenient reference and for available use in the departments of military science, politics, statistics, and history; a method which is plainly our inheritance from the Middle Ages, and which bears the marks of those days. Thus from this arbitrary arrangement, made without reference to any indwelling necessity, sprang the three groups with which we are familiar: Chronography and Topography forming the first, Ethnography and Anthropology the second, and Statistics and History the third, or Political Geography.

From these three groups our ordinary text-books compile their usual aggregate of facts, and each becomes after its own pattern a motley in miniature. They contain variable quantities of this triple mass of materials, and follow no law but the demands of the time when they see the light; they favor, like our light literature, the whim of the hour, and are political, military, or commercial, as the public may demand. A systematic exposition of geography is very seldom to be found in them. A harmony of parts, a true harmony, is very rarely attained in their pages. They are at the foundation only arbitrary and unmethodical collections of all facts which are ascertained to exist throughout the earth. They are arranged according to countries, or great natural divisions; but the relation of one great natural division to another, the mutual and immense influence of one country on another, is never mentioned. The description of Europe follows in them to-day the same order in which Strabo set the pattern. The facts are arranged as the pieces of a counterpane, as if every one existed in itself and for itself, and had no connections with others. The setting out of these facts follows the rubrical method of grouping, according to boundary, soil, mountains, rivers, products, and cities. The beginning is usually made with *boundaries* which are generally most unstable and uncertain, instead of being made with some rudimental fact around which all others arrange themselves as a center.

If we compare these geographical treatises with those made in the interest of any other great department, we shall speedily discover that they indicate knowledge rather than science; they form a mere aggregation and index of rich materials, a lexicon rather than a true text-book. And therefore ensues, despite the undenied interest of the subject and its high claims, the mechanical and unfruitful method only too common—the crowding of the memory without judgment, without thought; thence comes it that Geography has taken so low a place among our school studies, worthy only of the youngest of the pupils, and presenting little stimulus even to them.

It will be my effort, in the course of these lectures, to exhibit the subject of relations rather than to detain you with descriptions; in one word, to generalize rather than to add new details. In the lack of a thoroughly excellent text-book of geography, I shall presuppose an acquaintance on your part with the materials, so to speak, of which the science is to be constructed.

118. Ritter's "Remarks on Form and Numbers as Auxiliary in Representing the Relations of Geographical Spaces"

We call to our aid the auxiliary of form, in the use of well known geometrical figures, to bring into view before the mind, without the employment of measurements, the characteristic to which we would give prominence in the portion of the earth at any given time under our consideration. The right use of geometrical figures, and their intelligent application to the study of geographical forms, should be largely brought out in a scientific treatise, in order to conduct the student, in a very simple and direct manner, to well defined conceptions. He may be securely led to new applications of these figures, and thus to a constant succession of new views, which shall exhibit in the distinctest manner whatever features of the earth's surface are capable of representation in geometrical forms.

This use of geometrical figures was a long time since introduced into botanical science with very great advantage, and, while it has imparted clearness to it, it has been not at all wanting in accuracy. But in geographical science this coadjutor has, with very few exceptions, and those mostly of a light and playful sort, not been brought into use, because people have hitherto been content with the old-fashioned descriptions, and have made no attempt to reach scientific results, or even to grasp the central idea on which all special phenomena rest. In a method wide enough and consecutive enough to embrace the great system of countries that cover the globe, this use of geometrical figures, applied to superficial surfaces of very comprehensive or subordinate size, would lead to a clear view and definite comprehension of the earth's surface; provided only and always that these geometrical figures were not arbitrary and imaginary, but plainly existing in nature, and expressing natural forms, as in a mathematical formula. How speedily would such a method lead geography to entirely new phases, and rid it, whether in its elementary or in its scientific forms, of the vast mass of mere undigested descriptions which only burden the mind with endless details. And it is just because those materials have not been brought into system, and still lie in this crude state, that geography is yet in its rudimentary condition, and that the scholastic works and the treatises for schools are all in their primitive form. The free and full application of geometrical progress to this science would lead, as may easily be seen, to a reconstruction of it, to greater breadth of scope, and to a thorough digestion into a systematic shape of facts now loose and burdensome.

In the preceding paper the attempt was made to put to some use

---

From Carl Ritter, "Remarks on Form and Numbers as Auxiliary in Representing the Relations of Geographical Spaces," in *Geographical Studies* trans. William Leonard Gage, (Boston, 1863), pp. 242–246.

geometrical figures in the establishment of those general relations which spring from the grouping of the continents, their position on the earth, that respective length and breadth, their configuration, and brokenness or unbrokenness of outline.

This application of geometrical figures would lead to a clearer view, and therefore to a speedier and surer comparison of special countries, as well as to greater brevity and exactness in geographical terminology because a geometrical outline conveys at a glance what it would take many sentences to describe. But in order to reach a more perfect conception of geographical forms, since geometrical figures are only partially exact, not strictly coinciding with the outlines of countries or continents, here falling short, here having an excess, we must designate the countries which overrun or do not come up to the limits of the figure taken as the basis by the + and — signs. This simple arrangement proves very convenient and serviceable.

In the application of geometrical forms in the manner just indicated, we should find that some countries were measured by the square, as Spain and the Peloponnesus; others by the rectangle, as Thessaly and Epirus; others by the circle, the ellipse, the triangle, the pentagon, and so forth: that these things meet the eye at a glance, and that this variety of contact with the surroundings gives immediate and perceptible occasion to direct results. Yet it is not easy, it must be confessed, to reach, in the generality of instances, the geometrical figure which best expresses geographical forms, because the number and size of the irregularities prevent the observer from imagining its base type in his mind, and reaching that pure outline on which depend, however, a great many important things,—the climate, the productions, the river systems; these all have much more intimate relation with the geometrical figure which is the type than with the + and — excesses or fallings off from it.

After the analogies had been carefully traced between different countries and geometrical figures, it would be time to advance to deviations from the base types, peculiar as they are in the case of every separate district. Even in the individuality of these deviations does every country show that it has a specially designated place for itself, and for its peculiar productions, surroundings, and relations. The two great triangulated countries of North and South America (to speak only of the largest), when brought into contact with the triangle of hither India and the south half of Africa, will offer, upon careful consideration, points of resemblance, and yet again points of difference; they will at the same time show many other features less studied, in respect to size far inferior, but in which everything dependent upon conformation is subjected, though in less degrees, to the same general laws and influences.

In this way there will result from the use of geometrical forms as the types of geographical configuration certain classes and classifications, which, in reference to the type and the deviations from it, allow of a sharp scientific statement of the relations and characteristics which spring from the entire class, from the subordinate divisions, and from individual members.

This making intelligible and bringing into subjection of the varied, unwieldy, and almost unmanageable mass of material, by the simple employment of the element of form, seems to be the great improvement of the age in geographical science, which has always remained in the rear of her sister sciences of natural history; and so continues to-day a vast helpless mass, of very little service in the instruction of schools, or for yet higher needs, and whose very ponderousness and shapelessness has prevented her being reduced to compact and useful forms.

These geometrical figures are not only applicable in this broad way to the different continents and their natural or arbitrary subdivisions, but they may equally well serve to clearly indicate other characteristics of the earth, such as tracts of water, mountain and plateau districts, plains, lowlands, wastes, fruitful spots, connected forests, regions watered by river systems, grain lands, mineral locations, and the like; and here, where there is not palpable regularity, but only an approximation to the coincidence of geographical configuration with geometrical form, the deviations may, as before, be designated by the $+$ and $-$ signs.

119. Robert Dickinson on Ritter's main geographical concepts

First, Ritter conceived of geography as an empirical science rather than one based on deduction from rational principles or from *a priori* theories. He saw geographic study as proceeding from observation to observation, and although he was convinced that there were laws, he was in no hurry to establish them, as the enormous compilation of the *Erdkunde* shows.

Secondly, there is a coherence in the spatial arrangement of terrestrial phenomena which is described by both Ritter and Humboldt as *Zusammenhang*. Areal phenomena are so interrelated as to give rise to the uniqueness of areas as individual units. Ritter, therefore, considered that areal synthesis and description, must precede the world wide analysis of particular sets of phenomena.

"The earth and its inhabitants stand in the closest mutual relations, and one element cannot be seen in all its phases without the others. On this account history and geography must always go hand in hand. The country works upon the people, and the people upon the country."

Geography must rise above mere description. To avoid the "diversity of

From Robert E. Dickinson, *The Makers of Modern Geography* (London: Routledge & Kegan Paul, Ltd., 1969), pp. 42–44, by permission of the publisher.

origin of its materials" it must have a unique and guiding central principle. "It is to use the whole circle of sciences to illustrate its own individuality, not to exhibit their peculiarities. It must make them all give a portion, not the whole, and yet must keep itself single and clear." He sought for *relations* or *connections* between sets of phenomena in the same area and between one place and another. The task of geography is "to get away from mere description to the law of the thing described; to reach not a mere enumeration of facts and figures, but the connection of place with place, and the laws which bind together local and general phenomena of the earth's surface" and again "what then is the task imposed upon geography, in its work of analysis, but to reach the connection which exists between parts; or, in other words, to get at the relation between places and what fills and occupies them." The geographer traces "*causation* and *interdependence* of the (spatially distributed) phenomena, and the *relations* of every one to the country which supplies its conditions of being." These words give the keynote to Ritter's approach.

Thirdly, Ritter insisted that boundary lines whether wet or dry such as rivers or mountains, were "a means towards the real purpose of geography, which is the understanding of the content of areas." Geography, he maintained, is the study of *der irdisch erfüllten Räume de Erdoberfläche*.

Fourthly, geography according to both Humboldt and Ritter, was concerned with objects on the earth as they exist together in area. Hartshorne writes that Ritter considered "that as chronology provides the framework into which the multiplicity of historical facts are ordered, the area (*Raum*) provides the skeleton for geography; both fields are concerned with integrating different kinds of phenomena together, each in its respective frame." To pursue this end, Humboldt studied systematically particular sets of phenomena in their areal relations with other phenomena. Ritter studied areas synthetically, that is, in their totality. He deliberately deferred the formulation of laws till later.

Fifthly, Ritter held a holistic view with respect to the content and purpose of geographic study and the whole study was focussed on and culminated in man. He wrote that the purpose of the *Erdkunde* was "to present the generally most important geographic-physical conditions of the earth-surface in their natural coherent interrelation (*Naturzusammenhang*), and that (the earth-surface) in terms of its most essential characters and main outlines, especially as the fatherland of the peoples in its most manifold influence on humanity developing in body and mind." But this, Ritter claimed in the introduction to the *Erdkunde*, was not an exclusive concern with man. "Independent of man, the earth is also without him, and before him, the scene of the natural phenomena; the laws of its formation cannot proceed from man. In a science of the earth, the earth itself must be asked for its laws." The dilemma presented by this double viewpoint has already been noticed.

Sixthly, we turn to Ritter's concept of the unique and distinct geographical unit (*Individuen*). In his earlier years, Ritter, following Buache and Gatterer, adhered to the idea of the river basin and its bounding mountain ranges as natural units. It was later that he made the discovery of such units as the objective of study. In 1806 he expressed the notion of the organic unity of geographic areas. "Every naturally bounded area is a unity in respect of climate, production, culture, population and history." This deterministic view was changed later by the idea that the degree of cohesion of these phenomena is the objective of study. He came to regard the geographical unit not as something given by a natural framework, but as something to be discovered in the physiognomy of the earth.

The study of a geographic unit as a spatial concept was an essential purpose of Ritter's geography and is best illustrated in his volume on Africa. He recognised major physical divisions of each continent on a deductive basis. The earth he regarded as an organic entity (*organische Einheit*). In the further breakdown he arrived at distinct units (*Erdteile*) according to the major features of physical demarcation. The continent or *Erdteil* is not a generic complex. It is a complex of individual units (*Ländersysteme*). The geographer, he wrote, seeks to examine every earth area according to its characteristics, without approaching these on a one-sided logical system of division and classification. He is concerned with the total content of areas. This content, viewed geographically, embraced what he called the fixed forms, the mobile forms (atmosphere, hydrosphere, vulcanicity), and the material forms. This makes clear the concept of areal study (*räumliche Prinzip*) as opposed to systematic study (*sachlich systematisch*). In fact, he rejects the latter as not geographic (and this in spite of his frequent declaration of general laws of land-man relations). Ritter sought to build up systems of areas inductively. The ways in which these various sets of phenomena (for example, plant formations and climate) are spatially interdigitated and interdependent, he was never really clear about.

Finally, Ritter's philosophical viewpoint was teleological. Although he insisted that geography investigates the spatial variations of the total content of terrestrial areas, he also believed that the earth was designed to serve one end—namely, to be the abode of man. This viewpoint, expressed repeatedly in his writings, was based on his religious convictions, as well as on his primary interest in the history of man. It was his philosophic interpretation of what he could not understand. It had no effect on the method and substance of his work as contained in his masterpiece of geographic compilation, the *Erdkunde,* but this was the approach that dominated the geographic viewpoint of his successors, and brought the study into disrepute.

Chinese Geographical Writings

Joseph Needham's great work, *Science and Civilization in China,*[1] clearly established the accomplishments of Chinese science, including those in the earth sciences. This section includes three examples of early Chinese geographical writing: an excerpt from the oldest geographical document, *The Tribute of Yü;* two segments typical of the genre Needham calls "descriptions of southern regions and foreign countries," both written during the first millenium of the Christian era; and excerpts from a much later book (fourteenth century) on Chinese trade with lands around the Indian Ocean and beyond.

120. From The Tribute of Yü, an early Chinese work on geography

The following selection is taken from the *Shu Ching* ("Historical Classic") and dates from the fifth century B.C. Yü, a legendary emperor, was an early hydraulic engineer and therefore an important figure in the "hydraulic civilization" of China. This selection is a survey of the several provinces of China—of their soils and products and of the great rivers. It ends with a curious arrangement of the empire into what may have been concentric squares, ranging from the inner zone of the imperial dominations to the outermost zone

---

From James Legge, ed. and trans., *The Chinese Classics* (London, 1865) vol. III, pt. 1.

1. Joseph Needham is Reader Emeritus in Biochemistry at the University of Cambridge. Volumes of his *Science and Civilization in China* are still in the process of publication.

inhabited by barbarians. Needham calls this document "the first naturalistic geographical survey in Chinese history."[1] He believes it may well be contemporary with the work of the Ionian geographer-philosophers, and he emphasizes its long-lasting influence on subsequent Chinese geographical works.

James Legge (1815–1897), from whose translation these excerpts are taken, was a British sinologist who spent much of his life translating the Chinese classics.

___

The tribute of Yü, part 1

I. Yu divided the land. Following the course of the hills, he hewed down the woods. He determined the high hills and great rivers.

II. With respect to K'E-CHOW,—he did his work at Hoo-k'ow and took effective measures at Lëang and K'e. Having repaired *the words on* T'ae-yuen, he proceeded on to the south of *mount* Yŏ. He was successful with his labours on Tan-hwae, and went on to the cross-flowing stream of Chang.

The soil of this province was whitish and mellow. Its contribution of revenue was the first of the highest class, with some proportion of the second. Its fields were the average of the middle class.

The *waters of the* Hang and Wei were brought to their proper channels; and Ta-luh was made capable of cultivation.

The wild people of the islands brought dresses of skins. Keeping close on the right to the rocks of Këë, they entered the Ho.

III. *Between* the Tse and the Ho was YEN-CHOW.

The nine branches of the Ho were conducted by their proper channels. Luy-hea was formed into a marsh; *in which the waters of* the Yung and the Tseu were united. The mulberry grounds were made fit for silkworms, and then *the people* came down from the heights, and occupied the ground *below*.

The soil of this province was blackish and rich; the grass in it became luxuriant, and the trees grew high. Its fields were the lowest of the middle class. Its revenues just reached what could be deemed the correct amount; but they were not required from it as from the other provinces, till after it had been cultivated for thirteen years. Its articles of tribute were varnish and silk; the baskets from it were filled with woven ornamental *fabrics*.

They floated along the Tse and T'ă, and so reached the Ho.

IV. The sea and the Tae *mountain* were *the boundaries of* TS'ING-CHOW.

___

1. Joseph Needham, *Science and Civilization in China* (Cambridge. At the University Press, 1959), vol. 3, *Mathematics and the Sciences of the Heavens and the Earth*, p. 502.

The territory of Yu-e was defined; and the Wei and Tsze were conducted by their proper channels.

The soil of this province was whitish and rich; near the sea were wide tracts of salt land. Its fields were the lowest of the first class, and its contribution of revenue the highest of the second.

Its articles of tribute were salt, fine grass-cloth, and the productions of the sea, of various kinds; with silk, hemp, lead, pine-trees, and strange stones, from the valleys of the Tae. The wild tribes of Lae were taught tillage and pasturage, and brought in their baskets the silk from the mountain mulberry.

They floated along the Wăn, and reached the Tse.

V. The sea, the Tae *mountain,* and the Hwae were the *boundaries of* Ts'EU-CHOW.

The Hwae and the E *rivers* were regulated. The *hills of* Mung and Yu were brought under cultivation. The *lake of* Ta-yay was confined within its proper limits. The *country of* Tung-yuen was successfully brought under management.

The soil of this province was red, clayey, and rich. The trees and grass became more and more bushy. Its fields were the second of the highest class; its contribution of revenue was the average of the second.

Its articles of tribute were earth of five different colours; with the variegated feathers of pheasants from the valleys of the Yu; the solitary dryandra from the south of *mount* Yih; and the sounding stones that *seemed to* float near the banks of the Sze. The wild tribes about the Hwae brought oyster-pearls and fish; and their baskets full of deep azure silks, and other silken fabrics, chequered and pure white.

They floated along the Hwae and Sze, and so reached the Ho.

The tribute of Yü, part 2

I. *Yu* surveyed and described the hills, beginning with K'ëen and K'e, and proceeding to mount King; then, crossing the Ho, Hoo-k'ow and Luy-show, going on to T'ae-yo. After these came Te-ch'oo and Seih-ching, from which he went to Wang-uh; then there were Ta-hang, and mount Hăng, from which he proceeded to Këe-shih, where he reached the sea.

South from the Ho, he surveyed Se-k'ing, Choo-yu, and Neaou-shoo, going on to T'ae-hwa; then Heung-urh, Wae-fang. T'ung-pih, from which he proceeded to Pei-wei.

He surveyed and described Po-ch'ung, going to *the other* mount King; and Nuy-fang, from which he went to Ta-pëë.

He *did the same with* the south of mount Min, and then went on to mount Hăng. From this he crossed *the lake of* Kew-këang, and went on to the plain of Foo-tsëen.

II. He surveyed the Weak-water as far as Hŏ-le, from which its super-fluous waters went away among the Moving sands.

He surveyed the Black-water as far as San-wei, from which it went away to enter the southern sea.

He surveyed the Ho from Tseih-shih as far as Lung-mun; and thence, southwards, to the north of mount Hwa; eastward then to Te-ch'oo; east-ward again to the ford of Măng; eastwards still he passed the junction of the Lŏ, and went on to Ta-pei. From this the course was northwards, past the Keang-water, on to Ta-luh; north from which the stream was distributed and became the nine Ho, which united again and formed the meeting Ho, when they entered into the sea.

From Po-ch'ung he surveyed the Yang, which, flowing eastwards, be-came the Han. Farther east, it became the water of Ts'ang-lang; and after passing the three great dykes, went on to Ta-pëe, southwards from which it entered the Keang. Eastward still, and whirling on, it formed the marsh of P'ang-le; and from that its eastern flow was the northern Këang, at which it entered the sea.

From mount Min he surveyed the Këang, which branching off to the east formed the T'o; eastward again it reached the Le; after this it passed the nine Këang; and flowing eastward and winding to the north, it joined *the Han* in its eddying movements; from that its eastern flow was the middle Këang, as which it entered the sea.

He surveyed the Yen water, which flowing eastward became the Tse, and entered the Ho. *Thereafter* it flowed out, and became the Yung *marsh*. Eastward, it issued forth on the north of Taou-k'ew, and flowed further east to the *marsh of* Ko. North-east from this it united with the Wăn, and after flowing north went eastwards on to the sea.

He surveyed the Hwae from *the hill of* T'ung-pih. Flowing east, it united with the Sze and the E; and with an eastward course *still* entered the sea.

He surveyed the Wei from Neaou-urh-tung-heŭ. Flowing eastwards it met with the Fung, and eastwards again with the King. Farther east still, it passed the Tseih and the Tseu; and entered the Ho.

He surveyed the Lŏ from Heung-urh. Flowing to the north-east, it united with the Këen and the Ch'ëen; eastwards still, it united with the E; and then on the north-east entered the Ho.

III. *Thus,* throughout nine provinces a similar order was effected: — the grounds along the waters were everywhere made habitable; the hills were cleared of their superfluous wood and sacrificed to; the sources of the streams were cleared; the marshes were well banked; access to the capital was secured for all within the four seas.

A great order was effected in the six magazines *of material wealth;* the different parts of the country were subjected to an exact comparison, so that contribution of revenue could be carefully adjusted according to their resources. The fields were all classified with reference to the three characters of the soil; and the revenues for the Middle region were established.

IV. He conferred lands and surnames. *He said,* "Let me go before the empire with reverent attention to my virtue, that none may act contrary to my conduct."

Five hundred *le* constituted THE IMPERIAL DOMAIN. From the first hundred *le* they brought, as revenue, the whole plant of the grain; from the second, they brought the ears; from the third, they brought *only* the straw, but had to perform other services; from the fourth, *they gave* the grain in the husk; and from the fifth, the grain cleaned.

Five hundred *le beyond* constituted THE DOMAIN OF THE NOBLES. The first hundred *le* was occupied by the cities and lands of the high ministers and great officers; the second, by the principalities of the Nan; the other three hundred were occupied by the various princes.

Five hundred *le still beyond* formed the PEACE-SECURING DOMAIN. *In*

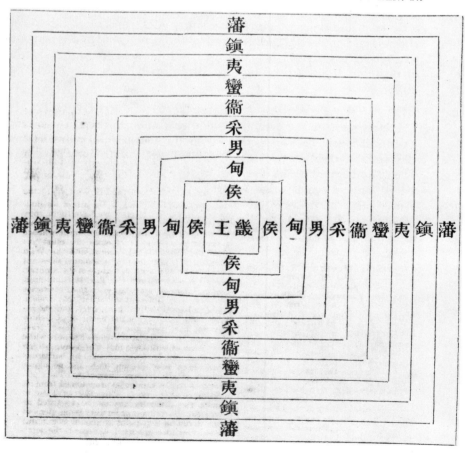

the first three hundred *le* they cultivated the lessons of learning and moral duties; in the other three hundred they showed the energies of war and defence.

Five hundred *le, remoter still,* constituted THE DOMAIN OF RESTRAINT. *The first* three hundred *le* were occupied by the tribes of the E; *the next* two hundred by criminals undergoing the lesser banishment.

Five hundred *le, the most remote,* constituted THE WILD DOMAIN. Three hundred *le* were occupied by the tribes of the Man, two hundred by criminals undergoing the greater banishment.

121. Fa-Hsien, a Chinese Buddhist, travels to the land of the Buddha

For Chinese Buddhists India, the home of the Buddha, was as important a goal of pilgrimages as the Holy Land was to Christians, or Mecca to Moslems. In the following excerpts Fa-Hsien, or Fa-Hian, a Chinese Buddhist of the fifth century A.D., describes his travels and relates some of his adventures on the way to India, a long route that led across the dry heart of Asia and the great mountain barrier of northern India, to the land of "the firm believers in the law of Buddha."

The rainy season being over, they again pressed on to reach Tun-hwang. The fortifications here are perhaps 80 li in extent from east to west, and 40 li from north to south. They all stopped here a month and some days, when Fa-hian and others, five men in all, set out first, in the train of an official, and so again parted with Pao-yun and the rest. The prefect of Tun-hwang, called Li-ho, provided them with means to cross the desert (*sand-river*). In this desert are many evil demons and hot winds; when encountered, then all die without exception. There are no flying birds above, no roaming beasts below, but everywhere gazing as far as the eye can reach in search of the onward route, it would be impossible to know the way but for dead men's decaying bones, which show the direction . . .

Fa-hian and the others, grateful for the presents they received of Fu Kung-sün, forthwith journey to the south-west. On the road there were no dwellings or people. The sufferings of their journey on account of the difficulties of the road and the rivers (*water*) exceed human power of comparison. They were on the road a month and five days, and then managed to reach Khotan.

From "The Travels of Fa-Hsien," in Samuel Beal, *The Si-Yu-Ki, Buddhist Records of the Western World* (London: Kegan Paul Trench Trübner, 1884).

This country is prosperous and rich [*happy*]; the people are very wealthy, and all without exception honour the law (*of Buddha*). They use religious music for mutual entertainment. The body of priests number even several myriads, principally belonging to the Great Vehicle. They all have food provided for them (*church-food, commons*); the people live here and there. Before their house doors they raise little towers, the least about twenty feet high. There are priests' houses for the entertainment of foreign priests and for providing them with what they need ...

From this going onwards towards North India, after being a month on the road, we managed to cross Ts'ung-ling. In Ts'ung-ling there is snow both in winter and summer. Moreover there are poison-dragons, who when evil-purposed spit poison, winds, rain, snow, drifting sand, and gravel-stones; not one of ten thousand meeting these calamities, escapes. The people of that land are also called Snowy-mountain-men (Tukhâras?). Having crossed (Ts'ung)-ling, we arrive at North India. On entering the borders there is a little country called To-li, where there is again a society of priests all belonging to the Little Vehicle. There was formerly an Arhat in this country who by magic power took up to the Tušita heaven a skilful carver of wood to observe the length and breadth (*size*), the colour and look, of Maitrêya Bôdhisattva, that returning below he might carve wood and make his image (*that is,* carve a wooden image of him). First and last he made three ascents for observation, and at last finished the figure. Its length is 80 feet, and its upturned foot 8 feet; on fast-days it ever shines brightly. The kings of the countries round vie with each other in their religious offerings to it. Now, as of yore, it is in this country.

Keeping along (Ts'ung)-ling, they journeyed southwest for fifteen days. The road was difficult and broken, with steep crags and precipices in the way. The mountain-side is simply a stone wall standing up 10,000 feet. Looking down, the sight is confused, and on going forward there is no sure foothold. Below is a river called Sin-t'u-ho. In old days men bored through the rocks to make a way, and spread outside-ladders, of which there are seven hundred (*steps?*) in all to pass. Having passed the ladders, we proceed by a hanging rope-bridge and cross the river. The two sides of the river are something less than 80 paces apart, as recorded by the *Kiu-yi;* but neither Chang-kin nor Kan-ying of the Han arrived here ...

Following down the river Ganges in an easterly direction for eighteen *yôjanas,* we come to the great kingdom of Chen-po (Champâ) on its southern shore. In the place where Buddha once dwelt, and where he moved to and fro for exercise, also where the four previous Buddhas sat down, in all these places towers have been erected, and there are still resident priests. From this continuing to go eastward nearly fifty *yôjanas,* we arrive at the kingdom of Tâmralipti. This is at the sea-mouth. There are twenty-four *sanghârâmas* in this country; all of them have resident priests, and the law of Buddha is generally respected. Fa-Hian remained here for two years, writing out copies of the sacred books (*sûtras*) and drawing image-pictures. He then shipped himself on board a great merchant vessel. Putting to sea, they proceeded in a south-westerly direction, catching the first fair wind of the winter season. They sailed for fourteen days and nights, and arrived at the country of the lions (Simhala, Ceylon). Men of that country (Tâmralipti) say that the distance between the two is about 700 *yôjanas*. This kingdom (*of lions*) is situated on a great island. From east to west it is fifty *yôjanas,* and from north to south thirty *yôjanas.* On every side of it are small islands, perhaps amounting to a hundred in number. They are distant from one another ten or twenty li and as much as 200 li. All of them depend on the great island. Most of them produce precious stones and pearls. The *mâni*-gem is also found in one district, embracing a surface perhaps of ten li. The king sends a guard to protect the place. If any gems are found, the king claims three out of every ten.

This kingdom had originally no inhabitants, but only demons and dragons dwelt in it. Merchants of different countries (*however*) came here to trade. At the time of traffic, the demons did not appear in person, but only exposed their valuable commodities with the value affixed. Then the merchantmen, according to the prices marked, purchased the goods and took them away. But in consequence of these visits (*coming, going, and stopping*), men of other countries, hearing of the delightful character of the place, flocked there in great numbers, and so a great kingdom was formed. This country enjoys an agreeable climate, without any differences in winter or summer. The plants and trees are always verdant. The fields are sown just according to men's inclination; there are no fixed seasons. Buddha came to this country from a desire to convert a malevolent dragon. By his spiritual power he planted one foot to the north of the royal city, and one on the top of a mountain, the distance between the two being fifteen *yôjanas*. Over the foot-impression (*on the hill*) to the north of the royal city, is erected a great tower, in height 470 feet. It is adorned with gold and silver, and perfected with every precious substance. By the side of this tower, moreover, is erected a *sanghârâma*, which is called Abhayagiri, containing 5000 priests. They have also built here a hall of Buddha, which is covered with gold and silver engraved work, conjoined with all precious substances. In the midst of this hall is a jasper figure (*of Buddha*), in height about 22 feet. The entire body glitters and sparkles with the seven precious substances, whilst the various characteristic marks are so gloriously portrayed that no words can describe the effect. In the right hand it holds a pearl of inestimable value. Fa-Hian had now been absent many years from the land of Han; the manners and customs of the people with whom he had intercourse were entirely strange to him. The towns, people, mountains, valleys, and plants and trees which met his eyes, were unlike those of old times. Moreover, his fellow-travellers were now separated from him—some had remained behind, and some were dead. To consider the shadow (*of the past*) was all that was left him; and so his heart was continually saddened. All at once, as he was standing by the side of this jasper figure, he beheld a merchant present to it as a religious offering a white taffeta fan of Chinese manufacture. Unwittingly (Fa-Hian) gave way to his sorrowful feelings, and the tears flowing down filled his eyes . . .

Beyond the deserts are the countries of Western India. The kings of these countries are all firm believers in the law of Buddha. They remove their caps of state when they make offerings to the priests. The members of the royal household and the chief ministers personally direct the food-giving; when the distribution of food is over, they spread a carpet on the ground opposite the chief seat (the president's seat) and sit down before it. They dare not sit on couches in the presence of the priests. The rules relating to the almsgiving of kings have been handed down from the time of Buddha till now. Southward from this is the so-called middle-country (Mâdhyadesa). The climate of this country is warm and equable, without frost or snow. The people are very well off, without poll-tax or official restrictions. Only those who till the royal lands return a portion of profit of the land. If they desire to go, they go; if they like to stop, they stop. The kings govern without corporal punishment; criminals are fined, according to circumstances, lightly or heavily. Even in cases of repeated rebellion they only cut off the right hand. The king's personal attendants, who guard him on the right and left, have fixed salaries. Throughout the country the people kill no living thing nor drink wine, nor do they eat garlic or onions, with the exception of Chandâlas only. The Chandâlas are named "evil men" and dwell apart from others; if they enter a town or market, they sound a piece of wood in order to separate themselves; then men, knowing who they are, avoid coming in contact with them. In this country they do not keep swine nor fowls, and do not deal in cattle; they have no shambles or wine-shops in their market-places. In selling they use cowrie shells. The Chandâlas only hunt and sell flesh. Down from the time of Buddha's *Nirvâna*, the kings of these countries, the chief men and householders, have raised *vihâras* for the priests, and provided for their support by bestowing on them fields, houses, and gardens, with men and oxen. Engraved title-deeds were prepared and handed down from one reign to another; no one has ventured to withdraw them, so that till now there has been no interruption. All the resident priests having chambers (*in these vihâras*) have their beds, mats, food, drink, and clothes provided without stint; in all places this is the case. The priests ever engage themselves in doing meritorious works for the purpose of religious advancement (*karma*—building up their religious character), or in reciting the scriptures, or in meditation. When a strange priest arrives, the senior priests go out to meet him, carrying for him his clothes and alms-bowl. They offer him water for washing his feet and oil for rubbing them; they provide untimely (*vikâla*) food. Having rested awhile, they again ask him as to his seniority in the priesthood, and according to this they give him a chamber and sleeping materials, arranging everything according to the *dharma*.

122. Hsüan Chang, a Chinese pilgrim, on Indian cosmography and on the lands and peoples of southern Asia

The following excerpt from the travels of Hsüan-Chuang, a Buddhist pilgrim of the seventh century A.D., is introduced by his description of the traditional system of Indian cosmography. This is one of the most interesting segments of his report since it describes the basis for early Indian maps showing the shape and divisions of the world— maps that are preserved only in Japanese copies of a much later date. The remainder of the excerpt is a matter-of-fact survey of the land, vegetation, and monuments of India.

This Sahalôka (Soh-ho) world is the three-thousand-great-thousand system of worlds (*chiliocosm*), over which one Buddha exercises spiritual authority (*converts and controls*). In the middle of the great chiliocosm, illuminated by one sun and moon, are the four continents, in which all the Buddhas, lords of the world, appear by apparitional birth, and here also die, for the purpose of guiding holy men and worldly men.

The mountain called Sumêru stands up in the midst of the great sea firmly fixed on a circle of gold, around which mountain the sun and moon revolve; this mountain is perfected by (*composed of*) four precious substances, and is the abode of the Dêvas. Around this are seven mountain-ranges and seven seas; between each range a flowing sea of the eight peculiar qualities. Outside the seven golden mountain-ranges is the salt sea. There are four lands (countries or islands, *dvîpas*) in the salt sea, which are inhabited. On the east, (Pûrva) vidêha; on the south, Jambudvîpa; on the west, Gôdhanya; on the north, Kurudvîpa.

A golden-wheel monarch rules righteously the four; a silver-wheel monarch rules the three (excepting Kuru); a copper-wheel monarch rules over two (excepting Kuru and Gôdhanya); and an iron-wheel monarch rules

---

From "The Travels of Hsüan-Chuang," in Samuel Beal, *The Si-Yu-Ki Buddhist Records of the Western World* (London: Kegan Paul Trench Trübner, 1884), pp. 9 10–15, 70, 88, 89, 188, 231, 232.

over Jambudvîpa only. When first a wheel-king is established in power a great wheel-gem appears floating in space, and coming towards him; its character—whether gold, silver, copper, or iron—determines the king's destiny and his name.

In the middle of Jambudvîpa there is a lake called Anavatapta, to the south of the Fragrant Mountains and to the north of the great Snowy Mountains; it is 800 li and more in circuit; its sides are composed of gold, silver, lapis-lazuli, and crystal; golden sands lie at the bottom, and its waters are clear as a mirror. The great earth Bôdhisattva, by the power of his vow, transforms himself into a Nâga-râja and dwells therein; from his dwelling the cool waters proceed forth and enrich Jambudvîpa (Shen-pu-chau).

From the eastern side of the lake, through the mouth of a silver ox, flows the Ganges (King-kia) river; encircling the lake once, it enters the south-eastern sea.

From the south of the lake, through a golden elephant's mouth, proceeds the Sindhu (Sin-to) river; encircling the lake once, it flows into the south-western sea.

From the western side of the lake, from the mouth of a horse of lapis-lazuli, proceeds the river Vakshu (Po-tsu), and encircling the lake once, it falls into the north-western sea. From the north side of the lake, through the mouth of a crystal lion, proceeds the river Sîtâ (Si-to), and encircling the lake once, it falls into the north-eastern sea. They also say that the streams of this river Sîtâ, entering the earth, flow out beneath the Tsih rock mountain, and give rise to the river of the middle country (China).

At the time when there is no paramount wheel-monarch, then the land of Jambudvîpa has four rulers.

On the south "the lord of elephants;" the land here is warm and humid, suitable for elephants.

On the west "the lord of treasures;" the land borders on the sea, and abounds in gems.

On the north "the lord of horses;" the country is cold and hard, suitable for horses.

On the east "the lord of men;" the climate is soft and agreeable (*exhilarating*), and therefore there are many men.

In the country of "the lord of elephants" the people are quick and enthusiastic, and entirely given to learning. They cultivate especially magical arts. They wear a robe thrown across them, with their right shoulder bare; their hair is done up in a ball on the top, and left undressed on the four sides. Their various tribes occupy different towns; their houses are built stage over stage.

In the country of "the lord of treasures" the people have no politeness or justice. They accumulate wealth. Their dress is short, with a left skirt. They cut their hair and cultivate their moustache. They dwell in walled towns and are eager in profiting by trade.

The people of the country of "the lord of horses" are naturally (*t'ien tsz'*) wild and fierce. They are cruel in disposition; they slaughter (*animals*) and live under large felt tents; they divide like birds (*going here and there*) attending their flocks.

The land of "the lord of men" is distinguished for the wisdom and virtue and justice of the people. They wear a head-covering and a girdle; the end of their dress (*girdle*) hangs to the right. They have carriages and robes according to rank; they cling to the soil and hardly ever change their abode; they are very earnest in work, and divided into classes . . .

Extent of India, climate, etc.

The countries embraced under this term of India are generally spoken of as the five Indies. In circuit this country is about 90,000 li; on three sides it is bordered by the great sea; on the north it is backed by the Snowy Mountains. The north part is broad, the southern part is narrow. Its shape is like the half-moon. The entire land is divided into seventy countries or so. The seasons are particularly hot; the land is well watered and humid. The north is a continuation of mountains and hills, the ground being dry and salt. On the east there are valleys and plains, which being well watered and cultivated, are fruitful and productive. The southern district is wooded and herbaceous; the western parts are stony and barren. Such is the general account of this country . . .

Plants and trees, agriculture, food, drink, cookery

The climate and the quality of the soil being different according to situation, the produce of the land is various in its character. The flowers and plants, the fruits and trees are of different kinds, and have distinct names. There is, for instance, the Amala fruit (*Ngán-mo-lo*), the Âmla fruit (*Ngán-mi-lo*), the Madhuka fruit (*Mo-tu-kia*), the Bhadra fruit (*po-ta-lo*), the Kapittha fruit (*kie-pi-ta*), the Amalâ fruit (*'O-mo-lo*), the Tinduka fruit (*Chin-tu-kia*), the Udumbara fruit (*Wu-tan-po-lo*), the Môcha fruit (*Mau-che*), the Nârikêla fruit (*Na-li-ki-lo*), the Panasa fruit (*Pan-na-so*). It would be difficult to enumerate all the kinds of fruit; we have briefly named those most esteemed by the people. As for the date (*Tsau*), the chestnut (*Lih*), the loquat (*P'i*), and the persimmon (*Thi*), they are not known. The pear (*Li*), the wild plum (*Nai*), the peach (*T'au*), the apricot (*Hang or Mui*), the grape (*Po-tau*), &c., these all have been brought from the country of Kaśmîr, and are found growing on every side. Pomegranates and sweet oranges are grown everywhere.

In cultivating the land, those whose duty it is sow and reap, plough and harrow (*weed*), and plant according to the season; and after their labour they rest awhile. Among the products of the ground, rice and corn are most plentiful. With respect to edible herbs and plants, we may name ginger and mustard, melons and pumpkins, the *Heun-to* (*Kaṇḍu?*) plant, and others. Onions and garlic are little grown; and few persons eat them; if any one uses them for food, they are expelled beyond the walls of the town. The most usual food is milk, butter, cream, soft sugar, sugar-candy, the oil of the mustard-seed, and all sorts of cakes made of corn are used as food. Fish, mutton, gazelle, and deer they eat generally fresh, sometimes salted; they are forbidden to eat the flesh of the ox, the ass, the elephant, the horse, the pig, the dog, the fox, the wolf, the lion, the monkey, and all the hairy kind. Those who eat them are despised and scorned and are universally reprobated; they live outside the walls, and are seldom seen among men.

With respect to the different kinds of wine and liquors, there are various sorts. The juice of the grape and sugar-cane, these are used by the Kshattriyas as drink; the Vaiśyas use strong fermented drinks; the Śramaṇs and Brâhmans drink a sort of syrup made from the grape or sugar-cane, but not of the nature of fermented wine.

The mixed classes and base-born differ in no way (*as to food or drink*) from the rest, except in respect of the vessels they use, which are very different both as to value and material. There is no lack of suitable things for household use. Although they have saucepans and stewpans, yet they do not know the steamer used for cooking rice. They have many vessels made of dried clay; they seldom use red copper vessels: they eat from one vessel, mixing all sorts of condiments together, which they take up with their fingers. They have no spoons or cups, and in short no sort of chopstick. When sick, however, they use copper drinking cups . . .

On the east of the Yamunâ, going about 800 li, we come to the Ganges river. The source of the river (*or* the river at its source) is 3 or 4 li wide; flowing south-east, it enters the sea, where it is 10 li and more in width. The water of the river is blue, like the ocean, and its waves are wide-rolling as the sea. The scaly monsters, though many, do no harm to men. The taste of the water is sweet and pleasant, and sands of extreme fineness border its course. In the common history of the country this river is called Fo-shwui, the *river of religious merit,* which can wash away countless sins. Those who are weary of life, if they end their days in it, are borne to heaven and receive happiness. If a man dies and his bones are cast into the river, he cannot fall into an evil way; whilst he is carried by its waters and forgotten by men, his soul is preserved in safety on the other side (in the other world).

Mo-lo-kiu-ch'a (Malakuta)

This country is about 5000 li in circuit; the capital is about 40 li. The land and fields are impregnated with salt, and the produce of the earth is not abundant. All the valuables that are collected in the neighboring islets are brought to this country and analysed. The temperature is very hot. The men are dark complexioned. They are firm and impetuous in disposition. Some follow the true doctrine, others are given to heresy. They do not esteem learning much, but are wholly given to commercial gain. There are the ruins of many old convents, but only the walls are preserved, and there are few religious followers. There are many hundred Dêva temples, and a multitude of heretics, mostly belonging to the Nirgranthas.

Not far to the east of this city is an old *saṅghârâma* of which the vestibule and court are covered with wild shrubs; the foundation walls only survive. This was built by Mahêndra, the younger brother of Aśôka-râja.

To the east of this is a *stûpa,* the lofty walls of which are buried in the earth, and only the crowning part of the cupola remains. This was built by Aśôka-râja. Here Tathâgata in old days preached the law and exhibited his miraculous powers, and converted endless people. To preserve the traces of this event, this memorial tower was built. For years past it has exhibited spiritual signs, and what is wished for in its presence is sometimes obtained.

On the south of this country, bordering the sea, are the Mo-la-ye (Malaya) mountains, remarkable for their high peaks and precipices, their deep valleys and mountain torrents. Here is found the white sandal-wood tree and the *Chan-t'an-nip'o* (*Chandanêva*) tree. These two are much alike, and the latter can only be distinguished by going in the height of summer to the top of some hill, and then looking at a distance great serpents may be seen entwining it: thus it is known. Its wood is naturally cold, and therefore serpents twine round it. After having noted the tree, they shoot an arrow into it to mark it. In the winter, after the snakes have gone, the tree is cut down. The tree from which *Kie-pu-lo* (*Karpûra*) scent is procured, is in trunk like the pine, but different leaves and flowers and fruit. When the tree is first cut down and sappy, it has no smell; but when the wood gets dry, it forms into veins and splits; then in the middle is the scent, in appearance like mica, of the colour of frozen snow. This is what is called (in Chinese) *long-nao-hiang,* the dragon-brain scent.

123. Chau Ju-kua on Chinese overseas trade

Western views of classical China have long tended to ignore the flourishing maritime trade of that country and the great overseas expeditions undertaken by Chinese, reflected in detailed reports on distant lands. The following selections from a thirteenth century

From Chau Ju-Kua, *His Work on the Chinese and Arab Trade in the Twelfth and Thirteenth Centuries,* trans. Friedrich Hirth and W. W. Rockhill (Tokyo, 1914), pp. 47, 48, 114, 115, 116, 119, 142, 143.

work on Chinese trade by Chau Ju-Kua represent this genre, which may well be called that of the geographical encyclopedia.

———

The sea-route to the east of Chan-ch'öng leads to Kuang-chóu; to the west it borders on Yün-nan; to the south it reaches to Chön-la; to the north it confines on Kiau-chï, whence it communicates with Yung-chóu. From Ts'üan-chóu one can make this country in twenty days' sailing with a favourable wind.

The country extends from east to west 700 *li;* from north to south 3000 li. The capital is called Sin-chóu. They use the designations "district city" (*hién*) and "market town" (*chön*).

The (capital) city walls are of brick and are flanked with stone towers.

When the king shows himself in public he is seated on an elephant or is carried in a kind of cotton hammock (or *juan-pu-tóu*) carried by four men. On his head he wears a golden cap and his body is ornamented with strings of pearls. Whenever the king holds his court, encircling his throne are thirty women attendants carrying swords and bucklers or his betel-nut. At audiences the officials present make one prostration and stop. When the business has been concluded, they again make a prostration and retire. The forms of prostration and salutation are the same for women as for men.

In cases of adultery both the man and the woman are put to death. Theft is punished by cutting off the fingers and the toes.

In battle they bind five men together in one file; if one runs, all who belong to the same file are doomed to death. If a Chinese should be left by a native while lying dangerously wounded, the latter is treated as a murderer and put to death.

The people of this country are fond of cleanliness, they bathe from three to five times daily. They rub themselves with a paste made of camphor and musk and perfume their clothes with fumes of various scented woods.

During the whole year the climate is agreeably warm; there is neither extreme cold nor heat.

Every year on New Year's day they lead a chained elephant through the city, after which they turn it loose. This ceremony is called "driving out evil." In the fourth moon they play at boat-sailing, when they have a procession of fishing boats and look at them.

The full-moon day of the eleventh moon is kept as the winter solstice. At that time cities and towns all bring the king the products of the soil and of their industry.

The people usually plough their fields with two buffaloes. Among the various kinds of cereals they have no wheat, but they have millet, hemp and beans. They do not cultivate tea, neither do they know how to make fermented liquors. They only drink the juice (or "wine") of cocoanuts. As to fruits, they have the lotus, sugar-cane, bananas and cocoanuts. The country also produces elephants' tusks, the *tsién, ch'ön* and *su* (varieties of gharu wood), yellow wax, ebony, white rattans, *ki-peï* cotton, figured cotton stuffs, silk, damasked cotton gauzes, white muslins (or *po-t'ié*) fine bamboo matting, peacocks, rhinoceros horns and parrots.

The cutting of scented wood in the mountains is conducted under government control; the tax paid the government is known as "the scented wood poll-tax", just like the Chinese "salt poll-tax". Once the full amount due has been paid, the people may trade in it on their private account.

Money is not used in trade; they barter with wine, rice and other food substances; with these they settle their accounts yearly ...

The Ta-shï are to the west and north (or north-west) of Ts'üan-chóu at a very great distance from it, so that the foreign ships find it difficult to make the voyage there direct. After these ships have left Ts'üan-chóu they come in some forty days to Lan-li where they trade. The following year they go to sea again, when with the aid of the regular wind they take some sixty days to make the journey.

The products of the country are for the most part brought to San-fo-ts'i, where they are sold to merchants who forward them to China.

This country of the Ta-shï is powerful and warlike. Its extent is very great, and its inhabitants are pre-eminent among all foreigners for their distinguished bearing.

The climate throughout a large part of it is cold, snow falling to a depth of two or three feet; consequently rugs are much prized.

The capital of the country, called Mi-sü-li (Note: Some make it to be Ma-lo-pa), is an important centre for the trade of foreign peoples. The king wears a turban of silk brocade and foreign cotton stuff (buckram). On each new moon and full moon he puts on an eight-sided flat-topped headdress of pure gold, set with the most precious jewels in the world. His robe is of silk brocade and is bound around him with a jade girdle. On his feet he wears golden shoes. In his residence the pillars are of cornelian stone, the walls of *lü-kan* stone (Note: It is as transparent as crystal), the tiles of rock-crystal, the bricks of green stone (jasper?), and the mortar of *huo* stone. The curtains and screens are of brocade with rich designs woven in all kinds of colour in silk and pure gold thread.

The king's throne is set with pearls and precious stones, and the steps of the throne are covered with pure gold. The various vessels and utensils around the throne are of gold or silver, and precious pearls are knotted in the screen behind it. In great court ceremonies the king sits behind this screen, and on either side, protecting him, "the ministers of state surround him" bearing golden bucklers and helmets and armed with precious swords.

His other officers are called *T'ai-weï;* each of them has the command of some twenty thousand horsemen. The horses are seven feet high and are shod with iron. His army is brave and excels in all military exercises.

The streets (of the capital) are more than fifty feet broad; in the middle is a roadway twenty feet broad and four feet high for the use of camels, horses, and oxen carrying goods about. On either side, for the convenience of pedestrians' business, there are sidewalks paved with green and black (or blueish black) flagstones of surpassing beauty.

The dwellings of the people are like those of the Chinese, with this difference that here thin flagstones (slates?) are used instead of tiles.

The food consists of rice and other cereals; mutton stewed with fine strips of dough is considered a delicacy. The poor live on fish, vegetables and fruits only; sweet dishes are preferred to sour. Wine is made out of the juice of grapes, and there is also the drink (called) *ssï*, a decoction of sugar and spices. By mixing of honey and spices they make a drink (called) *meï-ssï-ta-hua*, which is very heating.

Very rich persons use a measure instead of scales in business transactions in gold or silver. "The markets" are noisy and bustling, and "are filled with great store of gold and silver damasks, brocades, and such like wares. The artisans have the true artistic spirit."

The king, the officials and the people all serve (or revere) Heaven. They have also a Buddha by the name of Ma-hia-wu. Every seven days they cut their hair and clip their finger nails. At the New Year for a whole month they fast and chant prayers. Daily they pray to Heaven five times.

The peasants work their fields without fear of inundations or droughts; a sufficiency of water for irrigation is supplied by a river whose source is not known. During the season when no cultivation is in progress, the level of the river remains even with the banks; with the beginning of cultivation it rises day by day. Then it is that an official is appointed to watch the river and to await the highest water level, when he summons the people, who then plough and sow their fields. When they have had enough water, the river returns to its former level.

There is a great harbour (or anchorage) in this country, over two hundred feet deep, which opens to the south-east on the sea, and has branches connecting with all quarters of the country. On either bank of the harbour the people have their dwellings and here daily are held fairs, where crowd boats and wagons, all loaded with hemp, wheat, millet, beans, sugar, meal, oil, firewood, fowls, sheep, geese, ducks, fish, shrimps, date-cakes, grapes and other fruits.

The products of the country (of the Ta-shï) consists in pearls, ivory, rhinoceros horns, frankincense, ambergris, putchuck, cloves, nutmegs, benzoin (*an si hiang*), aloes, myrrh, dragon's-blood, asa-foetida, *wu-na-ts'i*, borax, opaque and transparent glass, *ch'ö-k'ü* shell, coral, cat's-eyes, gardenia flowers, rose-water, nut-galls, yellow wax, soft gold brocades, camel's-hair cloth, *tóu-lo* cottonades and foreign satins . . .

A foreign trader by the name of Shï-na-weï, a Ta-shï by birth, established himself in the southern suburb of Ts'üan-chóu. Disdaining wealth, but charitable and filled with the spirit of his western home, he built a charnel house in the south-western corner of the suburb (or outside the city in the south-west direction) as a last resting-place for the abandoned bodies of foreign traders. The Customs Inspector Lin Chï-k'i has recorded this fact.

Murabit, southern coast of Spain

The country of Mu-lan-p'i is to the west of the Ta-shï country. There is a great sea, and to the west of this sea there are countless countries, but Mu-lan-p'i is the one country which is visited by the big ships of the Ta-shï. Putting to sea from T'o-pan-ti in the country of the Ta-shï, after sailing due west for full an hundred days, one reaches this country. A single one of these (big) ships of theirs carries several thousand men, and on board they have stores of wine and provisions, as well as weaving looms. If one speaks of big ships, there are none so big as those of Mu-lan-p'i.

The products of this country are extraordinary; the grains of wheat are three inches long, the melons six feet round, enough for a meal for twenty or thirty men. The pomegranates weigh five catties, the peaches two catties, citrons over twenty catties, salads weigh over ten catties and have leaves three or four feet long. Rice and wheat are kept in silos for tens of years without spoiling. Among the native products are foreign sheep, which are several feet high and have tails as big as a fan. In the spring-time they slit open their bellies and take out some tens of catties of fat, after which they sew them up again, and the sheep live on; if the fat were not removed, (the animal) would swell up and die.

If one travels by land (from Mu-lan-p'i) two hundred days journey, the days are only six hours long. In autumn if the west wind arises, men and beasts must at once drink to keep alive, and if they are not quick enough about it they die of thirst.

Misr (Egypt)

The country of Wu-ssï-li is under the dominion of Pai-ta. The king is fair; he wears a turban, a jacket and black boots. When he shows himself in public he is on horseback, and before him go three hundred led horses with saddles and bridles ornamented with gold and jewels. There go also ten tigers held with iron chains; an hundred men watch them, and fifty men hold the chains. There are also an hundred club-bearers and thirty hawk-bearers. Furthermore a thousand horsemen surround and guard him, and three hundred body-slaves bear bucklers and swords. Two men carry the king's arms before him, and an hundred kettle-drummers follow him on horseback. The whole pageant is very grand.

The people live on cakes, and flesh; they eat no rice. Dry weather usually prevails. The government extends over sixteen provinces, with a circumference of over sixty stages. When rain falls the people's farming (is not helped thereby, but on the contrary) is washed out and destroyed. There is a river (in this country) of very clear and sweet water, and the source whence springs this river is not known. If there is a year of drought, the rivers of all other countries get low, this river alone remains as usual, with abundance of water for farming purposes, and the people avail themselves of it in their agriculture. Each succeeding year it is thus, and men of seventy or eighty years of age cannot recollect that it has rained.

An old tradition says that when Shï-su, a descendant in the third generation of P'u-lo-hung, seized the government of this country, he was afraid that the land would suffer from drought on account of there being no rain; so he chose a tract of land near the river on which he established three hundred and sixty villages, and all these villages had to grow wheat; and, so that the ensuing year the people of the whole country should be supplied with food for every day, each of these villages supplied it for one day, and thus the three hundred and sixty villages supplied enough food for a year.

Furthermore there is a city called Kié-yé on the bank of this river. Every two or three years an old man comes out of the water of the river; his hair is black and short, his beard is hoary. He seats himself on a rock in the water so that only half his body is visible. If he is thus seen taking up water in his hands, washing his face and cutting his nails, the strange being is recognized, and they go near him, kneel before him and say: "Will the present year bring the people happiness or misfortune?". The man says nothing, but if he laughs, then the year will be a plenteous one and sickness and plagues will not visit the people. If he frowns, then one may be sure that either in the present year, or in the next, they will suffer from famine or plague. The old man remains a long time seated before he dives down again.

In this river there are water-camels (cranes?), and water-horses which come up on the bank to eat the herbs, but they go back into the water as soon as they see a man.

Index